Systemic Crises of Global Climate Change

Sociological literature tends to view the social categories of race, class, and gender as distinct and has avoided discussing how multiple intersections inform and contribute to experiences of injustice and inequity. This limited focus is clearly inadequate.

Systemic Crises of Global Climate Change is an edited volume of 49 international, interdisciplinary contributions addressing global climate change (GCC) by intentionally engaging with the issues of race, gender, and class through an intersectional lens. The volume challenges and inspires readers to foster new theoretical and practical linkages and think beyond the traditional, and oftentimes reductionist, environmental science frame by examining issues within their turbulent political, cultural, and personal landscapes. Varied media and writing styles invite students and educators to reflexively engage different, yet complementary, approaches to GCC analysis and interpretation, mirroring the disparate voices and viewpoints within the field. The second volume, *Emergent Possibilities for Sustainability*, will take a similar approach but will examine the possibilities for solutions, as in the quest for global sustainability.

This book is a valuable resource for academics, researchers and both undergraduate and post-graduate students in the areas of Environmental Studies, Climate Change, Gender Studies, and International Studies as well as those seeking a more intersectional analysis of GCC.

Phoebe Godfrey is an Assistant Professor-in-Residence at UCONN in sociology. She co-founded the non-profit CLiCK dedicated to a local sustainable food system.

Denise Torres is a doctoral candidate at the Graduate Center, City University of New York. The unifying theme of her work and publications is the authentic inclusion of silenced and marginalized groups in the systems that affect them.

Routledge Advances in Climate Change Research

Inspiring, life affirming, poignant. This densely packed volume of wisdom from the hearts and minds of people most intimately targeted by climate change illustrates what is at stake, provides a vision of what must be done, and cogently illustrates an intersectional analysis that includes the environment. There is no other book like it.

Kari Marie Norgaard, *Associate Professor of Sociology and Environmental Studies, University of Oregon, USA*

At one level climate change seems so simple: too much carbon in the atmosphere. But as this provocative volume makes clear, the real truth is far more complicated and interesting. There's never been a better lens than global warming for looking at the ways that power is wielded on our planet, and these essays, poems, and images do much to bring that picture into sharp focus.

Bill McKibben, *founder of* 350.org, *USA*

Passionately mixing scholarly, activist, and poetic analysis of global climate change, this volume takes anti-racist and queerfeminist intersectional analysis to new levels of critical transgression and global understanding. It makes a strong case for using intersectionality to deconstruct and decolonialize borders between struggles for climate, environmental, social, and reproductive justice.

Nina Lykke, *Professor of Gender Studies and co-director of the Centre for Gender Excellence (GEXcel), Linkoeping University, Sweden*

The work offers an innovative synthesis of two key sociological approaches – "intersectionality" and "just sustainabilities". Challenging both media denialisms and the silo methodologies of academia, Godfrey, Torres, and their community of authors are relentless in their diagnosis of exploitive social prejudices that drive the Anthropocene, globalization, climate change, food scarcity, and violence.

Ariel Salleh, *author of* Ecofeminism as Politics

Systemic Crises of Global Climate Change

Intersections of race, class and gender

Edited by Phoebe Godfrey and
Denise Torres

LONDON AND NEW YORK

First published 2016 by Routledge

2 Park Square, Milton Park, Abingdon, Oxfordshire OX14 4RN
711 Third Avenue, New York, NY 10017

Routledge is an imprint of the Taylor & Francis Group, an informa business

First issued in paperback 2017

British Library Cataloguing-in-Publication Data
A catalogue record for this book is available from the British Library

Library of Congress Cataloging-in-Publication Data
Names: Godfrey, Phoebe (Phoebe Christina), editor. | Torres, Denise (Sociologist), editor.
Title: Systemic crises of global climate change : intersections of race, class, and gender / edited by Phoebe Godfrey and Denise Torres.
Description: Abingdon, Oxon ; New York, NY : Routledge, 2016.
Identifiers: LCCN 2015040964| ISBN 9781138830066 (hb) | ISBN 9781315737454 (ebook)
Subjects: LCSH: Climatic changes–Philosophy. | Anthropology. | Human ecology–Cross-cultural studies.
Classification: LCC QC903 .S97 2016 | DDC 363.738/74–dc23
LC record available at http://lccn.loc.gov/2015040964

ISBN: 978-1-138-83006-6 (hbk)
ISBN: 978-0-8153-5917-3 (pbk)

Typeset in Goudy
by Wearset Ltd, Boldon, Tyne and Wear

This book is dedicated to the memories of our fathers
Leonard Edward Andrew Godfrey (1928–2015)
and
Peter 'Jimmy' Torres Gutierrez (1941–2010)

Contents

Illustrations

Figures

Table

Contributors

Editors

Phoebe Godfrey, PhD, is an Assistant Professor-in-Residence at UCONN in Sociology. She is an activist teacher, community organizer, gardener, and artist. She co-founded the non-profit CLiCK dedicated to a local sustainable food system. She lives with her wife and three cats in their creative house and garden in Willimantic, CT.

Denise Torres, LCSW, MPhil, is a clinician, evaluator, and educator. The unifying theme of her work and publications is the authentic inclusion of silenced and marginalized groups in the systems that affect them. She is a New York native and doctoral candidate at the Graduate Center, City University of New York.

Authors

Marcella Ahumada received her Bachelor of Arts in English at Florida Atlantic University, and intends to pursue a Master of Arts degree in English – before her young children enter college themselves. A south Florida native, she currently works for a furniture retailer.

Jacquelyn Anderson is currently an undergraduate student at Florida Atlantic University.

Imna Arroyo's research is focused on African spirituality, and philosophical and aesthetic legacy in the African Diaspora. She traveled to Ghana and Nigeria, West Africa; Salvador de Bahia, Brazil; and made subsequent trips to Cuba to continue her research. Her artwork has been reviewed and exhibited nationally and internationally.

Parvez Babul is a journalist, author, and human rights activist in Bangladesh. He has written about 200 articles and four books on gender equality, media, climate change, sustainable development, poverty, food security, and nutrition. He is a regular contributor to the mainstream newspapers in Bangladesh and abroad.

Julie Bacon is a PhD candidate in Environmental Sciences, Studies, and Policy at the University of Oregon.

Subhankar Banerjee is a photographer, writer, activist, and environmental humanities scholar. His photographs have been exhibited in more than 50 museum exhibitions, including the *Rights of Nature: Art and Ecology in the Americas* at the Nottingham Contemporary. He is editor of *Arctic Voices: Resistance at the Tipping Point*, and received a 2012 Lannan Foundation Cultural Freedom Award.

Khalil Bendib, a Muslim, Progressive Arab American residing in Berkeley, brings a non-Eurocentric perspective focused on equity, racial injustice, labor/class struggles, imperialism, and environmental degradation. Through his cartoons, public speeches, and radio program, he offers social critiques and commentary, noting "The Pen is Funnier than the Sword" (slogan at www.bendib.com).

Chantal Bilodeau is a playwright and the Artistic Director of The Arctic Cycle – an organization created to support the writing, development, and production of eight plays that examine the impact of climate change on the eight countries of the Arctic. She is also founder of the blog "Artists And Climate Change."

Toban Black is a co-editor of the collection *A Line in the Tar Sands: Struggles for Environmental Justice* (Between the Lines/PM Press), and an associate editor for *Upping the Anti*. As an activist, he has focused on extreme energy projects and community-based alternatives.

Priyanka Borpujari is an independent journalist and photographer who has been reporting for various Indian and international publications about human rights abuses from different corners of India. She has won several awards and fellowships, and has widely exhibited her photographs. For a glimpse of her work, visit www.priyankaborpujari.contently.com and www.kaliwritesproject.wordpress.com.

Jane Caputi is Professor of Women, Gender and Sexuality Studies at Florida Atlantic University. She has written three books and the 2015 documentary, *Feed the Green: Feminist Voices for the Earth*.

Soraya Cardenas, PhD, teaches Sociology and Social Sustainability at Cascadia College. Her accolades include a Dissertation Fellowship, Presidential Fellowship, Garcia-Robles Fulbright Scholarship, Researcher of the Year, and Trustee Professorship. She has taught Sustainability and Environmental Sociology for over 14 years. She is currently working on her book, *The Environmental Wetback*.

Lisa M. Corrigan, (PhD, University of Maryland) is an Associate Professor of Communication and Director of the Gender Studies Program in the J. William Fulbright College of Arts and Sciences at the University of Arkansas.

Steve Cutts is an illustrator and animator working in London. He previously spent several years as an illustrator at London creative agency Glueisobar working on digital projects for a wide range of clients, including Coca-Cola, Google, Reebok, Magners, Kellogg's, Virgin, 3, Nokia, Sony, Bacardi, and Toyota. See www.stevecutts.com.

Antonia Darder, PhD, is an internationally recognized Freirian scholar. She holds the Leavey Presidential Chair of Ethics and Moral Leadership at Loyola Marymount University, Los Angeles, and is Professor Emerita of Education Policy, Organization and Leadership at the University of Illinois, Urbana Champaign. She is the author of *Culture and Power in the Classroom*, *Reinventing Paulo Freire: A Pedagogy of Love*, and *Freire and Education*.

Al Duvernay is a retired paleontologist with a constant focus on the human condition and societal inequities. He is a regular volunteer and contributor to initiatives aimed at improving the biosphere.

Prince Ea is the stage name of Richard Williams, an American rapper and activist, known for discussing topics like politics, sociality, environmentalism, and life issues. For more about his work see www.princeea.com.

Martha Eichloff graduated from Florida Atlantic University with a Bachelor of Arts degree in Criminal Justice and a minor in Sociology. She is currently a legal assistant for a law firm in Fort Lauderdale, FL.

Ana Marie Estrada is a former undergraduate of Education and Sociology at Florida Atlantic University. She is currently working for the Broward County School Board.

Mr. Fish (Dwayne Booth) has been a freelance artist for 25 years. His work has appeared in *Harper's* magazine, the *Los Angeles Times*, *The Village Voice*, and *The Nation*. His books include *Go Fish: How to Win Contempt and Influence People* (Akashic Books, 2011) and *WARNING! Graphic Content* (Annenberg Press, 2014).

Albert S. Fu is an Associate Professor of Sociology at Kutztown University of Pennsylvania. His research examines the intersection between built and natural environments as it pertains to natural disaster and urban growth. His work has appeared in journals such as: *Cities*, *Critical Sociology*, *Environment and Planning A*, the *International Journal of Urban and Regional Studies*, and *Urban Studies*.

José G. González is a K-12 public educator in formal and informal education in the arts, education, conservation, and the environment. He contributes to "Green Chicano" working on diverse environmental issues with an emphasis on the intersection of art, education and conservation. Find him on Twitter @JoseBililngue, @Green_Chicano and www.josegagonzalez.com.

Coco Gordon. Her Italian birth delivers through the no-time tunnel, a NYC Susan B. Anthony Award. As SuperSkyWoman, 200+ books or Sustainable

Futures Commission exploring all-species rights, letting go is as important as getting things done. Permaculture observation mode transforms reversals found or pursued. Experiencing 53 Nature senses, learning = teaching. Forget Dominance!

Bert Gunn, a clinical social worker and handyman, traveled with Tlakaelel, an indigenous Mexican Elder, for 22 years up until Tlakaelel's death in 2012, helping to deliver Tlakaelel's message around the world. He is the Tecutli (Executive Director) of Kalpulli Chaplin (and the Zihuakoatl (Administrative Director of both the International Confederation of Kalpultin and in Kaltonal).

Rebecca Hall is an attorney, a teacher, a scholar, and an activist. After receiving her law degree from UC Berkeley she worked with low-income tenants and homeless families for eight years. She received her PhD in 2004 from UC Santa Cruz and has published on racialized gender and resistance.

Dylan Harris has worked and lived in various mountain communities and is drawn to their stories. He is a Geography PhD student at Clark University where he is studying comparative climate change narratives among high-mountain cultures in Bolivia and Nepal.

Eyob Hassen, MD, is former head of the Addis Ababa Regional Health Bureau, Ethiopia. Prior to his appointment he worked as a medical director of one of the largest community health centers in Addis-Ababa.

R.D.K. Herman is Senior Geographer for the Smithsonian National Museum of the American Indian. He is currently Co-chair of the Indigenous Peoples Specialty Group of the Association of American Geographers. His research engages with the representation of Indigenous cultures and the importance of Indigenous epistemologies and ontologies.

David C. Jackson has been an active artist for the majority of his life. A graduate of UCONN, his lifetime goals are reflected in his artworks that are filled with social consciousness, charisma, and an unprecedented desire to express empathy for the subjects he chooses.

Cherese Mathews is a surprisingly optimistic, wild forager, artist, living the questions, healer, sacred being, fire keeper, agitator, dreamer, tree hugger, passionate, gardener, recycler, dancer, and truth seeker. She is destroying and rebuilding in her own small way in Northeast CT.

Isis Mattie is an attorney, writer, and entrepreneur who lives in New York with her husband and children where she pursues broad interests, including law, social and environmental activism, art, architecture, herbalism, and the study of Earth-based spiritual traditions.

Gabrielle Maughan is a poet in England. She was born in London and remembers growing up among houses still showing the signs of bomb damage from the Second World War. She now lives near one of the largest man-made lakes in Europe, Rutland Water, with some chickens, quails, and tortoises and an unparalleled view over the English countryside. She has worked to protect the natural environment since collecting caterpillars as a child on her way home from school.

Cara Murray earned a BA from Brown University and a MA in Marine Policy from the University of Rhode Island. Her poetry and prose have appeared in various journals, including *Platte Valley Review*, and aired on Rhode Island's NPR station.

Julianne Norton is a 22-two-year-old Connecticut-based painter and illustrator. Her work focuses on the long-term effects of genocide, cultural trauma, and environmental destruction on post-memory (trans-generational cultural memory following trauma). She is currently working on a graphic novel about Holocaust post-memory.

Andrew Pattison, PhD, is a Visiting Assistant Professor in the Department of Policy Studies at California Lutheran University. He studies public policy, local climate change policy and sustainable urban infrastructure.

Juliet Pinto, PhD, is an Associate Professor at Florida International University. Dr. Pinto studies environmental communication in Spanish- and English-language media. She developed classes in environmental journalism, has taken students to the Galapagos and to watch NASA astronauts train underwater, and produced the documentary, *South Florida's Rising Seas*.

Molly Rawn (BA, University of Arkansas) is Director of Development and Communications at the Amazeum in Bentonville, Arkansas.

Anthony J. Richardson is an Associate Professor and climate change ecologist at the University of Queensland and CSIRO. He collects and analyzes time series data to understand the response of marine systems to climate change and other human impacts. He also builds models to develop strategies to maintain healthy natural systems.

Carmen Rowe is currently a PhD student in Sociology at Boston University. She completed her BA degree in Sociology at Florida Atlantic University in 2011. Her research interests include gender, sexuality, race and ethnicity, and environmental, economic, and cultural sociology. In the future, she plans to examine the rise of female farm operators in the USA.

Chandra Russo is a PhD candidate in Sociology at the University of California, Santa Barbara. She studies US-based grassroots activism for peace, racial justice, and transnational solidarity. Her current research is a comparative ethnography of three groups seeking to enact solidarity with the targets of US National Security Policy.

Devin Samuels is an artist-activist from Providence, Rhode Island who has been performing for the past nine years. He has designed and hosted workshops on intersectionality, identity, and emotional knowledge around New England, including the 2014 Climate Change March in New York City and the 2014 Brave New Voices International Youth Poetry Festival.

Defne Sarsilmaz is enrolled in the Global and Sociocultural Studies Department at Florida International University. She specializes in political geography and feminist theory. Her research interests are ethnic/religious minority struggles and women's empowerment in the Middle East. Her dissertation investigates politicization and activism among Arab Alawite women in Antakya, Turkey.

Jade S. Sasser is an Assistant Professor in the Department of Gender and Sexuality Studies at the University of California, Riverside. Her PhD is in Environmental Science, Policy, and Management from the University of California, Berkeley. She researches race, class, and gender in environmentalism, global population advocacy, and international development.

Tina Shirshac is an artist, gardener, and practicing reiki master and healing facilitator. She lives in northeastern Connecticut with her wife Phoebe and their three cats. She has hopes that if we are grounded in the present moment we will be able to see that all things are connected.

Hamad Sindhi is a doctoral student in Sociology at the CUNY Graduate Center. His research is situated at the intersection of environmental disasters and citizenship, specifically examining post-disaster state response and claims to benefits and protection by vulnerable communities. He teaches Introduction to Sociology and Research Methods at CUNY.

Bruno Takahashi is an Assistant Professor at Michigan State University with a joint appointment with the School of Journalism and the Department of Communication. He is also Research Director of the Knight Center for Environmental Journalism. His research focuses on environmental/science communication, environmental journalism practices, and links between media/policy making.

Victoria Team, MD, MPH, DrPH, is a Research Fellow in the School of Social Sciences, Monash University. Her research interests are in the area of women's health. She is involved in a research project on contraceptive technologies among immigrant women in Australia. She previously worked in General Practice in Addis-Ababa.

Lewis Vande Pallen is a New York City poet of Ghanian origins; his worldview has been shaped by living on three continents. Author of *I am Beautiful* and *www.sayiambeautiful.com*, he adds his voice to the growing chorus of citizens raising awareness about global climate change.

Patricia Widener is an Associate Professor of Sociology at Florida Atlantic University. She is the author of *Oil Injustice: Resisting and Conceding a Pipeline in Ecuador*. She continues to study the social impacts of the oil industry, and her current research explores how community resistance and climate activism developed in response to deep-water oil proposals off the coast of Aotearoa, New Zealand.

Chris Williams is the author of *Ecology and Socialism: Solutions to Capitalist Ecological Crisis* (Haymarket Books, 2010), Adjunct Associate Professor at Pace University, and an environmental activist/educator. He has published widely on climate change. He was awarded the Lannan Cultural Freedom Fellowship in 2014 and is currently writer-in-residence at Truthout.

Foreword

As someone who has become steeped in the climate justice movement through its myriad intersections with other areas of work I have engaged in over time, I am honored to be asked by the editors to write this Foreword. My life's work has been as a researcher, policy analyst, program manager, activist, etc. on public health, women's rights, economic justice, racial justice, disability rights, housing rights, environmental justice, and so on throughout the USA, as well as in Sub-Saharan Africa, Latin America, the Caribbean, and beyond. Particularly in my work as a women's rights advocate in Sub-Saharan Africa, the intersection with gender justice was clear as I worked with women who experienced sexual assault in the course of climate-forced migration when they found their home countries unlivable due to drought stripping their livelihoods, or as the result of resource wars. I worked with girls who could no longer go to school because their days were consumed with walking twice as far to perform their duty of retrieving water for the household. Therefore, by force of intersectionality, I have increasingly become engaged in addressing climate change progressively up until now, when it has become the frame of a continued focus on all of these interconnected issues.

I extend warm and hearty congratulations to the editors of this volume for their vision in conceptualizing a book that weaves a narrative of the linkages between societal norms, political dynamics, communities, populations, institutions, and movements, but with an undercurrent of the spirit and soul of the collective climate justice movement and all of its actors and nodes. All hail to the authors who combine to form a robust tapestry of activists, community organizers, culture workers, academics, policy analysts, and more, bringing rich perspectives, experiences, analyses, and reflections on this critical topic of how intersecting 'isms' both contribute to, and are exacerbated by, climate change.

Climate change, as a crisis of our time, is a manifestation of the convergence of historic global patterns of domination, extraction, and oppression. The predicament in which we find ourselves in terms of this, the unfettered slide toward catastrophic climate change, is rooted in our commoditization of labor and natural resources toward the end of amassing wealth by a powerful few. Colonization had the central aim of the acquisition of natural resources (spices, minerals, gold, etc.), land, and people, and those patterns of racism, sexism, and classism have persisted over time and become institutionalized in systematic oppression.

As an illustrative example, burning coal is the number one culprit in advancing climate change, being the top contributor to carbon dioxide emissions. The average coal company CEO earns compensation of 289 times the average wage of the company's workers. The inextricable link between policies, policy makers, and corporations means that the 'rules' are written by corporations in such a way as to ensure that the wealth remains heavily concentrated among the rule makers. For example, just 1 percent of the US population has 35.6 percent of all private wealth, which is more than all of the people in the bottom 95 percent combined. We see how this has fatal consequences for our planet and its people in the fact that, for decades, the coalmining industry has focused millions of dollars in antiregulatory lobbying against measures that can protect workers' health. As a result, 76,000 coalminers had died of black lung disease since 1968 until the first-ever regulation on coal dust finally went into effect on August 1, 2014. Meanwhile, coal-based energy production erodes the Earth's natural balance that maintains stable temperatures and climate. The extreme subsidization of the fossil fuel industry, with an average $550 billion in subsidies per year, and the lack of regulations governing its practices that are destroying the environment and violating human rights, is a self-propelling cycle.

Entities such as the American Legislative Exchange Council (ALEC) and Americans for Prosperity (Koch Brothers), which are mostly invested in the fossil fuel industry, spend millions of dollars annually to influence research produced by academic institutions and think tanks, as well as to influence elected offices and those who preside over the courts and the decisions they make. The Koch brothers alone spent $61 million between 1997 and 2010 to support climate change-denying institutions who lobby against clean air and clean energy legislation. They have already signaled the intention to pump $889 million into the 2016 elections to maintain the status quo of decisions and rule making that holds wealth and power within a privileged few. While pushing for policies that repeal voting rights, the decisions they drive include blocking climate-saving regulations protecting air, water, and land, legislation on energy efficiency and clean energy, as well as furthering the disenfranchisement of those who would advance systems changes, such as low-income communities and communities of color.

All we have to do is pick up a newspaper and read the latest headlines to find real-time evidence of this worldwide dynamic of groups of people and nations that have been commoditized, criminalized, disenfranchised, and otherwise oppressed while the system is designed to protect the profits and power building of a select elite. The Black Lives Matter Movement, aided by the digital age, has brought to light what African American communities have been experiencing for decades in terms of racial profiling, police brutality, and mass incarceration. The Arab Spring illuminated the discontent with the authoritarian regimes that advance systematic violation of human rights, while Global North nations prop up these governments to maintain their oil supplies. The Greek

Austerity crisis revealed practices that have been going on for decades in Global South nations where international finance institutions such as the World Bank and the International Monetary Fund impose measures to restrict government spending toward the end of mandating debt servicing. This is usually at the expense of the well-being of the citizenry of focal nations, as services and systems that support basic human needs shrink. La Via Campesina highlights the intersection between global trade policies governed by the world trade organizations and the dissolution of national and local food systems while corporate agribusiness bloats and thrives. The Occupy Movement highlighted the fact that capitalism and corporate domination of our political systems is stealing our democracy and rendering the voice of the masses unheard, irrelevant, and powerless.

Trade policies which effectively prohibit some countries from manufacturing or restrict their ability to hold patents; predatory agribusinesses which develop seeds that do not regenerate and act as a poison that spreads to neighboring farms; energy production processes which pollute in order to maintain profits; national and global policy-making structures which exclude communities and countries from decision making that most impacts their lives while power is held by a select few; individuals, population groups, communities, and nations which experience grinding poverty amidst abundance while others enjoy extreme wealth through exploitation – all of these circumstances are connected to each other and linked to the proliferation of climate change. It is not possible to effectively reform one aspect without addressing the whole, given the inextricable connections.

Yet measures to address these global challenges, including climate change, fail to comprehensively tackle the fundamental systemic racism, sexism, classism, and xenophobia – all bolstered by capitalism – that maintains, and even strengthens, the status quo. Research, philanthropy, policy making, and programming all suffer from the fragmentation of siloes that only allow for a focus on one or two variables/issues/interventions. This keeps us in an endless cycle of ineffective band-aids while domination, extraction, and oppression persist.

As such, this book starts with an acknowledgment of the increasing state of chaos, and then, using a frame of the elements of air, water, fire, and earth, goes deep into exploring the myriad connections between climate change and the very fabric of our existence, including the systems that govern society. Several authors explore an analysis of gender, including population discourse, from a personal reflection perspective, as well as examining societal trends, incorporating intersections with race and class. Relatedly, unpacking power, privilege, and capitalism in the context of climate change politics is an important foundational analysis that this book delivers. Some authors examine public perceptions and attitudes that accept or deny climate change, including media and messaging and strategies for educating different populations about climate change. One chapter explores climate adaptation planning to build resilience, including the complexity of these interventions in urban contexts where gentrification resulting in displacement is a risk that can have counter-productive results for communities. Given the relationship between how we produce food as a driver of climate change and the impact of climate change on the food supply, there are several authors who examine farming and food security. As previously noted, climate-forced migration brings the issue of climate change into conjunction with immigrant rights issues that catalyzes/compounds situations ranging from the Syrian refugee crisis to displaced Haitians in the Dominican Republic and beyond. Disasters, from hurricanes to wildfires to oil spills to extreme heat, and how race and social class are overlaid with these situations to exacerbate impact, are another major topic of this volume. The vast majority of people worldwide hold some form of faith, and, in fact, the moral movement of faith leaders on climate is growing in volume and impact, so it is fitting that this volume also includes content on spirituality.

Interspersed throughout the chapters of the book there are moving and grounding reflections through culture work – such as poems, photos, and paintings – a critical component of movement building. Examples include "Dear Future Generations," an ode and an apology to future generations; visual arts such as "Small Extinction," a representation of an all-too-common disaster that defiles our waterways, and "El Agua es La Vida" (Water is Life), which affirms a precious resource that is threatened by poverty and privatization, by polluting practices, and by climate-driven drought; and poems such as "Crude," which depicts the effects of oil reliance.

"To change everything, we need everyone" was a very apt slogan for the People's Climate March, and it is a key theme that is foundational in the connections illustrated throughout this book. It makes the compelling case for the transformation we must collectively manifest, each of us playing our respective, essential roles. In this rich volume, we have a guidebook that provokes visionary analysis and demonstrates models of how we can organize ourselves as we advance the movement for climate justice. We cannot just tweak an existing system that is so deeply flawed and fundamentally built on principles and practices of domination and extraction with devastating impact. We must be individually and universally transformational to restructure relationships, institutions, and the threads and pathways that tie systems together. A fundamental shift, to a society that values all of its members and the earth upon which we rely for our existence, is necessary. The collective voices gathered here are clearly telling us that we must have "systems change, not climate change" for the survival of the planet and for the well-being of all living beings within it. Throughout this transformation, the authors implore us to keep in mind that we must be just as deliberative about cherishing and treasuring the heart, soul, and cultures of the movement, and its individuals, communities, and nations, as we are about strategies and tactics. A Luta Continua.

Jacqui Patterson
Director of NAACP Environmental Program
and Co-Founder of Women of Color United, serves on the Boards of Directors
for US Climate Action Network, GRID Alternatives, Center for StoryBased
Strategy, and Interfaith Moral Action on Climate Change, as well as on the
Leadership Body for the Climate Justice Alliance

Preface

A Zen story: accommodating the water

A Taoist story tells of an old man who accidentally fell into river rapids that led to a high and dangerous waterfall. Onlookers feared for his life. Miraculously, he came out alive and unharmed, downstream at the bottom of the falls. People asked him how he managed to survive. "I accommodated myself to the water, not the water to me. Without thinking, I allowed myself to be shaped by it. Plunging into the swirl, I came out with the swirl. This is how I survived."

Acknowledgments

This project has been three years in the making, and during that time many people have provided critical support. While it is impossible to capture all the individuals or the ways they have given meaning to this volume, we want to nevertheless acknowledge key sources of inspiration and perspiration.

A joint "thank you" to:

Jean Ait Belkhir and Christiane Charlemaine, editors of the journal *Race, Gender and Class*, for your courage in accepting a special edition on RGC and Global Climate Change and your collaboration with Phoebe Godfrey, the Guest Editor;

Rachel Hallum-Montes, Medani Bhandar, and Shangila Joshi for acting as co-editors at different points throughout this long journey;

Malaena Taylor, our Assistant Editor, for your eagle eye, organizational skills, and professionalism;

Louisa Earls, Margaret Farrelly, and Annabelle Harris at Routledge/Taylor & Francis for your guidance and structure; and

All our generous and patient contributors, both newcomers and those who have been with us since the inception.

From Phoebe:
I would also like to thank my wife, Tina Shirshac, for her unlimited support as an intellectual, creative, emotional, and spiritual companion. I would also like to thank my loving mother for always enthusiastically believing in my endeavors.

From Denise:
Thank you to my circle of support, those kin and fictive kin whose belief in and unwavering support of me have made a world of difference: the Barquet, Torres-Davila, Nerio-Mamayes, Zambello-Roiz, and Zambrana families. And, a special thank you to my son, James Elias Hamue, for being the air that I breathe, the fire that gets me up in the morning, the ground that steadies me, the water that refreshes me, and, truly, the chaos that makes it all worth it.

Introduction

Locating ourselves within the Anthropocene: applying intersectionality to anthropogenic climate change

Phoebe Godfrey and Denise Torres

> A human being is a part of the whole, called by us "Universe," a part limited in time and space. He experiences himself, his thoughts and feelings as something separated from the rest – a kind of optical delusion of his consciousness.
>
> (Einstein, 1951, in Caprice, 2005)

> Earth my body,
> Water my blood,
> Air my breath and
> Fire my spirit.
>
> (Chant, author unknown)

This volume is born out of our realization that, as our species enters the Anthropocene (denoting the present time in which many geologically significant conditions and processes are being shaped by human activities), we can no longer continue to destructively enact Western 'white' patriarchal capitalist society's (henceforth society, unless otherwise denoted) conceptualization of a clear boundary between the ideological and the material, the social and the environmental, the human and the non-human, the observer and the observed, the word and the flesh. Within these perceived oppositions there are no clear divides other than what our distorted thinking has divined. With the advent of anthropogenic Global Climate Change (GCC) and the concurring acts of ecological-social destruction, the vast conceptual veil perpetuating society's ultimate illusion that the fate of the human species is somehow separate from the fate of the Earth is finally tearing. Binaries and fossilized, immutable categories belie the fact that, as Einstein noted, we are part of a whole – one that is alive (Lovelock, 2000), permeable (Tuana, 2008), and continually co-constructed with and through space-time (Barad, 2007). Such a realization is, of course, not new; indigenous peoples, and non- and pre-Western cultures have long recognized the intra-action between the material and spiritual, the seen and unseen, and the felt and formulated long before Western physicists have come to 'discover' and 'confirm' it. Still, we maintain the 'optical delusion' even as the price of living in denial is paid by the least empowered, across all realms.

The indisputable existence of GCC (IPPC, 2007) "changes everything" (Klein, 2014), demanding that *we* change everything if we as a species desire, hence choose, to survive.

In an ongoing effort to project and protect the mistaken notion of human separation from, superiority to, and ownership of all life, society sacrifices a livable planet, and ultimately itself, echoing Freud's Todestrieb or death drive (also called Thanatos). Todestrieb, however, is in opposition to Eros, not just in the sense proposed by Freud as that of a survival instinct, but as conceived by Jung, in terms of being intrinsically bound by a "psychic relatedness" (Jung, 1982, p. 65). The transformative potential of the Thanatos–Eros tension is exemplified by the metamorphosis of the caterpillar into the butterfly. When the caterpillar creates the chrysalis, previously dormant imaginal cells begin to appear and multiply even as the caterpillar perceives these as foreign, a threat. Nevertheless, the imaginal cells persist and connect with each other, overwhelming the immune response of the caterpillar, ultimately dissolving it into fuel which they use to create a new form – completely different from the old one – a butterfly. While others have used this metaphor to describe the cultural changes underway and the forces of resistance (see Slater, 2008), we extend this further and emphasize that these two are intimately entangled (Barad, 2007), even when they appear as separate. The birth of the butterfly requires the death of the caterpillar.

In using 'born' to describe how and why this volume has come into being, we see the coming together of seemingly disparate ideas into a tangible whole that is greater than the sum of its constituent parts. As critical feminists, we seek to de-gender the term by recognizing that birth is the essence of creativity and thereby defines the universe. While we draw from Karen Barad's (2007) more recent insights that all 'things' "intra-act" and are "entangled," making it "impossible to differentiate in any absolute sense between creation and renewal, beginning and returning, continuity and discontinuity, here and there, past and future" (p. ix), other physicists have historically made these connections (Capra, 1976). Barad elaborates: "The world is an open process of mattering through which mattering itself acquires meaning and form through the realization of different agential possibilities" (2007, p. 141). In other words, things do not exist in and of themselves but constantly gain form and meaning, hence 'mattering,' in direct and dynamic intra-action with one another and that this process does not in fact have a designated beginning or end. As such, "there is no 'I' that is separate from the intra-active becoming of the world" (Barad, 2007, p. 394) and therefore "We are of the universe – there is no inside, no outside. There is only intra-acting from within as part of the world in its becoming" (p. 396). Such an intimate recognition where the essence of subjectivity is inseparable from that which forms the subject consequently makes our intra-actions more bound to the totality and therefore more responsible for the "world's vitality" and all "that might help us and it flourish" (p. 396).

Thus, like the imaginal cells in the chrysalis, we and contributors to this volume are attempting to build a model completely structurally different from the previous one, connecting to the many individuals globally who are working toward a new, more related form (Hawken, 2007). Indeed, a primary goal in birthing this volume is to add much-needed diversity in voices and views to the expanding social science research on GCC (Dunlap and Bullard, 2015), to highlight how our systemic social ideologies and their corresponding oppressive practices are in conflict with and continue to resist transformation. We conceptualize the existing Systemic Crisis of GCC herein addressed as the liquefying, indefensible caterpillar. In our second volume, *Emergent Possibilities for Global Sustainability* (Godfrey and Torres, 2016), we present the practices emerging from the goop of the current crisis, myriad unfolding butterflies.

Locating ourselves: intersectionality as holography

In holography, interacting beams of light create an "interference pattern," reflecting the information to be recorded on a plate, known as the hologram (Morgan, 1997, p. 75), with the object beam directed to the subject and the reference beam directed to the recording medium. The recording is then an encoding of these patterns of light – seemingly random variations in density and intensity on the surface of the medium. Remarkably, the hologram provides a more three-dimensional representation of the matter under consideration and any portion may be used to view and reconstruct the whole. That is, splitting the portraiture in half will give you two whole scenes of the hologram, although in smaller versions. This distributed nature of holograms has been used, therefore, as a metaphor to communicate the complexity and simultaneity of the brain and organizations (Morgan, 1997) as well as individual identity (Holvino, 2012).

Holograms, however, are snapshots in time and place, an interference pattern that is situated within a particular context and is only indistinguishable from the original when lit similarly. As rapidly shifting social constructions, we must find the correct key matched to the original light source to recapture its unique nature, to 'see.' We offer that intersectionality is *the* holographic process. As Vivian May states, it is "heuristic in nature," enabling users to gain insight into that which was previously obscured, one-dimensional, or nebulous (May, 2015, p. 19). From this perspective the emphasis is on what intersectionality "does or can do, not simply towards its definitional status as a noun" (May, 2015, p. 19). For us, *doing* intersectionality means embracing the dual foci of theory and praxis as mutually constitutive and necessary to creating the interference pattern required to both render a deeper, more complete image and revisit the scene that has been frozen in time and place. While similar terms have been used, such as Crenshaw's description of intersectionality as a "prism" (May, 2015, p. 19), or Collins' "matrix of domination," here hologram is most appropriate as we have attempted to engage intersectionality heuristically, throughout all aspects of the book.

Intersectionality theory: locating ourselves in the Anthropocene

Through our use of intersectional theory and analysis we have sought to not only engage with the core social constructions of identity represented by race, ethnicity, social class, gender, sexuality, and nationality; we also extend it to the imagined divide between the material and the social worlds as manifested by the continual emergence and creation of GCC. Taking our cue from Carbado (2013), we recognize that as a theory, intersectionality "Is never done, nor exhausted by its prior articulations and movement; it's always already an analysis-in-progress" (p. 304). Therefore, like others before us, we seek to move intersectionality "to engage an ever-widening range of experiences and structures of power," including calling for matter "to matter" (Barad, 2008, p. 120).

Given that intersectionality is not committed to particular "subjects nor to identities" but "to marking and mapping the production and contingency of both" (Carbado, 2013, p. 815) and to recognizing the act of doing so from given perspectives, we believe the application of intersectionality theory to GCC is essential in order to begin to fully conceptualize the complexity of how the social and natural worlds intersect. For us, intersectionality theory is the object or illumination beam focused on the systemic ideologies and the corresponding structural inequalities and oppressive practices undergirding GCC. Exposing and illuminating them to "reveal how power works in diffuse and differentiated ways through the creation and deployment of overlapping identity categories" (Cho et al., 2013, p. 797), both in relation to and intersecting with GCC.

More specifically, in building upon Barad (2007) and applying intersectionality theory to the social–material divide so that we explicitly include matter – in terms of the climate, environment, our physical bodies, and other beings – we recognize it as inseparable to experiential reality, and hence of any intersectional analysis. While Anna Kaijser and Annica Kronsell (2014) in their pioneering work on intersectionality and GCC propose that "questions such as 'How is nature represented?'" or "How are relations between humans and the environment portrayed?' [should] be addressed in any intersectional analysis of climate change" (p. 426), we argue that, regardless of the topic under examination, humans *are* nature and are inseparable from the environment and therefore such an intersection should ideally always be taken into account.

Lykke (2009) also makes a link between Barad's work and intersectionality in her call for greater attention to be given to the "human/'nature' or earth–other axis" (p. 39). In fact she goes so far as to state that the "human/earth–other axis has been suspiciously neglected in current feminist debates on intersectionality" (p. 39), representing "a problematic anthropocentrism" (p. 40). Ultimately, we recognize that all aspects of the social and the material are always intersecting regardless of what term is used with the options being many as in "entangled" (Barad, 2007), "porosity" (Tuana, 2008), "cosynthesis" (Kwan, 1997), "interconnectivity" (Valdes, 1995, p. 26), "multidimensionality" (Mutua, 2006, p. 370), and "assemblages" (Puar, 2007) (see Carbado, 2013, pp. 815–816) although, for us, intersectionality more than suffices.

Illuminating the intra-action between the social and material is vital given that the human body in particular, and the material world in general, has been so theoretically deconstructed to the point of apparent invisibility and forgetting. This is alarmingly ironic given that at no other time in human history has the body's welfare and the welfare of all life been in a more precarious state. Obviously, the two issues – the theoretical erasure of matter and the ecological state of the planet – are not separate in that the seduction of technology and the 'cyborg-self' allows for the illusionary dissolution of the material by blurring the lines between where technology can take our minds and where our bodies physically remain with their ineradicable need for air, water, and food. It is also for this reason that we assert the primacy of intersectionality in analyzing the complexity of GCC and the concurring acts of ecological-social destruction (Godfrey, 2012; Kaijser and Kronsell, 2014).

Indeed, our physical bodies serve not just as cultural canvasses upon which are writ identity categories, but have tangible physical responses to all that is around us. Other social theorists, feminists in particular, are conceptually engaging anew with matter, with nature, with physical bodies, even though it has apparently become necessary to assert that "Women *have* bodies; these bodies have pain as well as pleasure" (Alaimo and Hekman, 2008, p. 4, italics in original). Of course, from an intersectional perspective such a statement is essentially meaningless without unpacking all the ways in which 'women' are socially constructed and all the corresponding ways in which identity categories such as race, ethnicity, social class, gender, and sexual orientation imbue a body with differential meanings, hence experiences including those of pain and pleasure.

We are mindful of the ways in which theories and their desired corresponding praxes can lose their original intention as they evolve and travel (Said, 1983; see also Carbado, 2013) or just become a

> buzzword … reified into a formula merely to be mentioned, being largely stripped of the baggage of concretion, of context and history…. By just mentioning other "differences" besides "gender", the work to be done continues to be delegated to the respective "others."
>
> (Knapp, 2005, p. 255)

This is especially concerning as the initial intention to create progressive social change through a commitment to the ideals of democratic social justice, to take up "the social problems that most affected those most harmed by inequalities – poverty, poor education, substandard healthcare, inadequate housing, and violence – all became rethought through a lens of intersecting power relations of race, class, and gender" (Collins, 2009, p. viii), has instead "been systematically depoliticized" (Bilge, 2013, p. 405), made "colorblind" and "whitened" (Carbado, 2013).

We do not wish to flatten identity, nor intersectionality itself. Neither do we seek to take away from the importance of intersectionality's genealogy as we recognize Crenshaw's (1989, 1991) initial articulation of the term to demonstrate how the founding social categories of race, gender, and social class intersect, are consequently transformed, and are, therefore, in fact, inseparable. We wish to honor intersectionality's lenses as being ground by the struggle in body and mind of a "broader women's movement where Chicanas and other Latinas, native women and Asian women ... [were] at the forefront of raising claims about the interconnectedness of race, class, gender and sexuality in their everyday lived experience" (Collins, 2011, p. 91) by including the physical places and spaces their bodies inhabit as salient for analysis.

By including our environments as a 'category' or 'phenomena' for consideration – in calling for matter to matter – and explicitly linking to GCC works, we seek to add to intersectionality's interdisciplinary application and push "the theoretical boundaries" (Carbado, 2013, p. 841). We are, as Carbado (2013) proposed in building off Said's work, encouraging "movement [that] might radicalize and reinvigorate" while maintaining connection to the "initial articulation" (p. 812). The disparities, disproportionalities, and vulnerabilities we experience – those seemingly random variations between and among us – are, we offer, the holographic imprints and encoding generated by macro-systemic forces that include the physical world. Nevertheless, as with any theoretical lens, what is looked at is always partial, situated from a particular social position, hence perspective. These perspectives, as Peter Grzanka (2014) says of Haraway's situated knowledges (Haraway, 1988), "are not disinterested: they are knowledges made for doing something-truths with a purpose" (Grzanka, 2014, p. xxiii). Our purpose here is to address the creation and maintenance of inequalities as they intersect with the production and perpetuation of GCC.

Intersectionality praxis: recording the systemic crisis

This volume and its companion are acts of political intervention and intersectional praxis. Carbado (2013) describes intersectionality as "a method and a disposition, a heuristic and analytic tool" and that most fundamentally "intersectionality is what intersectionality does" (p. 312). Alternately stated, intersectionality is not just a theory, it is praxis: at once "a *practice* and a *political intervention*" (Luft and Ward, 2009, p. 10, italics in original). In a loosely holographic manner, we have attempted to be 'what intersectionality does' by bringing the most compelling issue humanity has ever created and resultantly encountered into the heart of intersectional analysis: global climate change.

There is no longer any doubt that the global climate is changing and that this change is as a result of human activity, and that the warming of the planet is occurring at a much faster rate than has previously been predicted or than what has naturally occurred over the last 1000 years (Bauer and Beckman, 2015). From the Pope, to the United Nations, to the Intergovernmental Panel on Climate Change (IPCC), to world leaders, to new scientific studies, to news headlines, to citizen movements, there is a growing global awareness and discussion that GCC, as well as all the destructive activities that contribute to and result from its creation such as the burning of fossil fuel, deforestation, industrial agriculture, ocean pollution, ecosystem destruction, species extinction, and high rates of consumption unevenly distributed, must be addressed (McNall, 2011).

Yet, other than mostly talk, the only apparent changes are those that are increasingly criminalizing dissent by attempting to "silence any who resist" (Hedges, 2014, p. 2) society's profit-driven agenda, whether that resistance be directed specifically at GCC as in the action taken by environmentalist Tim deChristopher, founder of Peaceful Uprising, or whether it be directed at structural racism and police brutality as in #BlackLivesMatter. These examples are emblematic of the systemic crisis of the Anthropocene. We recognize that progressive change will not happen without a struggle that is in essence a "battle for ideas as well as a battle for power" (Hedges, 2014, p. 2). As Marx observed, the production of ideas is inseparable from the economic means of production that manifests as social power and control. Society's narrative separation between the social and natural worlds – what Klein (2014) refers to as our "cognitive dissonance" (p. 2) and Einstein called our "delusion of consciousness" – persists despite mounting evidence. Indeed, we as a society not only continue in the same manner but are doing so with increased gusto given that a capitalist economy demands ever-increasing growth.

In the midst of the metamorphic goop and the battle between the dying caterpillar and awakening imaginal cells of the butterfly, we birth this volume. We focus on the 'systemic crisis of GCC' and the intersectional ways in which GCC is being caused, addressed, denied, understood, and experienced on the macro levels, as well as on the micro levels, and how such 'situated knowledges' can ultimately be applied to the proliferation of imaginal cells. From its very conception, our purpose has been to increase knowledge, understanding, and insight into the topics at hand as well as to inspire awareness of, engagement with, and commitment to social justice projects and practices around the world in particular as they intersect with issues of GCC.

Patricia Hill Collins, social theorist and contributing developer of intersectionality, observes that intersectional analysis "is both staunchly interdisciplinary and committed to claiming the much-neglected space of praxis" (Collins, 2009, p. vii). Thus, a core concern for us as editors was to gather voices from across countries, disciplines, ideologies, cosmologies, and myriad socially constructed identities in what has been termed "anthology-making" (Keating, 2002, p. 9) and "spiritual activism" (p. 19), seeing the volumes as simultaneously "an alliance making tool" (p. 16) and mechanisms for building a "transformative, coalitional consciousness" (p. 6) among dispersed imaginal cells.

As the recording medium upon which we have captured the subject, these volumes intentionally link the ideological, hence the socially constructed worlds, with the material. The elemental segmentation of this volume into Chaos, Air, Earth, Fire, and Water represents this connection, echoed by the Pagan chant "Earth my body, Water my blood, Air my breath, and Fire my Spirit." Furthermore, rather than place ideas or contributions in hierarchal relationship to one another and create chapter 'ghettos' segregating people of color, the disabled/unwell, the poor or other 'vulnerable groups' into special sections, these are distributed throughout. It is an explicit choice to demonstrate the material world's intra-action with the social world, and to confront and confound the taxonomic dichotomies typically used to grapple with GCC. Drawing from our holographic metaphor, we have explicitly attempted to capture the interference patterns we experience in a manner that problematizes statuses and their relationship to each other. As a political intervention we embrace the fact that "there is no hierarchy of oppressions" (Lorde, 1983, p. 9) and our use of the elements may be said to reflect an anti-categorical approach to intersectionality (McCall, 2005).

Similarly, our desire to inspire social justice action and intervention meant employing an "intracategorical" approach whereby contributors attempted to illuminate the experiences and perspectives of "a single category at a neglected point of intersection" (McCall, 2005, p. 1780). Our decision to include voices and perspectives not normally seen in either a social science volume nor in works addressing GCC reflects our recognition that it is through a more holistic, embodied engagement with ourselves, each other, and our world that we will see more clearly how our actions defy both democratic social justice and a livable planet. As well, such inclusion better positions us to imagine what we might do collectively and in solidarity to create authentic radical change.

As a truly interdisciplinary work, contributors approach GCC from various standpoints – as artists, activists, cartoonists, dancers, educators, journalists, playwrights, poets, researchers, and survivors – and contributors necessarily had varying comfort with explicating climate injustice intersectionally. This variability re-creates tensions inherent in *doing* intersectionality: how many categories should one examine and does increasing the number to the point of using 'et cetera' diminish the efficacy of the theory or the ability to engage in a coherent analysis (Cho et al., 2013, p. 787; Chang and Culp, 2002)? It has been noted that engaging multiple categories simultaneously is complex (McCall, 2005) and must reflect the issues under consideration, including their potential prominence in our individual lives and our potential 'blindness' (Carbado, 2013). While there are differences within and across the contributions in terms of how each has chosen to apply intersectionality and how intricately the matrices of oppression and privilege have been unpacked and thickly described, our goal as editors has been to ensure that all pieces are in dynamic relationship with the elements as an indication of their presence.

In addition, we wanted to include topics that were not only important but that have been given inadequate attention elsewhere or that had been given attention in a way that was not as accessible as it could have been. Hence, an explicit criterion for inclusion was whether an intersectional analysis was directly engaged or at least engaged in ways that we would both want to teach and that we felt students and others would want to read. Indeed, as the project developed, our mantra to contributors who wrote pieces became 'tell a story,' as in 'take your reader by the hand' in terms of 'intersecting with them' as opposed to 'writing at them.' This emphasis sought to further invite contributors to explore their own intersections and to do so in a way that was transparent and engaging. Such use of reflexive praxis helped us as editors and the contributors better understand how political-institutional forces and history give meaning to race–gender–class and other axes of identity, difference, and power and, therefore, how they are understood, perceived, and performed (Yuval-Davis, 2014; Knapp, 2005). That is, to actually do intersectionality in ways that are less depoliticized and blind.

In our praxis, we wrestled with intersectionality in terms of how our identities have intersected and how they have intersected with those of our contributors, as well as to practicing what we have called *contemplative editing*, seeking to gently 'intra-act' our editorial voices with the voices of others. In each case the goal has been to bring together, hence intersect, that which has previously been conceptualized and treated as separate in order to allow for seemingly mutually beneficial transformations. However, our focus here has not only been on the individual pieces of the volume but also on the whole and the ways in which the parts intersect with each other metaphorically, creating in the aggregate the holographic level of intersectional analysis we have sought. This level could be categorized as methodologically representing "intercategorical complexity" (McCall, 2005, p. 1773) in that as a book we have had to use categories, chapters, and sections as organizational tools but we have still attempted to recognize their relational engagement even as the contributors themselves use a range of methodologies or at the very least "adopt an intersectional way of thinking about the problem of sameness and difference and its relations to power" (Cho et al., 2013, p. 795).

The difficulties of doing intersectionality even when there is an explicit commitment to environmental justice – of examining the whole and its constituent parts simultaneously – is illustrated in a statement from the documentary *Fierce Green Fire* (Kitchell, 2013). In it, sociologist and environmental justice advocate Robert Bullard appeals to the universality of air by stating: "There is no Hispanic air. There is no African-American air. There's air!" (see www.afiercegreenfire.com/about.html). As one of our sectional elements, air represents attitudes, philosophy, pedagogy, resilience, and theory, and the ways in which these abstract concepts shape the larger discussions of GCC and how they intersect with the social constructs of race, social class, gender, and other identities in very material ways. Although, in the abstract, Bullard is correct in that there is only air as a universal, when we locate it materially, in time, place, and history, and examine the situated differences, all air is not equal. In point of fact, from a public health perspective as well as one of environmental justice – as Bullard himself has ironically shown in conceptualizing environmental racism (Bullard, 1993, 1994) – we may in fact experience 'Hispanic,' 'African-American,' or even 'Los Angles' air, just as internationally daily reports are given on Mexico City and Beijing air, in terms of air quality and the exposure to pollution. In the USA, then, these embodied experiences may result in higher rates of asthma for African-Americans and Latinos whose racialization intersects with social class, which influences housing access. Although one could add the category of age, as children are more likely to suffer from asthma than adults (hence poor, African-American, and Hispanic children suffer the highest rates along with all the other contributing and resulting consequences), we would be doing nothing new in terms of applying an adequate intersectional analysis, even as we would still be exceeding the complexity of the Centers for Disease Control and Prevention statistics which list each category separately. Thus, the use of intersectionality and the intra-actional ideal we sought addresses the tension between the materiality of air by unpacking the 'baggage of concretion' that engenders disparities.

Where we have attempted in the structure of the volume to break new ground is that we see air, as in the material realm, as an intersecting agent that should in this case be highlighted and analyzed as part of a comprehensive intersectional analysis. Air can no longer be seen as apart from the social realm in the Anthropocene, as it denotes the current geological age wherein human activity is the dominant force in 'the environment.' To fully analyze asthma or any study of human well-being, air and place must be taken into account as an intersection that is negatively transformed by the social, as in this case racism and social class inequality, just as water and place would be salient if one were doing an analysis on water-borne illnesses in developing countries. Similarly, race is always present even when it acts as a privilege, as with 'whiteness,' and is therefore often not seen as necessary to a given analysis even though it should be. Such an obscuring of racial privilege is referred to by Carbado (2013) as "colorblind intersectionality" (p. 817), just as she also identifies "Gender-blind intersectionality" (p. 817), meaning that the gender of men may not be seen as influential in a given analysis when in contrast the gender of women would be. To this concept we would like to add that of 'nature-blind intersectionality,' which includes the physical body, in that bodies as well as air and water are always present in any analysis even though they may only be highlighted when connected with negative human consequences. This point links back to the discussion mentioned above as to how many intersections may or should be used, and which ones. In our role as editors we have aimed to have contributors engage with at least three – if not directly, then in terms of their thinking (this applies in particular to our artists and poets) – but which three and how they have framed their intersectional analysis in relation to GCC and the concurring acts of ecological-social destruction we have left up to them.

Concluding thoughts

While in *Systemic Crises of Global Climate Change: Intersections of race, class and gender* we have focused on the problems, and in *Emergent Possibilities for Sustainability: Intersections of race, class and gender* (2016) we focus on solutions, we nevertheless recognize that these are not separate and are, not surprisingly interdependent. Like the old man in the Preface who survived falling into the river by becoming one with it, if we are to survive these current systemic crises then we must seek harmony with the whole web of life in a manner that ensures balance and union. That said, we struggled with trying to make clear divisions between the two volumes, while accepting that this division is dissonant. Ultimately, for us, inequality and injustice are the problems, and hence equality and justice are the solutions. Yet, as history has shown, and we hope to further illuminate, these ideals have remained unrealized and so the time has come to, as the name of the current Native American environmental movement states, 'Idle No More'!

References

Alaimo, S. and Hekman, S. eds, 2008. *Material Feminisms*. Bloomington: Indiana University Press.

Barad, K., 2007. *Meeting the Universe Halfway: Quantum physics and the entanglement of matter and meaning*. Durham, NC: Duke University Press.

Barad, K., 2008. Posthumanist performativity: Toward an understanding of how matter comes to matter. In *Material Feminisms*, ed. S. Alaimo and S. Hekman. Bloomington: Indiana University Press, pp. 120–156.

Bauer, S. and Beckman, M., 2015. The climate is starting to change faster. Pacific Northwest National Laboratory, US Department of Energy. Available at: www.pnnl.gov/news/release.aspx?id=4186 (accessed August 30, 2015).

Bilge, S., 2013. Intersectionality undone: Saving intersectionality from feminist intersectionlity studies. *Du Bois Review*, 10, 2, pp. 405–424.

Bullard, R.D. ed., 1993. *Confronting Environmental Racism: Voices from the grassroots*. Boston, MA: South End Press.

Bullard, R.D., 1994. *Dumping in Dixie: Race, class and environmental quality*. Boulder, CO: Westview Press.

Capra, F., 1976. *The Tao of Physics*. New York: Bantam Books.

Caprice, A., 2005. *The New Quotable Einstein*. Princeton, NJ: Princeton University Press.

Carbado, D.W., 2013. Colorblind intersectionality. *Signs: Journal of Women in Culture and Society* 38, 4, pp. 811–845.

Centers for Disease Control and Prevention. Available at: www.cdc.gov/nchs/data/databriefs/db94.htm (accessed August 23, 2015).

Chang, R.S. and Culp, J.M., 2002. After intersectionality. *University of Missouri-Kansas City Law Review*, 71, pp. 485–491.

Cho, S., Crenshaw, K., and McCall, L., 2013. Toward a field of intersectionality studies: Theory, application and praxis. *Signs: Journal of Culture and Society*, 38, 4, pp. 785–810.

Collins, P.H., 2000. *Black Feminist Thought: Knowledge, consciousness, and the politics of empowerment*, 2nd edn. New York: Routledge.

Collins, P.H., 2009. Foreword: merging intersections – Building knowledge and transforming institutions. In *Emerging Intersections: Race, class, and gender in theory, policy, and practice*, ed. Bonnie T. Dill and Ruth E. Zambrana. Piscataway, NJ: Rutgers University Press, pp. vii–xiv.

Collins, P.H., 2011. Piecing together a genealogical puzzle: Intersectionality and American pragmatism. *European Journal of Pragmatism and American Philosophy*, 3, 2, pp. 88–112.

Crenshaw, K., 1989. Demarginalizing the intersection of race and sex: A black feminist critique of antidiscrimination doctrine, feminist theory, and antiracist politics. *University of Chicago Legal Forum*, 139.

Crenshaw, K., 1991. Mapping the margins: Intersectionality, identity politics, and violence against women of color. *Stanford Law Review* 43, 6, pp. 1241–1299.

Dunlap, R. and Bullard, R., 2015. *Climate Change and Society: Sociological perspectives.* New York: Oxford University Press.

Godfrey, P., 2012. Introduction: Special issue on race, gender and climate change. *Race, Gender and Class*, 19, 1–2, pp. 3–8.

Grzanka, P.G. ed., 2014. *Intersectionality: A foundations and frontiers reader.* Boulder, CO: Westview Press.

Haraway, D., 1988. Situated knowledges: The science question in feminism. *Feminist studies*, 14, 4, pp. 579–599.

Hawken, P., 2007. *Blessed Unrest: How the largest social movement in history is restoring grace, justice and beauty to the world.* New York: Penguin Books.

Hedges, C., 2014. The last gasp of climate change liberals. truthdig, August 13. Available at: www.truthdig.com/report/item/the_last_gasp_of_climate_change_liberals _20140831 (accessed August 23, 2015).

Holvino, E., 2012. The "simultaneity" of identities: Models and skills for the twenty-first century. In *New Perspectives on Racial Identity Development: Integrating emerging frameworks*, ed. Charmaine Wijeyesinghe and Bailey W. Jackson. New York: New York University Press, pp. 161–191.

IPCC, Climate Change, 2007. *Synthesis Report.* Available at: www.ipcc.ch/publications_ and_data/ar4/syr/en/spms2.html (accessed August 20, 2015).

Jung, C., 1982. *Aspects of the Feminine.* Princeton, NJ: Princeton University Press.

Kaijser, A. and Kronsell, A., 2014. Climate change through the lens of intersectionality. *Environmental Politics*, 23, 3, pp. 417–433.

Keating, A., 2002. Charting pathways, marking thresholds…. A warning, an introduction. In *This Bridge We Call Home: Radical visions for transformation*, ed. G.E. Anzaldúa and A. Keating. New York: Routledge, pp. 6–20.

Kitchell, M., 2013. *Fierce Green Fire.* Bullfrog Films.

Klein, N., 2014. *This Changes Everything: Capitalism vs. the climate.* New York: Simon & Schuster.

Knapp, G.A., 2005. Race, class, gender: Reclaiming baggage in fast travelling theories. *European Journal of Women's Studies*, 12, 3, pp. 249–265.

Kwan, P., 1997. Jeffery Dahmer and the cosynthesis of categories. *Hastings Law Journal*, 48, 6, pp. 1257–1292.

Lorde, A., 1983. There is no hierarchy of oppressions. *Bulletin: Homophobia and Education* 14, 3/4, p. 9.

Lovelock, J., 2000. *Gaia: A new look at life on Earth.* London: Oxford University Press.

Luft, R.E. and Ward, J., 2009. Toward an intersectionality just out of reach: Confronting challenges to intersectional practice. *Advances in Gender Research*, 13, pp. 9–37.

Lykke, N., 2009. Non-innocent intersections of feminism and environmentalism. *Women, Gender and Research*, 18, 3–4, pp. 36–44.

McCall, L., 2005. The complexity of intersectionality. *Signs*, 30, pp. 1771–1800.

McNall, S., 2011. *Rapid Climate Change: Causes, consequences and solutions.* New York: Routledge.

May, V., 2015. *Pursuing Intersectionality, Unsettling Dominant Imaginaries.* New York: Routledge.

Mutua, A.D., ed., 2006. The rise, development and future directions of critical race theory and related scholarship. *Denver University Law Review*, 84, 2, pp. 329–394.

Morgan, G. 1997. *Images of Organization.* Thousand Oaks, CA: Sage.

Nash, J.C., 2008. Re-thinking intersectionality. *Feminist Review*, 89, pp. 1–15.

Puar, J.K., 2007. *Terrorist Assemblages: Homonationalism in queer times*. Durham, NC: Duke University Press.

Said, E.W., 1983. *The World, the Text, and the Critic*. Cambridge, MA: Harvard University Press.

Said, E.W., 2000. *"Reflections on Exile" and Other Essays*. Cambridge, MA: Harvard University Press.

Slater, P.E., 2008. *The Chrysalis Effect: The metamorphosis of global culture*. London: Sussex Academic Press.

Tuana, N., 2008. Viscous porosity: Witnessing Katrina. In *Material Feminisms*, ed. S. Alaimo and S. Hekman. Bloomington: Indiana University Press, pp. 188–213.

Valdes, F., 1995. Sex and race in queer legal culture: Ruminations on identities and inter-connectivities. *Southern California Review of Law and Women's Studies*, 5, 1, pp. 25–74.

Yuval-Davis, N., 2014. Dialogical epistemology – An intersectional resistance to the "oppression olympics." *Gender and Society*, 26, 1, pp. 45–54.

Part I

Chaos

Figure I.1 Chaos by Tina Shirshac.

1 Worlds turning; worlds colliding?

Phoebe Godfrey and Denise Torres

In Greek mythology, Chaos is the formless, primal void from which was born all existence, including Gaia, the personification of Earth, and the other primordial deities. Meaning 'yawning' or 'gap,' Chaos is thought to have referred to the original separation between Heaven and Earth, whereas our modern use of the word – to connote confusion and disorder – derives from the Elizabethan era. We play on both the Greek and English meanings in Part I, delivering our first selections birthed from the confusion and disorder of global climate change (GCC) and emerging from the perceived 'gap' between Heaven (the realm of society, hence ideology) and Earth (the realm of the physical and natural world).

Having developed as a means to analyze the matrixes of domination and privilege, intersectionality provides ample opportunity to examine gaps and intersections across the micro, meso, or macro levels. Thus, we explore the ways in which not all identities are internally or externally harmonious or integrated, and in fact can and do create conflicts and discords in relation to "within-group differences and inequalities" as well as "between-group power asymmetries" (May, 2015, p. 4). In fact, we see GCC as a perfect example of how these can collide in ways that are potentially extreme, disruptive, and destructive. The question as to whether or not 'chaos' is interpreted as 'positive' or 'negative' of course depends on how it is being interpreted and in relation to whom and what.

To illustrate the complexity and temporal quality of an intersectional analysis, David Jackson's 'Pulled from all angles … with strings attached' places a pregnant young woman of color in motion amid the chaos of the social and natural worlds. Jackson's dynamic image may be said to epitomize what the Anthropocene looks like: an age of human dominion even as most struggle with a multiplicity of simultaneous demands. In Jane Caputi's piece, 'Mother Earth meets the Anthropocene: an intersectional ecofeminist analysis,' she wittingly applies cutting feminist critical analysis to explore and unpack how terms such as 'the Anthropocene,' 'Mother Earth,' and GCC are used and abused, by whom, and for what purposes. Spoken word poet Devin Samuels spins an imaginative dialogue between 'Mother Earth' and her male child who, as the architect of the Anthropocene, boastingly declares his independence from and domination of the Earth. This satirical hubris gains empirical validity in Julie Bacon's piece, 'The Rush Limbaugh Show and the expanding the culture war: whiteness, masculinity, and conservative media denials of climate change and sexism,' as she applies an intersectional analysis to climate change denial in the conservative media. Phoebe Godfrey reflects on recent events to highlight the concept of *hegemonic masculinity* and offers a provocative visual interpretation. Coco Gordon's poem, 'Supersky Woman,' invites critical reflection as to what exactly we are doing and why, with a still from Steve Cutts' animated work, MAN, capturing the maniacal delight in the cacophony and chaos of his actions. Jade Sasser's 'Population, climate change, and the embodiment of environmental crisis' takes on the popular racist and sexist – hence, colonial – view that overpopulation, in particular on the part of those in the underdeveloped world, is responsible for GCC, and calls for an embodied and intersectional engagement with reproductive rights. Finally, to close the section and fill the 'gap' between Heaven and Earth is Antonia Darder's poem and contextualizing thoughts in 'The Great Mother Wails.'

Reference

May, V., 2015. *Pursuing Intersectionality, Unsettling Dominant Imaginaries*. New York: Routledge.

2 Pulled from all angles ... with strings attached

David C. Jackson

Figure 2.1 Pulled from all angles by David Jackson, oil and acrylic on canvas, 4'×5'. This is a literal, mixed-media painting examining the dualities of humans, how we're labeled in our environments, and the consequences in occupying space in these environments. It's an exploration on how we build and destroy; physically, emotionally, and spiritually.

3 Mother Earth meets the Anthropocene

An intersectional ecofeminist analysis

Jane Caputi

> Enter the Anthropocene – Age of Man. It's a new name for a new geologic epoch – one defined by our own massive impact on the planet.
>
> (*National Geographic*, Kolbert, 2011)

> [T]he old people laugh when they hear talk about the 'desecration' of the Earth. Because humankind, they know, is nothing in comparison to the earth. Blast it open, dig it up, or cook it with nuclear explosions: the Earth remains. Human desecrate only themselves. The Earth is inviolate.
>
> (Leslie Marmon Silko (Laguna), 1989, p. 124)

At the People's Climate March, on September 21, 2014 in New York City, the iconic figure of Mother Earth, probably humanity's oldest religious (Roach, 2005, p. 1107) – and scientific and philosophical – idea was everywhere. One poster, created for the event by Favianna Rodriguez, depicted a brown-skinned, child-carrying woman, her face superimposed over a radiant Earth. "*Defienda Nuestra Madre*," its words demanded – Defend Our Mother. Marchers carried signs emblazoned with a blue-green, black-haired woman, the Earth at her womb level, and issuing this warning: "Don't Mess with Our Mama: Defend Gaia." One group carried a large puppet, a dark-skinned grandmother in traditional peasant dress, with accompanying signs reiterating the call: "*Defienda a nuestra Madre Tierra*" (Defend our Mother Earth).

This defense is particularly imperative, as the Earth, according to prominent European and US spokespersons, has entered into a new geological era, the *Anthropocene*. This term has been proposed since the 1980s as the most accurate designation of an unprecedented time when "human activity" has become a "globally potent biogeophysical force" (Revkin, 2011, n.p.) and "humans" have become capable of "overwhelming the Great Forces of Nature" (Steffen et al., 2007, p. 614).

My interest here is in the meeting of these two gendered, racialized, sexualized, and classed terms, one ancient and vernacular, the other new and issued from on high. But is the Anthropocene really so new? I think not. Rather, it extends a centuries-long Western paradigm, one structured, to put it crudely, along the lines of 'The White Man fucks the Dark Mother.' 'The White Man' does not mean all white men or even only men, but is a metonymy for what is known as 'civilization' and 'rationality,' with the 'Dark Mother' standing as and for what is excluded and devalued: "the emotions, the body, the passions, animality, necessity" (Plumwood, 1993, p. 19), and for what is said to be matter without spirit. Matter thus becomes an object, something inferior to be mastered, exploited, made to serve, and manipulated. Nature, thus defined, includes not only dis-spirited matter and the entire non-human world, but all those humans deemed inferior (sexually, racially, bodily) because they are supposedly closer to an inferiorized 'nature' (Griffin, 1989).

This civilization's objectification of Mother Earth/Nature makes possible the devastation of the Anthropocene. My plan here is to critique this and related assumptions undergirding the Anthropocene, while also arguing for that defense of Mother Earth. The name *Mother Earth*, according to the *Preamble to the International Declaration of Mother Earth Rights* (Rights of Mother Earth, 2010, n.p.), means an "indivisible, living community of interrelated and interdependent beings with a common destiny." The name Mother Earth further signifies the Earth/Nature as a force including humans, but as Silko (1989) attests, ultimately inviolate and upon which humankind utterly depends. Defense of Mother Earth then is not only defense of the planet; it is defense of ourselves.

The 'age of *humans*'?

The designation of the Anthropocene may seem helpful in countering ongoing denial that human activities have led to climate change. Still, the designation is also problematic, as it implies that *all* humans are doing this and, moreover, that this kind of behavior is intrinsic to human nature. But the 'humans' behind global warming and concomitant environmental damages are more or less privileged individuals (e.g., those commanding the world's most wealthy and powerful militaries, governments, families, classes, and corporations), and the affluent, who benefit in the short term from the devastation, while able to distance themselves from the damage (a privilege unavailable, for example, to climate refugees in the Arctic and the South Pacific and to poor people living in contaminated neighborhoods). Those living in affluence contribute through everyday acts simply contingent upon living in a consumer culture (e.g., driving, shopping, investing, eating industrial grains, vegetables, and meat, or flipping a switch).

Nobel laureate Paul Crutzen is the scientist most responsible for promoting adoption of the Anthropocene. He writes with journalist Christian Schwägerl (2011) of the need for a "new global ethos," which includes some necessary social changes: an end to hyper-consumption, factory farming of animals, and the use of private vehicles. However, they do not simultaneously demand equal access to food, health, space, air, and water. They do not call for international human rights, including women's rights. Nor do they recognize the need for what a global, indigenous-led movement calls *The Universal Declaration of the Rights of Mother Earth*. This document was adopted in 2010 at the World Peoples Conference on Climate Change and the Rights of Mother Earth in Bolivia

(Rights of Mother Earth, 2010). Its principles, based in an older, indigenous, and still vital ecological ethos, abjure mastery of the Earth, recognizing that: the Earth is alive and purposeful; all life, including human life, is dependent on the Earth; all humans are equal; humans are not exceptional or superior, but one life form among many intrinsically interrelated ones that form a community. Moreover, this ecological ethos affirms that human rights cannot be achieved without a simultaneous attainment of Mother Earth Rights. For those holding to Mother Earth Rights, Crutzen and Schwägerl's (2011) ethos only reinforces the Western, ecologically destructive paradigm. This becomes clear in statements such as the following:

> A long-held religious and philosophical idea – humans as the masters of planet Earth – has turned into a stark reality. It's no longer us against 'Nature.' Instead, it's we who decide what nature is and what it will be.... We should shift our mission from crusade to management, so we can steer nature's course symbiotically instead of enslaving the formerly natural world.
>
> (Crutzen and Schwägerl, 2011, n.p.)

This is corporate-speak – 'steering' and 'managing' – rather than plantation-speak – 'mastering' and 'enslaving' – but the underlying framework of domination remains unchanged. Moreover, their false universalization of terms such as 'human' and 'we' is what political scientist and feminist theorist Iris Marion Young (1990, pp. 58–61) refers to as *cultural imperialism*, where the dominant group universalizes its subjectivity, its experiences, and its worldview as the norm, while rendering everyone else's as deviant, or inferior, or simply non-existent. This contrived notion of 'humans' also functions to obscure not only who is responsible for the damage, but also who is best able to provide practical, conceptual, and spiritual/ethical leadership in solving the problem.

Best-selling US science writer Diana Ackerman's (2014) *The Human Age: The world shaped by us* is riddled with cultural imperialism. Ackerman adopts an optimistic view, impressed with the ways that 'humans' have proved so skilled at what she calls "making the planet our sandbox" (p. 11), even while acknowledging some resulting environmental scars. This juvenile trope continues as Ackerman minimizes all this damage as the result of a phase of 'human' childhood, where "without really meaning to, we have nearly emptied the world's pantry, left all the taps running, torn the furniture, strewn our old toys where they're becoming a menace, polluted and spilled and generally messed up our planetary home" (p. 307). Not only is this metaphor inaccurate – some of the most ancient human cultures have lived sustainably for millennia – but, again, it is not everyone doing all this scarring. In his review of her work, Nixon (2014, n.p.) gets right to the heart of the matter:

> In 2013, the world's 85 wealthiest individuals had a net worth equal to that of our planet's 3.5 billion poorest people. Since 1751, a mere 90 corporations, primarily oil and coal companies, have generated two-thirds of

humanity's CO_2 emissions. That's a serious concentration of earth-altering power.

Ackerman profiles a number of key thinkers and practitioners (all appear to be white, educated, and affluent men) and the beneficial emerging technologies they speak for, including synthetic biology, 3-D printing, and nanotechnology. On the latter, she quotes scientist, inventor, and post-human philosopher Ray Kurzweil, who predicts 'human' immortality and avows that "by the year 2030 we'll be putting millions of nanobots inside our bodies to augment our immune system, to basically wipe out disease" (Ackerman, 2014, p. 181). Even if this were possible, do any of us imagine that this will be accessible to any but the most privileged? Not much thought seems to be given to the capacity of the Earth to sustain a population of human immortals, nor to the ongoing industrial poisoning of earth, air, and water.

Most Anthropocene promoters accept that humanity as a species is exceptional, existing somehow outside of, over, and above 'nature,' and, moreover, that eco-destructive activity is something innately human. But the truth is that many human cultures have not done these types of things. Or, if they have, they have either wiped themselves out or, learning from their mistakes, developed ecological knowledge systems and life-ways (Sanchez, 1989). The damages marking the Anthropocene proceed not from human nature but from an imperial logic and practice, established over several millennia of hierarchical and patriarchal culture (Lerner, 1986), where the top tier wield powers of both domination and definition, naming themselves 'culture,' 'normalcy,' and 'progress,' while embarking on a mission to overwhelm 'savagery,' 'deviance,' and 'backwardness.' And they are not admitting any mistakes. Quite the contrary.

The 'mother-fucking paradigm'

Crutzen has declared: "What I hope is that the term 'Anthropocene' will be a warning to the world" (Kolbert, 2011, n.p.). But can the Anthropocene serve credibly as a warning? Or, as key words used by its proponents – 'potent,' 'overwhelming' – suggest, does it actually function as a boast, and a phallic one at that?

Consider some iconic imagery illustrating the February 2008 cover of *GSA Today* (Geological Society of America) and posing the question: "Are We Now Living in the Anthropocene?" Two skyscrapers dominate the built environment of the megacity of Shanghai. The same iconography – a tower dwarfing Dubai – appears in an article in *National Geographic*, "Enter the Anthropocene – Age of Man." These photos recall a *New Yorker* cartoon (Mankoff, 1999) depicting two white, male executives looking out of their office window and beholding the Empire State Building. Written down the face of the iconic skyscraper is the word VIAGRA. One man quips: "Now that's product placement."

Everyone gets the joke because skyscrapers, along with weapons, scepters, and other signs of power as supremacy, are well understood as phallic symbols.

Phallic symbols don't really have much to do with the biological penis, which is attached, embodied, fertile, and far more flower than tower. The detached, permanently hard, penetrating (but impenetrable), and implicitly white phallus is the emblem of white, male superiority and "the values of 'civilization' rather than 'nature'" (Bordo, 1999, p. 89).

Bordo is referencing and critiquing the Western worldview that severs 'nature' from what it then classifies as 'culture' or 'civilization.' This split is based in a type of binary thinking (Collins, 2000) that splits wholes and constructs two oppositional distortions placed in a hierarchy (e.g., mind over body, male over female, light over dark, life over death, high over low, normal over freak, clean over dirty). The qualities said to be inferior then are projected onto an 'other,' with the paradigmatic 'other' being Nature (the Dark M/Other) as well as all humans who become aligned with her side of things. For example, black men, in white supremacist stereotypes, are defined as having super-sized, animalistic penises, a stamp not of the human, mind, and 'culture,' but of the beast, of mindlessness, and of 'nature.'

Within this culture/nature binary, women *qua* women have long been defined in patriarchal Western cultures as inferior, more nature than culture with variations in proximity to nature ordained by intersections of such factors as race, ability, sexuality, class, and age. Heterosexually identified, feminized, 'clean,' and chaste white women gain status via association with privileged white men and their institutions, while 'othered' women (e.g., women of color, poor women, sexually autonomous women, prostitutes, lesbians, disabled and/or aged women) are cast as 'dirty,' contaminating, angry, mad, threatening, and the lowest of 'the low.'

This oppressive dynamic is not only steeped in normative gender, but also in sexualized violence. An intersectional feminist critique recognizes what is basically a sex–murder dynamic underlying the practices and ideologies of the Anthropocene, which one hip hop artist André 3000 (2003) characterizes pithily as *motherfucking*: "And when I say *motherfucker* I do mean *motherfucker/* Because Mother Earth is dying and we continue to fuck her to death." In the phallic universe, *fuck* literally means not only "to engage in heterosexual intercourse," but also "to harm irreparably; finish; victimize" (Sheidlower, 1999, p. 124). When Crutzen and others write about 'Man' being able to 'overwhelm' Nature, they may not be actually saying *motherfucking*, but they may as well be.

This dynamic of sexualized and gendered conquest of 'nature' informed the European genocide of indigenous peoples (Smith, 2005); the European and American slave trade (Collins, 2000; Farajaje-Jones, 2000); and the European Witch Craze (Merchant, 1980; Sjöö and Mor, 1991). In each of these cases, violators cast victims in gendered, sexual, dark, and obscene terms: female "witches" driven by their "insatiable ... carnal lust" (Kramer and Sprenger, 1971, p. 47); "savages," figured by the colonial "porno-tropical" imagination as females who were the very "epitome of sexual aberration and excess" (McClintock, 1995, pp. 21–24); and males who were gender deviants and sodomites (Carocci, 2013). 'Conquest' itself was openly figured via metaphors of white, masculine, sexual domination of a dark, feminine (sometimes promiscuous, sometimes virginal) land, forced to give way to the thrust of (white, male, phallic) penetration, including military invasion, occupation, and resource exploitation.

This same time period was marked by a 'scientific revolution' in Europe, an intellectual venture likewise characterized by a "metaphysics of gender violence" that defined the acquisition of knowledge itself as "the male penetration and exposure of a veiled, female interior" (McClintock, 1995, p. 23). Descartes spoke for this worldview when he celebrated the expansion of male knowledge as now able to make men "masters and possessors of nature" (cited in McClintock, 1995, p. 24). The ancient patriarchal definition of women as inferior by virtue of being closer to Nature (e.g., in Aristotle as in *Genesis*) made all this mastery of Nature seem just so, laying down the path that has led inexorably to the Anthropocene.

Again, many human cultures do not and have not entertained this egomaniacal, phallic fantasy that their destiny is to 'fuck with' Nature. US writer Barry Lopez (1986, p. 39) recounts a conversation he had with a Yup'ik hunter, who told him:

> what traditional Eskimos fear most about us is the extent of our power to alter the land, the scale of that power, and the fact that we can easily effect some of these changes electronically, from a distant city.... They call us, with a mixture of incredulity and apprehension, "the people who change nature."

This propensity to change Nature is not evidence of superiority, but of a brokenness, within and without. The land, as plant biologist and traditional ecological knowledge educator and practitioner Robin Wall Kimmerer (2013, p. 9) recognizes, "shows the bruises of an abusive relationship. It's not just land that is broken, but more importantly our relationship with land."

A healthy relationship with land and the kinship of all life is at the heart of *The Declaration of Mother Earth Rights* (2010). Naming Earth 'Mother' signifies a vast force of "continuous rebirth" (LaDuke, 2002, p. 98), via ongoing cycles of birth, growth, decline, death, decay, and transformation. But, in the Anthropocenic framework, connection, interrelationship, and dependency on Nature are denied, as is the animacy, the spirit of Earth, and of all matter.

Objectification/de-spiritualization

In several works, Australian ecofeminist theorist Val Plumwood (1993, 2002), acknowledging her debt to indigenous thought, critiques the Western 'master consciousness' that denies human dependency on Nature (including the body) and promulgates an alleged human 'hyper-separation' from and supposed mastery over Nature. This conceptual trick enables the illusion that humans are able to injure 'Nature' without somehow simultaneously injuring themselves. It refuses awareness that "spirit is not a hyper-separated extra ingredient but a certain mode of organisation of a material body" (2002, p. 223). Spirit is, to put it simply, what 'keeps it together.' Hierarchical and binary thinking dis-spirits; it severs wholes, results in brokenness and an ensuing objectification of all defined as Nature, including non-human as well as human beings.

In *Black Feminist Thought*, Patricia Hill Collins (2000, pp. 70–72) critiques the ways in which this paradigm underlies intersecting racism and sexism as well as environmental devastation. She cites the work of social theorist Dona Richards (1980), who draws upon indigenous African philosophy while observing that this calculated "separation of the 'knowing self' from the 'known object'" is characteristic of Western thought. This type of totalizing objectification is a "prerequisite for the despiritualization of the universe" (Richards, cited in Collins, 2000, p. 70) and the kinds of resulting destructions without the possibility of rebirth characterizing the Anthropocene.

In *The Death of Nature*, historian Carolyn Merchant (1980) identifies despiritualization and fragmentation as intrinsic to the worldview installed by the European scientific revolution in early modern Europe. This replaced an older ethos, an "organic worldview" (Merchant, 1992, p. 41) centered on the metaphor of the Earth as a living and enspirited being, a nurturing mother, a metaphor that worked to constrain violence against the Earth and elements. The scientific revolution replaced this with a 'mechanical worldview,' with Earth re-imagined as dead, an inert object, void of agency, purpose, will, and spirit, something able to be cut into parts, rearranged, ultimately even replaced with a new and improved model – and by men (and the women who identify with them) who see themselves as not only playing god, but now being 'God.'

Theology/geology

> The Western belief that god lives apart from earth ... is one that has taken us toward collective destruction. It is a belief narrow enough to forget the value of matter, the very thing that soul inhabits.
>
> (Linda Hogan (Chickasaw), 1995, p. 86)

In her 1973 book, *Beyond God the Father*, radical feminist theologian Mary Daly exposed the oppressive gender politics behind the Abrahamic religious projection of the (white) male God. This image of "God in 'his' heaven" as a "father ruling 'his' people" proclaimed symbolically that it was "in the 'nature' of things and according to divine plan and the order of the universe that society be male-dominated" (p. 13). Writer Linda Hogan also understands that the Western notion of God as an absent and immaterial father has led to "collective destruction" (1995, p. 86). Indeed, this hyper-separated and immaterial male God is said (in the Old Testament) to have made the Earth and it is also prophesied (in the New Testament) that he ultimately will destroy the Earth and eliminate all the Earth's "elemental spirits" in a glorious, global inferno (Daly, 1984, pp. 7–10). This same God then serves as a role model, his model of divinity assumed by some of the most arrogant of the largely straight-identified group of Eurocentric men boosting the Anthropocene. For example, techno-guru Stewart Brand (2010, epigraph), author of *The Whole Earth Catalog*, is given to such oft-quoted mantras as "We are as gods and might as well get good at it." More of the same shows up in *The God Species: Saving the planet in the age of humans*, where another arrogant white man, Mark Lynas (2011, p. 8), trumpets: "Nature no longer runs the Earth. We do."

The title of another religiously inflected Anthropocenic tract, *Regenesis: How synthetic biology will reinvent Nature and ourselves*, makes direct reference to Genesis, the first chapter of the Old Testament. Its authors, Harvard geneticist George Church and journalist Ed Regis (2012), place synthetic biology in the forefront of those techniques that they claim will allow "humans [to] remake nature according to their wishes" (p. 143). Proselytizing that field (particularly when fused with corporate capitalism), Church and Regis define *synthetic biology* as "the science of selectively altering the genes of organisms to make them do things that they wouldn't do in their original, natural, untouched state" (p. 2). This openly non-consensual model of inter-species relationship speaks to the mother-fucking paradigm underlying the types of sciences and technologies that have led to the Anthropocene. As implicit in that telling title, *Regenesis*, the synthetic biologist is now the creator and dominator of his own little sandbox/garden: the planet.

A bumper sticker available on the Internet reads: "Father God Made Mother Earth." The Genesis story does indeed feature an external Father God, who creates the Earth, reversing understandings of the Earth *as* the creator. The potency of (Mother) Earth is further denied and reversed when the human figure, Adam, effectively gives birth to Eve (Daly, 1973, p. 95). More reversals abound as the serpent (often depicted in European art as female and/or green) is identified in the story with evil. But that green serpent is an archetype of earthly life force, an avatar of the Earth Mother (Campbell with Moyers, 1988, p. 48). Eve's wise acceptance of the gift of knowledge or *conocimiento* (Anzaldúa, 2000, p. 266) from the serpent is condemned, identified with original sin. Consequently, Eve is condemned for bringing not only sin, but also 'nature,' into the world, all that patriarchal, hyper-separated "civilization" experiences as "dirtiness" (Delaney, 1988, p. 79) – sex, death, defecation, urination, menstruation, and sweat. God then mandates that men dominate the Earth (and women).

The *Regenesis* (literally rebirth) that Church and Regis prophesy as a result of synthetic biology is no longer just about 'The White Man' exerting dominion over 'The Dark Mother,' but is now about synthetic biologists actually being able to replace/Nature with an artificial substitute that they both create and control. Their vision of paradise has no serpent, no Mother, no tree, but, interestingly enough, in the logo of a much-beloved computer company, an artificial apple with a bite taken out.

Call your mother!

Mother Nature continues to appear in Anthropocene discourse, but in conde-scending ways. Ackerman (2014, p. 308) writes:

> Nature is still our mother, but she's grown older and less independent. We've grown more self-reliant, and, as a result we're beginning to redefine our relationship to her.... As we're beginning to see her limits as well as her bounty, we're trying to grow into the role of loving caregivers.

Reciprocity of care between humans and non-human Nature is certainly requisite, but Ackerman, typically, has to aggrandize 'humans' as the ones in charge.

Similar demeaning portrayals appear throughout US popular culture. For example, a series of ads for Tampax tampons (2012) is built around the theme "Outsmart Mother Nature." Nature is played by a middle-aged white woman dressed in a frumpy green suit. Reminiscent of the Genesis story, this green, Nature-identified figure attempts to deliver an apple-red gift package (menstrua-tion) to various Eves. These young women are shown as rightly refusing Nature's gift. In one of the print ads, a gigantic tampon dwarfs this now spurned and ineffectual Mother Nature. The copy reads: "Cut Mother Nature Down to Size" (to view, see Kissling, 2010).

Because of the profusion of such sexist stereotyping, many ecologically minded feminists (e.g., Gray, 1981) express discomfort with continuing to invoke Mother Nature or Mother Earth. But I disagree. I think that what we need to do is refute these stereotypes, expose their implicit sexist, racist, ageist underpinnings, and reclaim and rename Mother Nature/Earth outside of these. For example, we can understand the Earth beyond sexist and binary distortions, resulting in the caricatures of the 'good' mother – endlessly fertile, nurturing, and forgiving – and the 'bad' mother – treacherous, castrating, and deathly. If the whole is perceived, it becomes clear that "Living and dying," as Paula Gunn Allen (Laguna) writes, "are twin beings, gifts of our mother, the Earth" (1990, p. 52).

Some genderqueer eco-folks understand Earth in differently sexual and gen-dered ways. Deborah Taj Anapol (2014) writes: "What if Earth, what if Nature herself, was not mother, not even female, but male, or more likely both?" Eco-sexual artists, educators, and activists Annie Sprinkle and Beth Stephens recog-nize Earth as *Lover* (Stephens and Sprinkle, 2012). They ritualize this understanding in a manifesto, in theater performances, photography, and wedding ceremonies where ecosexual activists 'marry' the Earth. Of course, outside of sexist and sex-negative binaries, Mothers are also Lovers.

And mothering manifests, too, as a verb, as a "mothering power" (Standing Bear, 2004, p. 39) and not reductively gendered. Two Tewa scholars, Tito Naranjo and Rina Swentzell (1989, p. 260), deepen understandings of the Earth as Mother, explaining that the Tewa word *Gia* (Mother) is one "used pervasively to identify ideal behavior" as well as one "used to address the earth ... for the highest supernaturals who remain in the underworld, for males who are outstanding leaders, for strong community level females, and for biological mothers." *Mothers*, too, are "people who love and help."

Mothering power is taken up by womanist theory, based in social justice principles as well as in African and African-American spiritualties. For womanism, *mothering* is an essential means of social transformation (Phillips, 2006, pp. xxix–xxx), evidenced in activities of "other-mothers" (non-biological parents) who play essential, beneficial roles in sustaining any community. This expanded conception will recognize a mothering power as operative every time anyone, whatever their sex or gender, creates community, exercises egalitarian leadership, guidance, and community-oriented decision making.

It is also important to call the Earth 'Mother' because 'Mother Earth' remains a metonymy for a vital alternative ecological knowledge system and ethos. Ecofeminist and global activist Vandana Shiva (2005) contends that a core aim of European imperialism was the destruction and cooptation of the ecological knowledge systems encountered among indigenous peoples of Africa, Asia, and the Americas. But, these systems have survived, their principles embedded into the concept of *Mother Earth*, and explicitly gathered and articulated in *The International Declaration of Mother Earth Rights* (2010).

They also appear in what Santa Clara Pueblo educator Greg Cajete (2000) calls 'Native Science,' and what plant biologist Robin Wall Kimmerer (Potawatami) further identifies as 'traditional ecological knowledge' (TEK). TEK, Kimmerer (2002) explains, drawing on Berkes (1993), is the reliable, rational knowledge and practices honed over generations of observation, experimentation, and practices and concerning the relationship of living beings to one another and to the physical environment. TEK is the science held by "peoples in relatively nontechnological societies with a direct dependence upon local resources" (Kimmerer, 2002, p. 433). Kimmerer is Native American, but makes it clear that TEK is "not unique to Native American culture but exists all over the world ... born of long intimacy and attentiveness to a homeland and can arise wherever people are materially and spiritually integrated with their landscape" (Kimmerer, 2002, p. 433). Significantly, a core value of TEK (contra those malicious Tampax ads) is a responsibility to accept the gifts of Mother Nature. Then, "In return for the gifts of Mother Earth ... [our] duty is to care for those gifts. It is this reciprocity that allows the world to continue" (Kimmerer, 2007, p. 13).

In 1991, Cherokee poet Awiakta gave a poetry reading at the University of New Mexico concerning her understanding of reciprocity and the mothering power of the Earth. These themes cohered in her recitation of her poem, 'When Earth Becomes an It,' avowing that when people 'call' the Earth 'Mother,' they take from and give back to her with love, but that when they disrespectfully call the Earth 'it,' the Mother takes "all green" back, leading to their demise (Awiakta, 1993, p. 6). During an ensuing discussion, Navajo attorney and New Mexico state representative Leonard Tsosie made this comment:

One time in listening to an elderly Navajo man talking about the atom, one of the things that I remember him saying is that Mother Earth is good, but Mother Earth also has to protect herself and you can only go so far before you trigger Mother Earth's response. And one of the things that he was talking about – in Navajo we call it 'ant'iih – meaning that something that is bad. And if you keep digging and digging and digging, that elderly man said, eventually you will find Mother Earth's 'ant'iih. And then that's when we all go. When we dig that up. We may have gotten to the 'ant'iih, and, if that is the case, that may be the end.

(cited in Caputi, 1993, p. 187)

The Anthropocene discredits these and other powers of the Earth Mother, projecting her as enfeebled, as replaceable by a synthetic, controlled object, and as the ultimate victim of their phallic interference. But this is so wrong. However much the Anthropocenic culture tries to 'fuck' the Mother to death, it is not capable of this – though it can do enormous amounts of damage to other beings and ultimately may take out all of humanity as well. But Earth/Nature is no passive victim, unable to resist. Indeed, Earth herself is the ultimate 'Muthafucka.'

Writing from a black, queer feminist perspective, L.H. Stallings (2007, p. 25) defines *Muthafucka* as a "sacredly profane word," a favorite term of tricksters who turn oppressive structures upside down, "disrupting master narratives on Black womanhood and motherhood" and, I would add, Dark Earth Motherhood. Although the master narrative of The Anthropocene regales The White Man as having become capable of "overwhelming the Great Forces of Nature" (Steffen et al., 2007, p. 614), this is a reversal. The Dark Mother Earth/Nature *is* the vast force of ultimate creation/destruction/transformation, the one who actually will be doing any overwhelming that is going to take place.

Defienda A Nuestra Madre, Defendernos. Defense of Mother Earth is intrinsically *spiritual activism* (Anzaldúa, 2000, pp. 38, 178), one that mobilizes the Earth energy that connects all life to each other. Defense of Mother Earth is spiritual activism because it is, as well, "spirituality for social change, spirituality that recognizes the many differences among us yet insists on our commonalities and uses these commonalities as catalysts for transformation" (Keating, 2006, p. 11). Defense of Mother Earth is not a defense of a helpless and hapless planet: a damsel in distress who needs saving, a feeble grandmother meriting only a kindly condescension. Rather, it is a defense of the Earth community. *Defienda A Nuestra Madre Tierra, Defendernos.* Defend Our Mother Earth, Defend Ourselves (Caputi, 2015)

References

Ackerman, D., 2014. *The Human Age: The world shaped by us.* New York: W.W. Norton.

Allen, P.G., 1990. The woman I love is a planet the planet I love is a tree. In I. Diamond and G. Feman Orenstein, eds, *Reweaving the World: The emergence of ecofeminism.* San Francisco, CA: Sierra Club Books. pp. 52–57.

Anapol, D.T., 2014. Gender queering mother earth. *Love Without Limits*. Available at: www.lovewithoutlimits.com/articles/Gender_Queering_Mother_Earth.html (accessed June 6, 2015).

André 3000, 2003. Vibrate. OutKast, *Speakerboxx*.

Anzaldúa, G., 2000. *Interviews/Entrevistas*, ed. AnaLouise Keating. New York: Routledge.

Awiakta, M., 1993. *Selu: Seeking the Corn Mother's wisdom*. Golden, CO: Fulcrum Publishing.

Berkes, F., 1993. Traditional ecological knowledge in perspective. In J.T. Inglis, ed., *Traditional Ecological Knowledge*. International Program on Traditional Ecological Knowledge and International Development Research Centre, Ottawa, ON.

Bordo, S., 1999. *The Male Body*. New York: Farrar Straus & Giroux.

Brand, S., 2010. *Whole Earth Discipline: Why dense cities, nuclear power, transgenic crops, restored wildlands, and geoengineering are necessary*. New York: Penguin Books.

Cajete, G., 2000. *Native Science*. Santa Fe: Clear Light Publishers.

Campbell, J. with Moyers, B., 1988. *The Power of Myth*. New York: Doubleday.

Caputi, J., 1993. *Gossips, Gorgons, and Crones: The fates of the Earth*. Santa Fe: Bear and Company.

Caputi, J., 2015. *Feed the Green: Feminist voices for the Earth*. DVD, Women Make Movies, WMM.com.

Carocci, M., 2013. Sodomy, ambiguity, and feminization: Homosexual meanings and the male Native American body. In J. Fear-Siegal and R. Tillett, eds, *Indigenous Bodies*. Albany, NY: State University of New York Press, pp. 69–83.

Church, G. and Regis, E., 2012. *Regenesis: How synthetic biology will reinvent Nature and ourselves*. New York: Basic Books.

Collins, P.H., 2000. *Black Feminist Thought: Knowledge, consciousness, and the politics of empowerment*. New York: Routledge.

Crutzen, P.J. and Schwägerl, C., 2011. Living in the anthropocene: Toward a new global ethos. *Yale Environment 360*. Available at: http://e360.yale.edu/feature/living_in_the_anthropocene_toward_a_new_global_ethos/2363/ (accessed June 6, 2015).

Daly, M., 1973. *Beyond God the Father: Toward a Philosophy of women's liberation*. Boston, MA: Beacon Press.

Daly, M., 1984. *Pure Lust: Elemental feminist philosophy*. Boston, MA: Beacon Press.

Delaney, C., 1988. Mortal flow: Menstruation in Turkish village society. In T. Buckley and A. Gottlieb, eds, *Blood Magic: The anthropology of menstruation*. Berkeley: University of California Press. pp. 75–93.

Farajaje-Jones, E., 2000. Holy Fuck. In K. Kay, J. Nagle, and B. Gould, eds, *Male Lust*. New York: Harrington Park Press. pp. 327–336.

Gray, E.D. (1981). Green Paradise Lost. Available at: www.American-buddha.com/lit.greenparadiselost.toc.htm.

Griffin, S., 1989. Split culture. In J. Plant, ed., *Healing the Wounds: The promise of ecofeminism*. Philadelphia, PA: New Society Publishers, pp. 7–17.

Hogan, L., 1995. *Dwellings: A spiritual history of the living world*. New York: Touchstone Books.

Keating, A.L., 2006. From borderlands and new mestizas to nepantlas and nepantleras: Anzaldúan theories for social change. *Human Architecture: Journal of Sociology of Self-Knowledge*, 4, pp. 5–16.

Kimmerer, R.W., 2002. Weaving traditional ecological knowledge into biological education: A call to action. *BioScience*, 52(5), pp. 432–438.

Kimmerer, R.W., 2007. The sacred and the superfund. *Stone Canoe* (Syracuse University), 1, pp. 1–17.

Kimmerer, R.W., 2013. *Braiding Sweetgrass: Indigenous wisdom, scientific knowledge, and the teachings of plants*. Minneapolis, MN: Milkweed Editions.

Kissling, E., 2010. Is mother nature winning? *Menstruation Research*. Available at: http://menstruationresearch.org/tag/mother-nature/ (accessed June 6, 2015).

Kolbert, E., 2011. Enter the anthropocene: age of man. *National Geographic*. Available at: http://ngm.nationalgeographic.com/2011/03/age-of-man/kolbert-text (accessed June 6, 2015).

Kramer, H. and Sprenger, J., 1928. *The Malleus Maleficarum*. Translated by M. Summers, 1971. New York: Dover Press.

LaDuke, W., 2002. A society based on conquest cannot be sustained. In R. Hoffrichter and M. Gelobter, eds, *Toxic Struggles*. Salt Lake City: University of Utah Press, pp. 98–106.

Lerner, G., 1986. *The Creation of Patriarchy*. New York: Oxford University Press.

Lopez, B., 1986. *Arctic Dreams*. New York: Vintage Books.

Lynas, M., 2011. *The God Species: Saving the planet in the age of humans*. Washington, DC: National Geographic.

Mankoff, R., 1999. Now that's product placement. *New Yorker*, November 22. Available at: www.condenaststore.com/-sp/Now-that-s-product-placement-New-Yorker-Cartoon-Prints_i8474558_.htm (accessed June 5, 2015).

McClintock, A., 1995. *Imperial Leather: Race, gender and sexuality in the colonial context*. New York: Routledge.

Merchant, C., 1980. *The Death of Nature: Women, ecology and the scientific revolution*. San Francisco CA: Harper and Row.

Merchant, C., 1992. *Radical Ecology: The search for a livable world*. New York: Routledge.

Naranjo, T. and Swentzell, R., 1989. Healing spaces in the Tewa Pueblo world. *American Indian Culture and Research Journal*, 13(3), pp. 257–265.

Nixon, R., 2014. Future Footprints: Review of *The Human Age*, by Diane Ackerman. *New York Times Book Review*, September 7, p. BR 10.

Phillips, L., 2006. Womanism: On its own. In L. Phillips, ed., *The Womanist Reader*. New York: Routledge, pp. xiv–lv.

Plumwood, V., 1993. *Feminism and the Mastery of Nature*. New York: Routledge.

Plumwood, V., 2002. *Environmental Culture: The ecological crisis of reason*. New York: Routledge.

Revkin, A., 2011. Confronting the anthropocene. *New York Times*. Available at: http://dotearth.blogs.nytimes.com/2011/05/11/confronting-the-anthropocene/?_php=true&_type=blogs&_r=0 (accessed June 6, 2015).

Rights of Mother Earth. 2010. Preamble [pdf]. Available at: www.thealliancefordemocracy.org/pdf/AfDJR6117.pdf (accessed June 6, 2015).

Roach, C.M., 2005. Mother nature imagery. In B.R. Taylor, ed., *Encyclopedia of Religion and Nature*. New York: Thoemmes Continuum, pp. 1107–1110.

Sanchez, C.L., 1989. New world tribal communities: An alternative approach for recreating egalitarian societies. In J. Plaskow and C.P. Christ, eds, *Weaving the Visions*. San Francisco, CA: Harper and Row, pp. 344–356.

Sheidlower, J., ed., 1999. *The F-Word*, 2nd edn. New York: Random House.

Shiva, V., 2005. *Earth Democracy*. Boston, MA: South End Press.

Silko, L.M., 1989. The fourth world. *Artforum International*, 27(10), pp. 124–125.

Sjöö, M. and Mor, B., 1991. *The Great Cosmic Mother*, 2nd edn. San Francisco, CA: HarperSanFrancisco.

Smith, A., 2005. *Conquest*. Boston, MA: South End Press.

Stallings, L. H., 2007. *Mutha' Is Half a Word: Intersections of folklore, vernacular, myth, and queerness in black female culture.* Columbus: Ohio State University Press.

Standing Bear, L., 2004. Nature (1933). In R.S. Gottlieb, ed., *This Sacred Earth*, 2nd edn. New York: Routledge, pp. 39–42.

Steffen, W., Crutzen, P.J. and McNeill, J., 2007. The anthropocene: Are humans now overwhelming the great forces of nature? *Ambia*, 38(8), pp. 614–621 [pdf]. Available at: www.pik-potsdam.de/news/public-events/archiv/alter-net/former-ss/2007/05-09.2007/steffen/literature/ambi-36-08-06_614_621.pdf (accessed June 6, 2015).

Stephens, B. and Sprinkle, A., 2012. The ecosex manifesto. *SexEcology*, Available at: http://sexecology.org/research-writing/ecosex-manifesto/ (accessed June 6, 2015).

Vidal, J., 2011. Bolivia enshrines natural world's rights with equal status for Mother Earth. *Guardian*, April 10.

Young, I.M., 1990. *Justice and the Politics of Difference.* Princeton, NJ: Princeton University Press.

4 The environment in the twenty-first century

A play in two parts

Devin Samuels

1 *The mother disciplines her loudest son*
Sickness,
Might make you think twice about doing that pollution thing again
You know how many times I have told you to clean up your messes,
But somehow you think that if you can just move your waste to a place I can't see it
I won't get mad, or ask about it again.
Really?! I see everything.
You are the blind one.
This whole house is my body, did you really think I wasn't going to find out?

I remember your first words.
I held your hand and walked you two legged to the water
I taught your tongue its own name
Now, you speak it like it birthed itself
Human, progress, develop...

Industry? Are you joking? Don't swing at me unless you plan to finish the job.
Don't raise your hand against me and then ask for dinner.
Don't think you're getting away with this.
I will tsunami typhoon you till you don't know which way is up.
I'll choke you then drown your play cities.
I could swallow you whole with this ocean mouth

You are so young.
You know I do this out of love
If you can't listen I don't know what I'm going to do
I just don't want to see you keep hurting us.
Honestly, you are getting too old and too big to act like this.

2 *The child talks back*
I'm so smart I can make you better,
I'm so big you can't ground me,
I'm so strong I can fight you and walk away,
Don't try to stop me, *Mom*.

I own this house now.
I'm the *man* of this house now.
You keep acting like you know what's best but you don't.
I'm my own *man* now. I grow my own *food* now.
I build cities, FUCKING CITIES.
Since when have you done anything that, Metal.
Anything that, Permanent.
Well? That's what I thought.

And if you won't let me do what I want then I'll make you,
If you still don't, then I'll leave mom.
That's right, I'll run away.
I'll find a better mom under a different sun,
then I'll own that house too.
I'll be the man there too.
You'll see me write my city name in constellations
And you won't be able to threaten me anymore.

You El Niño Climate Change temper?
You have rain but I have steel houses,
You have floods but I have boats,
If you kill my food, *I'll build it out of fetus's.*
Didn't know I could do that? Yea, I figured it out yesterday.

I'm sick of you treating me like a kid
And I won't listen to you until you stop.

3 *In-script*
Stop it.
You're getting hurt.
I want to love you.
I think you're dying.

5 *The Rush Limbaugh Show* and the expanding culture war

Whiteness, masculinity, and conservative media denials of climate change and sexism

Julie Bacon

Introduction

As 2012 drew to a close, it would have been difficult for a person attentive to both environmental health and social justice in the USA to ignore the fact that the year had been one of record-breaking heat, devastating storms, economic uncertainty, and widespread political mobilization against the rights of women. As a high-stakes political year, it was perhaps not surprising that seemingly endless debates over women's reproductive control were amplified in media and public discourse, but what may have shocked was how sporadic references to climate change wove in and out, occasionally disappearing altogether, even as massive storms ravaged the eastern seaboard.

Regardless of one's level of certainty or skepticism about climate science, the intensity of discourse surrounding this issue remains apparent. Although the scientific community has drawn close to consensus regarding the occurrence and major cause of climate change, significant media outlets dedicatedly promulgate climate change denial. In the public and in mainstream media, few other scientific lines of inquiry have been so furiously attacked and hotly debated. Arguments about climate change, not unlike arguments surrounding reproductive rights, have become intensely emotional and partisan. They have also shaped the interests of those in particular gendered and racialized social locations.

As a queer person deeply invested in decolonial methodologies and suspicious of uncritical relationships with Western science, my own complicated relationships with climate change and ecofeminism render me perhaps an unusual candidate for addressing how sexism and climate denial co-exist in mainstream media. I feel that the debates over climate change as they have been framed in mainstream American media at best do nothing to address the impacts of climate change, and at worst relegate to obscurity a host of other environmental problems that also require immediate attention. As regards ecofeminism, I find that many of the arguments coming from this theory rely too heavily on essentialized images of women as heterosexual/reproductive subjects. Even so, I do find the suggestion that Western practices of domination over women and the planet have transpired under some strong cultural concepts that reductively link the identities of these two types – 'feminine' and 'natural' – compelling. As such, I believe that analysis of the relationship between media denials of sexism and climate change may reveal gendered and racialized cultural dimensions of what has been portrayed largely as political or scientific disagreement.

Media and climate change

Problems with framing and conveying climate science in the mainstream media have long been a topic of research. Significantly, analysis of US mainstream news from 1988 to 2004 indicates that adherence to journalistic norms such as personalization, novelty, and balance has led to "informationally deficient mass-media coverage" of climate change (Boykoff and Boykoff, 2007, p. 1190). Similarly in the United Kingdom, Anabela Carvalho's work reveals that the mutually constitutive nature of media discourse and ideology gives media particular power in "allowing or disallowing other social actors to advance their ideological standings" (Carvalho, 2007, p. 225). With regard to climate change, McCright and Dunlap suggest, "the industrial sector and the conservative movement ... defend the industrial capitalist order from critique by denying the significance of problems such as climate change" (McCright and Dunlap, 2011b, p. 155). I suggest that the conservative movement deploys a similar form of denial in response to issues of systemic inequality. These two forms of denial commingle clearly in the discourse of conservative mainstream media pundits who wield a great deal of influence over their audience's understanding of these issues (Akerlof et al., 2012).

Race, gender, and risk

The conservative media's sanction of climate change denial likely accounts for some of the increased political and demographic polarization, but there also exists the possibility that pre-existing forms of socio-political polarization have simply taken up climate change as another front in the culture wars. I use the term *denial* here in accordance with the findings of Peter J. Jacques' (2012) "General Theory of Climate Denial," in which he considers the similarities between Holocaust denial and climate denial. Jacques contends that, despite debate over which term most adequately describes those who consider climate change a "hoax," denial is the appropriate term in accordance with Lang's "General Theory of Historical Denial" and in accordance with Deborah Lipstadt's "architecture of reasoning in Holocaust denial: (1) It is a movement. (2) It is a defense of a threatened ideology. (3) Its true objectives are camouflaged. (4) Its tactics include sowing confusion by creating knowledge claims that appear legitimate to the general public" (Lipstadt, in Jacques, 2012, p. 10).

Some theorists have suggested that political polarization of climate change may in part result from conservative tendencies toward system-justifying behaviors

(Feygina et al., 2010), the elite cues hypothesis (Krosnick et al., 2000), and, at the Congressional level, conflicts not over science, but over the economic impacts of CO_2 regulation (Fisher et al., 2013). Yet, the partisan divide is not the only trend. In a series of articles, McCright and Dunlap report a predominance of conservative white men, both within the elite ranks of deniers and within the climate-denying general public.[1]

Scholars who study risk and risk perception refer to such demographic imbalance as the white male effect (Finucane et al., 2000; Flynn et al., 1994; McCright and Dunlap, 2011a). This term describes a trend observed in numerous risk perception studies often having nothing to do with politics. Attempts to understand the white male effect have been ongoing, but a 2007 study published in the *Journal of Empirical Legal Studies* proposes that the white male effect may be linked to cultural-identity-protective cognition (Kahan et al., 2007). Researchers conclude, "individuals ... adopt a posture of extreme skepticism, in particular, when charges of societal danger are leveled at activities integral to social roles constructed by their cultural commitments" (Kahan et al., 2007, p. 467). In other words, those whose worldviews are most threatened by an assertion that something is dangerous are those who are most likely to be skeptical of that danger. This force seems to be at work in white male conservative framings of both systemic inequality and global climate change.

The relationship between race, gender, and environmental concern is complex. The dominant perspectives of the climate "debate" are both epistemologically Eurocentric and masculinist, regardless of political inclination. The mainstream environmental movement itself emerges historically from the interests and practices of white male elites. Calls for environmental preservation and conservation, which sparked the environmental protections of the progressive era, grew out of the vanishing frontier myth and a sense that civilization – particularly immigrant-filled urban space – was corrupting masculine vitality in the white race (Cronon, 1995; Taylor, 1997; Dunaway, 2000). This legacy persists, albeit subtly, in the mainstream as groups like the Sierra Club and The Nature Conservancy have been critiqued for their ongoing patterns of organizational race and gender inequality (Park and Pellow, 2011; Smith, 2005; Taylor, 2014).

Alternatively, studies of extractive industry workers reveal the role of hegemonic masculinity in shaping an ethic of environmental domination (Bell and York, 2010; Miller, 2004; O'Shaughnessy and Krogman, 2011), and such concepts of masculinity likewise appear in studies of adaptations to new economic and ecological relationships in the face of environmental degradation (Anahita and Mix, 2006; Dunaway, 2000).

The conservative, white, male tendency toward climate denial must be seen not only in its relation to party politics but also as part of a complex sociopolitical trajectory in which race and gender have been instrumental in shaping environmental ideas and practices.

Data and methods

In order to explore the relationship between conservative media's climate change and gender inequality denial strategies, I chose to focus on the 2012 *Rush Limbaugh Show* transcripts. This decision was initially prompted by a significant study regarding mass-media dissemination of climate modeling information which indicates that since 2007, *Rush Limbaugh* has offered more explanations of climate models than any other mass-consumed news source regardless of political inclination, and has simultaneously generated the least sentences suggesting that such models may be accurate. Furthermore, *The Rush Limbaugh Show* is the top-rated news radio program in the USA, with an estimated weekly audience of 14 million people (Akerlof et al., 2012, p. 652). As a conservative program, it stands to reason that much of the audience shares similarities with the demographic least likely to see climate change as a concern. In addition, since *The Rush Limbaugh Show* covers a wide range of political and social issues, it is a rich source for a brief analysis of how climate change denial and denials of sexism intersect in conservative media.[2]

Using *The Rush Limbaugh Show* archive, I compiled all transcripts containing the phrases "climate change" or "global warming" between January 1 and December 31, 2012 (124 results). I also searched for transcripts containing "feminism" or "feminist" (53 results) and "sexism" (31 results) during the same time frame. Employing textual analysis of these transcripts in order to determine what patterns, if any, emerge in the denial strategies, I began by coding the climate articles, paying particular attention to the types of denial strategies employed in each. I noted four major denial strategies – absolute dismissal, party-line arguments, anti-intellectualism, and discrediting through association – recurring throughout the majority of the climate transcripts. These were my primary codes, each with numerous sub-codes. I then coded the "feminist" and "sexism" transcripts, noting the same four denial strategies.

Results

Generally, both claims of institutional oppression – in this case sexism – and claims of environmental/climate concern are met with some form of dismissal or denial in *The Rush Limbaugh Show* transcripts. By dismissal I mean that environmental or injustice claims are deemed unworthy of serious consideration because claimants themselves are deemed unworthy or unreliable, while denial consists of a more direct assertion that particular claims are untrue. In all, I noted four common denial or dismissal strategies.

First is a tendency toward absolute dismissal or the "wacko dismissal." In this mode, Limbaugh and his sympathetic callers refer to environmental and/or gender inequality issues as "hoaxes" and "lies" and to those who believe in them or raise any concern about them as "idiots" or "wackos." Limbaugh uses "wacko" to describe people as widely diverse as James Holmes – perpetrator of the 2012 Aurora movie theater shooting – to performance artist Karen Finley. But despite the widespread use of this appellation, the term "wacko" appears with tremendous frequency in conjunction with the terms "environment" and "climate change."

"Wacko" appears less frequently with the term "feminist" in part due to the existence of a special term: "feminazi." Limbaugh promoted the use of this term in his early days of rampaging against NOW and other mainstream feminist organizations. Loose comparisons with Nazism figure heavily into Limbaugh's dismissal projects; for example, in addition to "feminazi" as a term of derision for feminists, Limbaugh uses "the Big Lie" to describe a variety of issues including global warming. Like "wacko" comparisons to Nazism – or, less frequently, Stalinism – constitute a form of fear-based othering that seeks to render ideas incompatible with the status quo unthinkable and unworthy of contemplation.

One example of his "wacko" dismissal includes his coverage of activist responses to the 2012 hurricane season. Limbaugh comments on a *Boston.com* report, saying:

> [t]wo hundred environmentalist wackos "participated in a round-the-clock vigil since last Tuesday to protest the lack of discussion of climate change in the presidential debate and call on Scott Brown and [Princess] Elizabeth Warren to discuss the issue during their final scheduled debate" today. And then the storm hit and the environmentalist wackos dispersed and the climate change convention ended.
> (Bastardi: Global Warming Didn't Cause Sandy, October 30, 2012)

Not only does this report illustrate a special effort to mock female political leaders; this single report goes on to use the phrases "nuts," "dumb," "stupid," "insane," "faking," "fudging," "hoaxed-up," and "hoax," demonstrating the rapid-fire rhetorical intensity of Limbaugh's absolute dismissal tactic. Another episode embeds absolute dismissal in the title, "Enviro-Idiots: Two Earths Needed by 2030." This consistent pattern suggests that anyone concerned about gender equality or the environment are either stupid ("idiots"), insane ("wackos"), or totalitarians bent on destroying democracy ("feminazis").

Interestingly, Limbaugh does not employ this type of absolute dismissal when attacking conservatives, but in the case of conservatives, any sign of adherence to global warming or open advocacy of gender equality becomes an abnegation of conservative identity. These party-line dismissals are especially prominent early in the year when Limbaugh and his callers deployed belief in climate change to measure presidential candidates' conservative credibility. Throughout January and February, transcripts referring to Newt Gingrich repeatedly call into question his status as a "real" conservative because he acknowledged climate change. In an effort to suggest that Gingrich had become erratic, Limbaugh claims, "Newt's been all over the board. Newt believes in manmade global warming and is just trying to cover it up." Later in the month, callers echo this sentiment: "I was an avid supporter of Newt. But if Newt Gingrich changes his mind on global warming and he wrote about it and talked about it for five years, sat on the bench with Nancy Pelosi." The association with Pelosi also tarnishes Newt's credibility and recurs in later transcripts: "RUSH: We don't know why he'd sit on a couch with Pelosi for ANY reason, not to mention global warming." Caller responses to Gingrich's attempt at bipartisanism, especially bipartisanism with a woman addressing climate change, indicate that these efforts corrupt his conservative identity and suggest that such behavior is effeminate. Callers and Limbaugh frame Gingrich's belief in global warming, and his bipartisan work with Pelosi in terms of "Newt's weakness" and inability to be the "street fighter" Republicans need. Limbaugh reinforces this equation of belief in global warming or gender equality with relinquishing conservative identity in episodes that frame concern with women's rights and environmentalism as exclusively Democratic Party issues. This type of argument draws on the already-existing partisan polarity surrounding these issues, and simultan-eously reinforces the political dichotomy in the public imagination by assert-ing the impossibility of being both conservative and concerned about global warming or gender equality.

Another important trend related to both climate change and systemic injus-tice denial is Limbaugh's wide-reaching anti-intellectual project. When it comes to climate science, Limbaugh openly avows that

> It's insane. But college professors, high school teachers all over the country will likely pick this up, and it will become part of the daily lesson plan ... I want you to stop and think how literally ridiculous this is, but these people, and they are full-fledged liberals, leftists, whatever, they mean it.

The relationship between discrediting science and climate change denial further emerges through Limbaugh's use of a rhetorical association with "global warming" to debunk non-climate science claims. For example, Limbaugh attacks a University of California-San Francisco editorial published in *Nature*, which warns that increased sugar consumption contributes to chronic disease epidem-ics in the USA. The editorial calls for a "public health intervention" and Lim-baugh responds, "US scientists, the same kind of people that brought us manmade global warming, say sugar is now a 'substance.'"

Just as in the case of party-line arguments, this type of argument also works in reverse; if some older scientific study is proven wrong, its wrong-ness becomes associated with "global warming." For example, in a June episode, Limbaugh explores new claims that salt intake may not be as dangerous as was once sus-pected. He says,

> [i]f they tell you salt is killing you, salt is raising your blood pressure, you'll listen to them. Nobody wants to die. So all of these things get started. They become accepted conventional wisdom. And slowly but surely we're learn-ing, just like global warming and just like coconut oil, just like a bunch of other things that everybody presumed that were fact, they're not.

These discrediting-by-association moves are not limited to science claims but are deployed against any kind of claim Limbaugh wishes to disgrace. For example, in an October episode called "Polls Put Liberals on Suicide Watch," Limbaugh explores conflicts over polling methods and results. He says, "Everybody seems to be coming up with different numbers. But the consensus … [ahem] The 'consensus,' as in the global warming consensus…. The consensus among political scientists is that they oversampled Republicans by 5 percent."

These tendencies come together powerfully in episodes where both sexism and climate change are discussed. For example, in one particularly complex argument over contraceptives[3] that ranges from global warming to free guns and toothpaste, Limbaugh asserts that Democrats want health insurance to cover contraceptives because they "fear kids." He says, "[l]ook at what people lead to, global warming, overpopulation, starving, pestilence…. Overpopulation will lead to disease, starving, famine, thirst, squalor, poverty, can't have any of that. So gotta pass out the contraceptives." Limbaugh then returns to the issue of sexism:

> if you're an average Democrat couple, you fear that if you have a boy, he'll grow up to be a horrible man…. So you couldn't have that. "I don't want to have a child. If I have a little girl, she's just gonna grow up to be a slave to some man and if I have a little boy he's gonna grow up to be a man who enslaves some little girl. [crying] Give me my contraceptive."

This episode works to establish both feminism and environmentalism as particularly Democratic concerns, and then attempts to discredit both through dismissals such as "dumb asses," "freak," and of course "wacko." In this episode denials of climate change and sexism reinforce each other through mutual discrediting associations.

Similarly, in a later segment called "'War on Men' Column Causes Stir, Illustrates the Left-Wing Stuff We Laugh About But Lots of People Believe," Limbaugh addresses a column created by Suzzane Venker of *FoxNews.com*. The title of this episode alone echoes what this report has found through archival research, namely that Limbaugh uses mockery and dismissal to deny any knowledge claims which compete with his ideology or threaten the dominance of his group. In typical fashion for Limbaugh, this transcript, like all 2012 transcripts dealing with gender inequality, contains no direct mention of patriarchy, nor any acquiescence that sexism is (or ever was) a real oppression or social concern.

This transcript contains examples of all the major types of dismissal and denial noted in the larger dataset: absolute dismissal, party-line arguments, anti-intellectualism, and discrediting through association. Although the episode begins with Venker's assertions regarding men's waning interest in marriage, Limbaugh quickly moves into an anti-intellectually charged synopsis of how radical feminism has "chickified" colleges and universities.

I think back to previous years, in fact, eras of this program. And we did our feminist updates, and what were the feminist updates? We chronicled and laughed at what was being done in universities. We laughed at some of the radical, cockeyed ideas that radical feminists and feminazis were doing.... But they've become mainstreamed, and it's not just with feminism. It's a lot of other liberalism. The same kind of thing with environmentalism. The whole hoax of global warming.

This passage contains mockery of intellectualism, absolute dismissal through the terms "hoax" and "feminazi," party-line assertions of ending sexism and climate change as "liberalism," and a move to mutually discredit both feminism and environmentalism through association.

The transcript further asserts that those who believe in "these things" are "arrogantly cocky about it" and "have a moral superiority about their countenance." These assertions seem ironic if one considers the general tone Limbaugh and his callers take in their assaults on both environmentalists and reproductive rights, but it is more than that. Throughout the episode, Limbaugh repeats his dismissals of feminist and environmental concerns, even characterizing those dismissals as laughter. Limbaugh says, "we sit here and laugh ourselves silly at this," all the while bemoaning the "arrogant condescension" and "close-mindedness" of those with whom he disagrees. Perhaps more tellingly, Limbaugh says, "[w]hat they believe is morally superior to say what I believe, what they believe and what they live and how they live is morally superior. So they kind of look down their noses at people." In claiming that he feels he is considered morally inferior, Limbaugh reveals the conservative sense that feminism and environmentalism are deep threats to the established order in which conservative white male identity represent the ultimate source of intellectual, moral, and social legitimacy.

Conclusion

In the USA, elite white men have generally amassed privilege and power through social exclusion, political and economic disenfranchisement, and environmental plunder. As a result, white men (especially conservatives with system-justifying tendencies) are those citizens whose worldviews are most threatened by, and thus they are most skeptical of, any major shift in the ideological or material conditions that have so enriched them. In accordance with the concepts suggested by studies in cultural-identity-protective cognition, white conservative males generally have a stronger opposition to shifts such as any move away from fossil fuels and any move toward human rights and justice (including but not limited to eradicating sexism, racism, colonialism, homophobia). More succinctly, the demographic of climate denial is also the demographic most likely to openly dismiss any type of social mobilization for inclusion and equality. Not surprisingly, then, conservative mass-media rhetoric of denial and dismissal are comparable whether the topic is sexism or climate change. Similar to the findings of Anabel Carvalho in her study of media in the UK, I find that *The Rush Limbaugh Show* regularly discredits those who purvey "unwanted knowledge" (Carvalho, 2007, p. 237). Furthermore, like *The Times*, Limbaugh also does not hesitate to abuse the reputations of those whose "knowledge claims appear to constitute a threat to [his] ideological principles and arrangements in the political, social and economic realms" (Carvalho, 2007, p. 237). Common modes of denial and dismissal include a loss of political credibility for conservatives, suggestions of effeminacy for men, or outright claims of insanity for those most invested in resisting climate change or gender inequality.

Like McCright and Dunlap, I find that through climate denial the "culture wars have taken on a new dimension" (2011b, p. 180). The conservative media broadly, but in particular *The Rush Limbaugh Show*, works to discredit climate science and preserve the political and industrial status quo functions much like their work to maintain other types of hegemony. Beginning in the 1990s Limbaugh asserted his ridicule of feminism through use of the term "feminazi," a term which he continues to deploy to this day. *The Rush Limbaugh Show*'s dismissal of feminism – and its inherent denial of sexism – persists alongside various other identity- and status-conserving rhetorics of which climate denial is now one.

Critically, this expanded culture war, like all culture wars, has particular ramifications not only for those defending their identities through logics of denial, but also for those whose lives may be irrevocably harmed by continued American polarization and stagnation around these challenges. The intersection of climate and sexism denials is especially disconcerting, as studies continue to reveal how systemic inequalities produce increased vulnerability for women – in particular women of color – in the event of climate-related disasters (e.g., Arora-Jonsson, 2011; Denton, 2002; Terry, 2009). It will be difficult, if not impossible, to effectively address women's increased risk while so many people – enabled in part by media campaigns of denial – cannot even acknowledge the existence of those risks.

Notes

1 Some 29.6 percent of conservative white males believe that the effects of global warming will never happen, compared to 7.4 percent of all other adults, and 39.1 percent of conservative white males say they do not worry about global warming at all, compared to 14.4 percent of all other adults (McCright and Dunlap, 2011a, p. 5).
2 It is worth noting that Rush Limbaugh has plenty to say about race and racism as well (93 results for the word "racism"), and his patterns of denial and dismissal are nearly identical to those he employs in the refutation of sexism and climate change. This could be a rich area for future research.
3 Contraception was a major issue this year on the show (79 hits for contraceptives in 2012). Of these, five refer to global warming as well.

References

Akerlof, K., Rowan, K.E., Fitzgerald, D. and Cedeno, A.Y. 2012, Communication of climate projections in US media amid politicization of model science. *Nature Climate Change*, vol. 2, no. 10, pp. 648–654.

Anahita, S. and Mix, T. 2006, Retrofitting frontier masculinity for Alaska's war against wolves. *Gender and Society*, vol. 20, pp. 332–353.

Antionio, R.J. and Brulle, R.J. 2011, The unbearable lightness of politics: Climate change denial and political polarization. *The Sociological Quarterly*, vol. 52, pp. 195–202.

Arora-Jonsson, S. 2011, Virtue and vulnerability: Discourses on women, gender and climate change. *Global Environmental Change*, vol. 21, no. 2, pp. 744–751.

Bell, S.E. and York, R. 2010, Community economic identity: The coal industry and ideology construction in West Virginia. *Rural Sociology*, vol. 75, pp. 111–143.

Boykoff, M.T. and Boykoff, J.M. 2007, Climate change and journalistic norms: A case-study of US mass-media coverage. *Geoforum*, vol. 38, no. 6, pp. 1190–1204.

Carvalho, A. 2007, Ideological cultures and media discourses on scientific knowledge: Re-reading news on climate change. *Public Understanding of Science*, vol. 16, no. 2, pp. 223–243.

Cronon, W. 1995. *Uncommon Ground: Toward Reinventing Nature* (Vol. 95). New York: Norton.

Denton, F. 2002, Climate change vulnerability, impacts, and adaptation: Why does gender matter? *Gender and Development*, vol. 10, no. 2, pp. 10–20.

Dunaway, F. 2000, Hunting with the camera: Nature photography, manliness, and modern memory, 1890–1930. *Journal of American Studies*, vol. 34, pp. 207–230.

Feygina, I., Jost, J.T. and Goldsmith, R.E. 2010, System justification, the denial of global warming, and the possibility of "system-sanctioned change." *Personality and Social Psychology Bulletin*, vol. 36, no. 3, pp. 326–338.

Finucane, M.L., Slovic, P., Mertz, C.K., Flynn, J. and Satterfield, T.A. 2000. Gender, race, and perceived risk: The 'white male' effect. *Health, Risk and Society*, vol. 2, no. 2, pp. 159–172.

Fisher, D.R., Waggle, J. and Leifeld, P. 2013, Where does political polarization come from? Locating polarization within the US climate change debate. *American Behavioral Scientist*, vol. 57, no. 1, pp. 70–92.

Flynn, J., Slovic, P. and Mertz, C.K. 1994, Gender, race, and perception of environmental health risks. *Risk Analysis*, vol. 14, no. 6, pp. 1101–1108. doi:10.1111/j.1539–6924 .1994.tb00082.x.

Hoffman, A. 2011, Talking past each other? Cultural framing of skeptical and convinced logics in the climate change debate. *Organization and Environment*, vol. 24, no. 1, pp. 3–33.

Jacques, P.J. 2012, A general theory of climate denial. *Global Environmental Politics*, vol. 12, no. 2, pp. 9–17.

Jacques, P., Dunlap, R. and Freeman, M. 2008, The organisation of denial: Conservative think tanks and environmental scepticism. *Environmental Politics*, vol. 17, no. 3, pp. 349–385.

Kahan, D.M., Braman, D., Gastil, J., Slovic, P. and Mertz, C.K. 2007, Culture and identity-protective cognition: Explaining the white-male effect in risk perception. *Journal of Empirical Legal Studies*, vol. 4, no. 3, pp. 465–505.

Kimmel, M.S. 1993, Invisible masculinity. *Society*, vol. 30, pp. 28–35.

Kimmel, M.S. 2007, Racism as adolescent male rite of passage – ex-Nazis in Scandinavia. *Journal of Contemporary Ethnography*, vol. 36, pp. 202–218.

Knight, G. and Greenberg, J. 2011, Talk of the enemy: Adversarial framing and climate change discourse. *Social Movement Studies*, vol. 10, no. 4, pp. 323–340.

Krosnick, J.A., Holbrook, A.L. and Visser, P.S. 2000, The impact of the fall 1997 debate about global warming on American public opinion. *Public Understanding of Science*, vol. 9, no. 3, pp. 239–260.

McCright, A.M. and Dunlap, R.E. 2011a, Cool dudes: The denial of climate change among conservative white males in the United States. *Global Environmental Change*, vol. 21, no. 4, pp. 1163–1172.

McCright, A.M. and Dunlap, R.E. 2011b, The politicization of climate change and polarization in the American public's views of global warming, 2001–2010. *Sociological Quarterly*, vol. 52, no. 2, pp. 155–194.

Miller, G.E. 2004. Frontier masculinity in the oil industry: The experience of women engineers. *Gender, Work and Organization*, vol. 11, no. 1, pp. 47–73.

Olofsson, A. and Rashid, S. 2011, The white (male) effect and risk perception: Can equality make a difference? *Risk Analysis: An Official Publication of the Society for Risk Analysis*, vol. 31, no. 6, pp. 1016–1032.

O'Shaughnessy, S. and Krogman, N.T. 2011, Gender as contradiction: From dichotomies to diversity in natural resource extraction. *Journal of Rural Studies*, vol. 27, pp. 134–143.

Park, L.S. and Pellow, D. 2011, *The Slums of Aspen: Immigrants vs the environment in America's Eden*. New York: New York University Press.

Schmidt, C.W. 2010, A closer look at climate change skepticism. *Environmental Health Perspectives*, vol. 118, no. 12, pp. A537–A540.

Smith, A. 2005, *Conquest: Sexual violence and American Indian genocide*. Cambridge, MA: South End Press.

Taylor, D.E. 1997, American environmentalism: The role of race, class and gender in shaping activism. *Race, Gender and Class*, vol. 5, p. 1662.

Taylor, D.E. 2014, The state of diversity in the mainstream environmental sector. *Green 2 0*, Washington, DC.

Terry, G. 2009, No climate justice without gender justice: An overview of the issues. *Gender and Development*, vol. 17, no. 1, pp. 5–18.

Selected transcripts

'Conservatives Want a Street Fighter' 2012, *The Rush Limbaugh Show*, radio program transcript, Premiere Radio Networks, January 18, viewed June 18, 2015, www.rushlimbaugh.com/daily/2012/01/18/conservatives_want_a_street_fighter.

'Did Obama Blame the Movie for the Colorado Shooting?' 2012, *The Rush Limbaugh Show*, radio program transcript, Premiere Radio Networks, October 11, viewed June 18, 2015, www.rushlimbaugh.com/daily/2012/10/11/did_obama_blame_the_movie_for_the_colorado_shooting.

'Partisan Media Tries to Link Rush and Koran-Burner from Cape Girardeau' 2010, *The Rush Limbaugh Show*, radio program transcript, Premiere Radio Networks, September 10, viewed June 18, 2015, www.rushlimbaugh.com/daily/2010/09/10/partisan_media_tries_to_link_rush_koran_burner_from_cape_girardeau.

'Polls Put Liberals on Suicide Watch' 2012, *The Rush Limbaugh Show*, radio program transcript, Premiere Radio Networks, October 9, viewed June 18, 2015, www.rushlimbaugh.com/daily/2012/10/09/polls_put_liberals_on_suicide_watch.

'A Romney Girl on Newt's Weaknesses' 2012, *The Rush Limbaugh Show*, radio program transcript, Premiere Radio Networks, January 31, viewed June 18, 2015, www.rushlimbaugh.com/daily/2012/01/31/a_romney_girl_on_newt_s_weaknesses.

'Salt Doesn't Kill You After All' 2012, *The Rush Limbaugh Show*, radio program transcript, Premiere Radio Networks, June 4, viewed June 18, 2015, www.rushlimbaugh.com/daily/2012/06/04/salt_doesn_t_kill_you_after_all.

'"War on Men" Column Causes Stir, Illustrates the Left-Wing Stuff We Laugh About But Lots of People Believe' 2012, *The Rush Limbaugh Show*, radio program transcript, Premiere Radio Networks, November 30, viewed June 18, 2015, www.rushlimbaugh.com/daily/2012/11/30/war_on_men_column_causes_stir_illustrates_the_left_wing_stuff_we_laugh_about_but_lots_of_people_believe.

'What's So Crucial About Free Contraceptives, Why Not Free Toothpaste?' 2012, *The Rush Limbaugh Show*, radio program transcript, Premiere Radio Networks, February 14, viewed June 18, 2015, www.rushlimbaugh.com/daily/2012/02/14/what_s_so_crucial_about_free_contraceptives_why_not_free_toothpaste.

'Why Conservatives are Wary of Romney' 2012, *The Rush Limbaugh Show*, radio program transcript, Premiere Radio Networks, January 13, viewed June 18, 2015, www.rushlimbaugh.com/daily/2012/01/13/why_conservatives_are_wary_of_romney.

6 Hegemonic masculinity in three parts

Phoebe Godfrey

Conservative radio host Michael Savage in response to the Pope's Encyclical May 24th, 2015, focusing on climate change, declared on the June 16 edition of his show: "The pope is a danger to the world." He continued by calling the Pope a "great deceiver," "stealth Marxist," and "eco-wolf in pope's clothing," and comparing him to the false prophet in the book of Revelation "directing mankind to worship the Antichrist." Savage concluded that "we are living in global tyranny right now."

(Robbins, 2015)

Dylann Roof who shot nine Black church goers on June 17th 2015, told them as he shot into the crowd, "You rape our women, and you're taking over our country, and you have to go."

(Sanchez and Forster, 2015)

Chief Justice Scalia, who voted on June 26th, 2015 in opposition to the Supreme Court Ruling on gay marriage, identifies it as a "…threat to American democracy."

(Pavlich, 2015)

Social and gender theorist Rob Connell back in 1987 coined the term 'hegemonic masculinity,' based on the work of the Italian Marxist theorist Antonio Gramsci. Simply stated, hegemonic masculinity is a culturally based way of organizing society, everything from ideologies, to institutions, to identities – all to ensure male domination. What is added to the pre-existing idea of just plain old 'masculinity' and/or 'patriarchy' was the recognition that the concept of hegemonic masculinity recognizes that there are many 'masculinities' and that although an entire culture may aspire to the hegemonic version, few men may actually embody it. In fact, hegemonic masculinity is more of a cultural ideal that, although it is first and foremost about male domination of women, also intersects with ideologies, institutions, and identities based on race, social class, nationality, religion, body type/size, etc. For example, if we take popular action films and look at how a lead male character embodies masculinity, we will see that he is most likely white (and, if not, then still upholding the norms of white supremacy), athletic, tall, aggressive, arrogant, 'lone wolf'/'take-no-shit-from-anyone,' heterosexual/sexually dominant, patriotic to the USA, Christian, and even if he is a rebel he is still seen as fighting 'evil,' as in subordinate masculinities embodying all that he is not. We can see this hegemonic masculinity in everything from truck advertising (driving over/dominating nature, truck power as ego), to sports (needing to be number 1/aggressive/revering physical strength), to Wall Street (boasting a 'bull-market' and scoring high, wealth as the power to dominate others), to our foreign policy (USA number 1/'nuke 'em'/'these colors don't run'), to the interpretation of Christianity, ('Fags burn in hell'/'Women are to be silent in church'), to the rape culture on college campuses ('nice ass, baby'). As Connell and Messerschmidt state, writing years after Connell's first article, hegemonic masculinity exists "At a society-wide level" ... where "there is a circulation of models of admired masculine conduct, which may be exalted by churches, narrated by mass media, or celebrated by the state. Such models refer to, but also in various ways distort, the everyday realities of social practice" (2005, p. 838). What they mean by this is that, although most men in a given society regardless of their social status (including their race, class, body size, nationality, religion), admire hegemonic masculinity, at the same time it 'distorts' and even perverts their own experiences as individuals. Therefore, these models of hegemonic masculinity "express widespread ideals, fantasies, and desires" (Connell and Messerschmidt, 2005, p. 838) that a given society constructs as the social ideal.

Having now explained the concept of 'hegemonic masculinity,' I wish to turn to my epigraphs and apply it to them as a way to better theorize what they mean and why particular individuals think/act/speak as they do.

The Pope's Encyclical (2015) comprises 184 pages – much of which takes on global capitalism and the mantra of unlimited growth, consumption, and environmental destruction. For example, he states,

> The idea of infinite or unlimited growth, which proves so attractive to economists, financiers and experts in technology ... is based on the lie that there is an infinite supply of the earth's goods, and this leads to the planet being squeezed dry at every limit.
>
> (ch. 3, section II, para. 106)

In other words, the 'hegemonic masculinity' that we see embodied in capitalism (the pursuit of profit through unlimited consumption, hence environmental destruction) is being critiqued by one in the role of what is expected to also exercise, or at least support, the same hegemonic masculinity. In addition, the theme of 'our home' was expressed here when he states, "Never have we so hurt and mistreated our common home as we have in the last 200 years" (ch. 1, section VI, para. 53). Again, by owning our mistreatment of the Earth (usually gendered as 'female,' as in Mother Earth), the Pope is failing to exhibit hegemonic masculinity. Therefore, when we look at how many who have made their media careers out of practicing hegemonic masculinity, we can understand why they need to attack, insult, and ridicule the Pope, much like why socially decreed 'effeminate' men have had to endure the same treatment. By attacking the one who isn't toeing the line, they are reasserting their own dominance. For Michael Savage to refer to the Pope as if he were the one engaging in 'global tyranny' or that he is directing "mankind to worship the Antichrist," we can know that the Pope has threatened a sacred aspect of our society – and that is hegemonic masculinity, which is of course responding exactly in true character form.

The tragedy of Dylann Roof is not only that nine innocent people were slaughtered in their own church, but also that what Dylann did is no different than what we have recently seen so many 'boys in blue' do, or what we have seen our military do in other countries, or what we have seen throughout our history where white males (or again those males of other races who have agreed to support the status quo) enact a narrative of their white male supremacy by killing, maiming, torturing, and oppressing other males deemed inferior either due to race, and/or class and/or nationality and/or religion, and/or sexuality, etc. In addition, his words/opinions are ones that have been used as justification for racist violence, that keep the one acting out hegemonic masculinity as justified in his actions. Those nine innocent African Americans 'asked for it' because they (regardless of sex) "rape our women … are taking over our country." Time and time again, these ideas/words have been spoken by white males just as they are the ones who are 'raping women,' who are 'taking over a country,' but because they are the ones in hegemony this domination is seen as normal, as just; hence so is the violence they commit. When we look at images of Dylann Roof we see not strength but, rather, a pathetic young white man who clings to these empowered ideals as if they were his religion, which in many ways they are. Even if he were of large stature, his level of ignorance and distorted self-righteousness would still speak to his contemptible attempt to enact hegemonic masculinity.

Finally, we have the Supreme Court ruling making gay marriage legal in all US states and territories. As a women married to another woman, I am of course pleased that the absurdity of our previous situation (married in this state but not in that) has been resolved. However, if I step back, the whole idea that a government based on the constitutional premise of the separation of church and state previously didn't recognize people's right to marry whoever they chose is absurd. Of course, there never has been genuine separation of church and state – and probably never will be – but it should nevertheless remain an ideal. Homophobia is of course a key attribute of hegemonic masculinity, which is why women are allowed much more freedom in their gender expressions than men. I, and all women, can wear men's clothing with little fear of attack, whereas men who choose to wear women's clothing and/or fully 'pass' as women are in constant danger of attack. This is because masculinity is the valued gender – of course women want to be like men, but why would any man want to be like a woman? Therefore, when Chief Justice Scalia says the new ruling is a "threat to American democracy," what he is really saying is that it is a threat to

Figure 6.1 Hegemonic masculinity (source: photograph by Phoebe Godfrey, 2015).

52   P. Godfrey

the version of 'American democracy' that isn't in fact a democracy at all. Rather, it is the fantasy version that we claim has existed continuously since 1776. This would be the version that existed during 200 years of slavery, that didn't give women the vote until 1920, that still allows for race and class to be the determiners of our criminal justice system, and that has the highest poverty level in the industrialized world. In short, the 'democracy' he is talking about is that defined by hegemonic masculinity, and in this regard he is right – it is threatened, and thankfully so.

Connell and Messerschmidt (2005) state in their conclusion: "The making and contestation of hegemony in historically changing gender orders is a process of enormous importance for which we continue to need conceptual tools" (p. 854). Therefore, in applying hegemonic masculinity to these three recent social events, my hope is to show how there is not only a constant power struggle going on in terms of how societies are organized, but that it is never clear who/what ideologies/institutions/identities will prevail, and what transformations will occur in the process. The Pope has launched a much-needed global conversation on climate change, just as the killing of the nine 'martyrs' has heightened the need for national reform on race/racism beginning with the Confederate flag, and the Supreme Court ruling, although a direct victory for all same-sex couples seeking 'legally sanctioned' marriages should open up more discussion on the treatment of LGBTQ folks nationally/globally, as well as the need for national healthcare and other ways of accessing social benefit. To achieve all these socially progressive – hence positive in terms of authentic democracy – things, we must continue to deconstruct hegemonic masculinity and create many other ways to be male, female, and ultimately human.

References

Connell, R.W. and Messerschmidt, J.W., 2005. Hegemonic Masculinity: Rethinking the Concept. *Gender and Society*, 19(6), pp. 829–859.
Pavlich, K., 2015. June 26, Breaking: Supreme Court Rules 5–4 Gay Marriage is a Constitutional Right, Bans Struck Down. Available at: http://townhall.com/tipsheet/katiepavlich/2015/06/26/breaking-supreme-court-upholds-state-bans-on-gay-marriage-n2017268 (accessed September 29, 2015).
Pope Francis Encyclicals, 2015. Laudato Si, On Care for Our Common Home. Available at: http://w2.vatican.va/content/francesco/en/encyclicals/documents/papa-francesco_20150524_enciclica-laudato-si.html (accessed September 29, 2015).
Robbins, D., 2015. June 18, Conservative Media vs. The Pope: The Worst Reactions to Pope Francis' Climate Change Encyclical. Available at: http://mediamatters.org/research/2015/06/18/conservative-media-vs-the-pope-the-worst-reacti/204037 (accessed September 29, 2015).
Sanchez, R. and Forester, P., 2015. June 18, "You Rape Our Women and Are Taking Over Our Country" Charleston Church Gunman told Black Victims. *Daily Telegraph*. Available at: www.telegraph.co.uk/news/worldnews/northamerica/usa/11684957/You-rape-our-women-and-are-taking-over-our-country-Charleston-church-gunman-told-black-victims.html (accessed September 29, 2015).

7 Towards

Coco Gordon aka SuperSkyWoman

| |: Keep it simple and treed: | | Key line.
  Wild one.

Towards
  "the problem is the solution":
or how permaculture
hones a globally-warmed poet
  to see top-down wide-angle
differently

  A Sea one...

let's talk about weeds dynamic accumulators
  Telling acid or alkaline.
  before we got here
In a zone of alarm
useful healings, When we noticed

  taught what we noticed
tender subtle our destinations
what song are you singing??

kindom I beg of you.

hear the old stain? No
  too small
behind us
  that far,

let's not talk about the heart
get one mend one
  & seeds? Still eating them?
Are you eating the future, said Bill Mollison...
We had time then
  safe we thought
we will need all our seeds to plant.

Strip sambuco elderberry
to innerbark
    to make paper
while green still protects us
    that beta cellulose...

sharpen knife.
    Let's not talk about the heart
carved
writing on the wall
debris tumbled
trees
    running tsunamis
    indistinct un-named
    clean up after
the
    running water
where are you going?
    little beluga whale
acoustic wonder

fallacies:
post-colonial studies (of extinctions)
| |: keep it simple & treed: | |

"does a disease of resentment" Yes I said that infect
    global study
censor
    new historicism (disruptor police state
stoic)
tear-gas away at 'em
they sit to say
Oh say do you see?
    Flooding (up & away)
over timber
    degrees
    simplified
thinking,
Renewable fish ... (GMO)

substitute life
    after death
who says environmentalists?

Jailed without cause
    would have to shut up shop
who are the canaries
sit
    & feel

tremor
   wave
pressure
   from feeling to emotion
ozone at street level

   feet misplaced from protection
   blanket safe
up there
   down under
permafrost
   melt
ice
darken
   low lift.

Would a forest without cause

have to shut up shop?
   That's what we've done
death by plastic
   oil &
gas
   endocrine disruptor
species
   failure slow VS
brittle reefs
   fast showing.

   Do we dare to dive
   & tell
science has lasting
   elastic
tentacles
   & the contrasts feel

larger than life.

8 MAN still #63

Steve Cutts

Figure 8.1 MAN still #63 by Steve Cutts.

9 Population, climate change, and the embodiment of environmental crisis

Jade S. Sasser

At the start of the millennium, a topic that had previously garnered scant attention in the realm of global climate change (GCC) seemed to leap to prominence in public discussions: the role of population growth. For example, in December 2009, at the United Nations (UN) Climate Change Conference (COP 17) in Copenhagen, Zhao Baige, the Vice Minister of the National Population and Family Planning Commission, stated that "dealing with climate change is not simply an issue of CO_2 emission reduction but a comprehensive challenge involving political, economic, social, cultural and ecological issues, and the population concern fits right into the picture" (Hull, 2009). Referring to China's infamous one-child policy, the Minister then argued that 400 million births had been averted, which saw a saving of 1.8 billion tons of CO_2 each year. Claiming a moral authority, Zhao argued that China's population policy had provided benefits across sectors, and on a global scale. In this case, a state that has long been thought of as a pariah for its population policy was able to use well-worn global population anxieties to represent itself as environmentally and socially responsible.

Zhao was not the only public figure calling attention to population and GCC. When, during a 2009 meeting attended by Hilary Clinton and the Indian Environment Minister, a panelist noted that "it's rather odd to talk about climate change and what we must do to stop and prevent the ill effects without talking about population and family planning," Clinton concurred, adding, "and yet, we talk about these things in very separate and often unconnected ways" (Plautz, 2014). In an interview the following year, Gloria Steinem argued that "overpopulation is still the biggest reason for global warming, for all the pressure on the environment," adding that allowing women to make their own decisions about childbearing has direct and beneficial impacts on the environment. Mary Robinson, former President of Ireland and a prominent women's rights and climate justice advocate, was quoted at Durban's international climate change meetings speaking in favor of family planning, noting that "if we were to solve this problem we would not only help these women ... we would also do great work for the climate" (Goering, 2011).

These statements assert that global attention to climate change also offers an opportunity for Global North audiences to think about women's sexual and reproductive lives and needs. But which women? Who is included in that category, and under what conditions? The fact that the above statements were made by wealthy, 'white' Western women highlights the key differences between who can speak with authority on GCC and women's fertility, who can identify problems, and who is targeted for interventionist solutions. As the rest of this chapter will demonstrate, these differences are organized along the axes of race, class, and geography. This project joins the broader body of intersectional feminist scholarship in an effort to "harness the most effective tools … to illuminate how intersecting axes of power and inequality operate to our collective and individual disadvantage" (Cho et al., 2013, p. 795).

According to the logic linking population growth with GCC, broad access to contraceptives offers three key solutions: first, it facilitates women's control over reproductive decision-making, thereby enhancing their social empowerment; second, it helps reduce the future number of greenhouse gas-emitting resource users; finally, it reduces the burden on families and communities as they struggle to adapt to the already apparent effects of GCC, by ensuring that there are fewer demands on food, education, healthcare, and other basic social services. Tying this framework together is a concept that I refer to as women's *embodied environmental responsibility*: the perspective that, through contraceptive use, women fulfill a social responsibility to help slow population growth and the associated environmental pressures.

Climate change, like all things, is an embodied experience. As Bennett (2010) reminds us, all matter is alive in the sense of a complex, interwoven web of materials, all affecting each other, competing, forming alliances, initiating new processes, and dissipating others. Humans are inextricably enmeshed in these webs, or assemblages, which include organisms, matter, events, processes, and occurrences. Thus GCC is a living phenomenon, wherein shifts in temperatures, weather patterns, atmospheric concentrations of gases, and policy negotiations act on, with, and through human bodies in a variety of complex ways that are seen and unseen. The distinctions between bodies, objects, and contexts are blurred (Coole and Frost, 2010). One obvious way this happens is in the health impacts of GCC – through changing infectious disease patterns, increased deaths from heatwaves and intense storms, malnutrition due to decreased crop yields, and hygiene and sanitation problems caused by shortages of clean water (IWGCCH, 2009; Portier et al., 2010; Whitmee et al., 2014). A less obvious – though equally important – way in which explanations for climate change become attached to particular kinds of bodies and communities is through discourse. As Butler (1993, p. 10) argues,

> to claim that discourse is formative is not to claim that it originates, causes, or exhaustively composes that which it concedes; rather, it is to claim that there is no reference to a pure body which is not at the same time a further formation of that body.

Discourses positioning the fertility of poor women of color (WOC) as a driver of, and solution for, climate change provide a useful tool for analyzing the ways in which these bodies are always already marked as problematic and environ-mentally polluting.

It is also important to note that knowledge is developed within bodies as well – bodies whose perspectives are limited, partial, and governed by embodied experience (Haraway, 1988). Scientists, policy-makers, and others who link women's reproduction to climate change cite just one component of the body – its reproductive functions – as if they exist, separate and dislocated, from other components of women's lived realities. This fragmenting, facilitated by modern medicine, serves to separate women from their embodied experiences, treating "the person as a machine and assum[ing] the body can be fixed by mechanical manipulations" (Martin, 2001, p. 19). This chapter exposes this fragmentation and works to "reintegrate the whole person from the jigsaw of parts created by modern scientific medicine" (Martin, 2001, p. 159) – as well as, in this case, through environmentalist discourse. As DiChiro (2008, p. 285) argues,

> all environmental issues are reproductive issues; efforts to protect the health and integrity of natural systems – water, air, soil, biodiversity – are struggles to sustain the ecosystems that make all life possible and enable the produc-tion and reproduction processes upon which all communities (human and non-human) depend.

I assert that we will not find solutions to GCC, nor to questions around popula-tion, in technology focused contraceptive distribution campaigns, but rather in social movements guided by intersectional frameworks and practices such as reproductive justice (RJ) – which links women's reproductive capacities to issues of agency, political, economic, and social empowerment, and human rights (Ross, 2006).

In what follows, I further explore the concept of embodied environmental responsibility. First, I ground the concept in the complex history of how inter-national development actors have framed the links between population, gender, and environment, alternating between blame and empowerment while firmly defining women as responsible for environmental problems. I then explore two empirical examples, the first of which links climate change to population growth through narratives of impending environmental catastrophe and contraceptive necessity. The second case explores a blogger's approach to gendered embodied environmental responsibility, centered on childlessness and environmental ethics. Notably, its author uses an intersectional analysis to argue that, as an American, middle-class white woman, she and others like her should willingly shoulder the reproductive blame for environmental problems. I follow with an analysis of the ways in which RJ and climate justice (CJ) activists can forge movements foregrounding reproductive bodily integrity for poor women and WOC, and broader movements for a healthy population and planet.

Women and embodied environmental responsibility

International development narratives linking women, gender, and environment have usually operated along two lines: women as victims or potential solvers of environmental problems. Feminist scholars have demonstrated the ways in which WOC bodies have been located in images of the Earth, situated within these images as static, timeless, and closer to nature. These images circulated frequently in the 1980s and 1990s in environmental development materials, which represented women's environmental roles as both natural and universal (Leach, 2007). Persistent women–environment linkages (Resurreccion, 2013) have endured through the decades, despite sustained critique from feminist scholars and activists. One of the reasons for this persistence is the strategic benefits they offer: collapsing the complex conditions of women's lives, experiences, values, and environmental roles, regardless of racial, ethnic, economic, and geographic contexts offers women a particular kind of visibility in international development and environmental policy-making. Similarly, policy discourses linking population growth and environmental degradation have been pervasive in development circles from the mid-twentieth century, comprising a 'blueprint development' narrative used to explain a range of complex problems and standardize solutions (Roe, 1991). Population growth, and women's reproduction by extension, have consistently been blamed in mainstream development writing for a range of social, economic, and environmental problems (Connelly, 2008; Silliman and King, 1999; Hartmann, 1995).

For much of the twentieth century, policy-makers and activists advocated population control to slow population growth; however, after the 1994 International Conference on Population and Development (ICPD) in Cairo, the focus changed to supporting individual women's access to contraceptives for voluntary use. While the Cairo Program of Action did not eliminate the neo-Malthusian ideologies that identify population growth as a direct cause of food shortages, deforestation, air and water pollution, and a broad range of other environmental problems, it did center women's human rights via a new emphasis on sexual and reproductive health, including universal, voluntary access to contraceptives. The individual woman-centered approach in the document resulted from strategic alliances brokered by neo-Malthusian development practitioners, environmentalists, and feminist activists, who compromised their individual agendas to achieve policy consensus (Hodgson and Watkins, 1997; Hartmann, 2006). Since that time, population policy advocates have applied the Cairo Program approach to environmental problems including GCC, arguing that this framing aligns population interventions with progressive, social justice movements (Sasser, 2014; see also Mazur, 2010).

Thus, Western development agencies promote family planning as an embodied environmental responsibility for poor women – an 'ethical' way of behaving that manages fertility and reproduction, reduces environmental harm, and evades questions of resource consumption and the development of polluting technologies. Notably the Cairo agenda's emphasis on individual fertility and reproduction management reinforces the embodied responsibility of poor women in the Global South reproducing the argument that population is a key cause of resource depletion, extraction, and degradation at local and global scales. Thus, even as it calls for an end to coercive policies and top-down demographic targets, its linked focus on slowing population growth in the Global South and promoting sustainable development still suggests that women's reproduction is a direct cause of, and key solution for, environmental problems.

The climate–population linkage and the idea of women's embodied environmental responsibility is problematic. First, GCC is a complex, long-term process of atmospheric changes arising from the emission and accumulation of greenhouse gases (GG) in the atmosphere, as well as the destruction of carbon sinks – the forests, soils, and oceans that store carbon from the atmosphere. These factors are mediated by countless factors, including: modes of national economic production; the transportation and energy consumption habits of the relatively affluent; the extractive practices of corporations and militaries; industrial agriculture; production of meat for consumption; and the everyday livelihood practices of urban and rural communities. They are also influenced by activist and policy efforts to slow fossil fuel extraction and development, to develop cleaner alternative energy sources, and to shift diets away from methane-intensive beef, and toward more plant-based consumption (Klein, 2014). In other words, reducing the issue to a simple 'more people, more emissions' logic obscures the complex roles of powerful institutions, affluence-based consumption, and social and policy activism (Peet et al., 2011). It also sets up a logic whereby blaming women's fertility, particularly that of poor WOC, appears to be common sense.

I challenge this logic, since GCC does not occur in an 'environment' that is separate from human bodies, but rather is occurring in, and on, and around, and through the body at all times. As Clark (2010) argues, the unpredictable changes in the natural world – the "inherent variability and volatility of our planet" – are closely tied to the "no-less-inherent vulnerability and openness of human bodies – to each other and to the wider universe" (p. 34). In other words, human bodies are deeply embedded in, constantly acting on, and acted in by the many elements of the non-human world in which we live. What Alaimo (2010) describes as the "literal contact zone between human corporeality and more-than-human nature" (p. 2) is the body itself, and the many manifestations of the ways in which nature is inseparable from bodily processes, functions, illnesses, and possibilities. Her concept of transcorporeality is instructive in understanding how "the human is always intermeshed with the more-than-human world" (p. 2), such that the environment is "always the very substance of ourselves" (p. 4). In other words, binaries between human/non-human, the body/the environment simplify the complex interrelationships between the two (Oppermann, 2012, p. 104).

Thinking about bodies and environments through the lens of transcorporeality, and particularly the ways in which race, class, gender, and geography structure the ways in which bodies and environments are enmeshed with one another makes it possible to reimagine environmental issues like GCC in the context of our own embodied existences as intimately connected. It also opens up the possibility of asking questions about the extent to which humans and nonhuman climate and weather phenomena are mutually emergent and constitute each other (Neimanis and Walker, 2014). Scholars have proposed corporeal citizenship as a way of framing humans' inescapable embeddedness in both social and natural contexts and understanding the human body as porous but resistant, plural, and connected (Gabrielson and Parady, 2010). Considering human (and non-human) bodies and GCC this way challenges the unidirectional binary that posits bodies as distinct from, and solely acting on, the environment. It can also complicate embodied environmental responsibility by highlighting the ways in which marginalized women embody existing environmental crises – and all of the other socio-economic issues generating them.

RJ activists and scholars cite extensive histories of reproductive oppression in which women of color and immigrant women, including African Americans, Puerto Ricans, Mexicans, and Native Americans have been targeted for coercive sterilization, abusive state welfare policies, restricted access to comprehensive contraceptive options, and racialized campaigns characterizing their fertility as socially, politically, economically, and environmentally threatening (Nelson, 2003; Silliman et al., 2004; Solinger, 2007). These examples demonstrate the ways in which 'external' environments – whether policy-based or discursive – operate on and through bodies of color. Intersectionally, however, access to comprehensive health services is also a function of other social categories, given that poor and disabled white women have endured brutal histories of forced sterilization, institutionalization, and reproductive oppression. RJ exposes the historical patterning of reproductive racism – the ways in which access to reproductive self-determination has followed broader societal patterns of racial discrimination – to reject the color-blind language and rhetoric of individual privacy and choice promoted by the mainstream women's reproductive rights movement, and argues that privacy assumes access to resources and individual autonomy at a level that many women do not have (Luna and Luker, 2013). RJ acknowledges that women's reproductive lives are shaped in political, social, and economic context, in relationship to family and community (Asian Communities for Reproductive Justice, 2005). From an RJ perspective, climate population interventions continue a long, problematic history of scapegoating the bodies and fertilities of primarily poor WOC.

As an intersectional framework, RJ offers helpful ways of thinking about GCC and bodies because it challenges single-issue organizing, addresses multiple identities, and aligns well with other movements that critique race, gender, and class inequality (Ross, 2006). It also maps well with critical analyses of environmental racism and provides frameworks for understanding how toxic exposures disproportionately burden marginalized women. In her 2014 book, *This Changes Everything: Capitalism vs. the climate*, Naomi Klein writes evocatively about the disrupted cycles of fertility apparent in Gulf Coast wildlife and human populations following the 2010 BP oil spill, a key impact of the fossil fuel economy. Drawing direct connections between GG-emitting fossil fuel production and women's bodies, Klein explores the pervasive infertility, miscarriages, birth defects, and extreme birth ratio imbalances in African American and Native American communities located near oil and gas refinery and fracking sites. These embodied environmental crises in women's reproductive health are deeply linked to GCC and environmental pollution in ways that have little to do with mainstream framings of population and climate. In what follows, I analyze two examples of framings of climate change and women's embodied environmental responsibility, in order to highlight both the dominant ways these linkages are framed, as well as how intersectionality may be used to critique and undo these framings. They also provide opportunities to apply intersectional frameworks to other analyses of climate change impacts – from the development of new technologies, to the productive potential of forging alliances across RJ and climate-related social movements, a topic to which I will return in the conclusion.

(Dis)embodying and averting emissions

Contemporaneously, a British charity and think tank known as the Optimum Population Trust (OPT) launched a project website called "PopOffsets." With the tagline "smaller families, less carbon," PopOffsets (2015) juxtaposed the notion of shared environmental responsibility with a consumerist model of reducing emissions:

> We are all responsible for carbon dioxide and other greenhouse gases. Large amounts come from fossil fuels used for travel, power and heating and from making the goods and services we all use. Individuals and organizations can offset their carbon emissions by supporting activities that have either reduced or no emissions, or that capture the gases released by others.
>
> (PopOffsets, 2015)

OPT offers the opportunity to invest in family planning projects as a form of carbon offset – a trading scheme in which carbon dioxide or GG is reduced in one place to offset an emission made elsewhere. Users click a link to an online calculator tool which offers individuals or organizations the option of contributing various amounts based on a calculation of the number of tons of greenhouse gases emitted through their everyday activities, such as driving a car, traveling in an airplane, purchasing energy-guzzling household appliances, or taking a week-long vacation at an air-conditioned hotel. How is the offset calculated? According to the site, "We have chosen a figure of £10 – approximately $15 or €12 – which we feel comfortable that, if spent on family planning, after overhead, would save one tonne of carbon emissions" (PopOffsets, 2015). Population projects they have funded include a pop-up vasectomy clinic in New York; a nurse in Kenya who distributes contraceptive supplies from a backpack to families in rural areas; and a network of organizations in Ethiopia that integrate contraceptive distribution and family planning information into environmental conservation projects. Previous projects include providing heavily subsidized contraceptives to poor communities in Madagascar, and a youth sexual and reproductive services organization in the UK.

Climate population proponents often describe contraceptives in terms of their expected ecological usefulness, as a tool for reducing human numbers, a proxy for slowing resource consumption, and a technology that eliminates the messiness of complex human decision-making. This website extends these ideas by almost completely obscuring the role of women – with the possible exception of those women who are contributing funds to the offsets program – almost completely. All mention of women and gender is absent from their descriptions; fertility and reproduction are referred to largely in abstract terms, through GGE offsets; and consumers have the opportunity to engage in responsible environmental actions via their online consumption practices. But what is the role of contraceptive users? Do they figure into the offsets equation beyond the abstract?

Potential contraceptive users do figure into the equation – as potential targets, or recipients, of cost-effective aid. The website states:

> Carbon offsetting is a multinational, multibillion dollar market. PopOffsets is simply the most cost-effective and environmentally beneficial way of off-setting greenhouse gas emissions.... PopOffsets helps you to offset your footprint by helping others to avoid unplanned pregnancies with no environmental downside: it's win – win! Incidentally, supporting PopOffsets also sends a strong message to decision-makers that it's time to address the population issue – wisely, democratically and compassionately.
>
> (PopOffsets, 2015)

Contraceptives, and online investments in them, link individual consumers with specific women's bodies and reproductive lives through notions of responsibility, consumption, and choice. By advocating for private investment in contraceptive programs, OPT suggests that not only can environmental responsibility be embodied, it is also a shared responsibility in which the embodied fertility of abstract others is predicated on the consumerist responsibility of the privileged. Transforming humans into potential emissions, and potentially averted humans into potentially averted emissions, seemingly erases the role of social conditions in which women's reproductive behaviors are enabled, constrained, negotiated, and enacted. At the same time, these reductive transformations symbolically interpret poor WOC as (dis)embodied vessels for the environmental concerns of male and female consumers many miles away. While these consumers drive gas-guzzling vehicles, fly around the globe, and click and tap at the latest power-sucking devices, their very consumerism acts as a means of erasing the impact of their daily consumption practices. Their own embodiment of environmental responsibility is mediated by polluting technologies – and free of the discourses of blame leveled at marginalized women. This lays bare the problems with characterizing 'women' as a monolithic group (Mohanty,

1984) of actors, given that many of these privileged consumers are women. Thus, if race, class, and geography mediate the discourses linking women's bodies and climate change in the Global South, are affluent women in the Global North discursively erased? Do they embody environmental responsibility? Some would answer yes, and that there is in fact a greater responsibility for wealthier women to limit their childbearing for environmental reasons. The following case explores an example of this approach in depth.

Embodying responsibility

In August 2011, blogger Lisa Hymas wrote an attention-grabbing headline for the RH Reality Check blog: "I Am the Population Problem." Her article centered on a personal narrative in which she argued that, rather than the "Africans and Asians" who are often targeted in discussions of global population growth, women like herself should shoulder the reproductive blame: "the population problem is all about *me*: white, middle-class, American me. Steer the blame right over here" (Hymas, 2011, n.p., emphasis in original). Hymas detailed a long list of gadgets, technologies, cars, homes, foods, and household water consumed daily by what she describes as average Americans, ending with the argument that the average middle-class white woman is "just the sort of person who should *not* have kids." Her blog goes further to state that, regardless of individual environmentalist values and actions, such as limiting resource and energy consumption, taking public transportation, or keeping a low heating bill, simply living in the USA as a middle-class person makes one's environmental impact unsustainable.

One of the interesting elements of Hymas' article is that she argues that it is her specific position as a middle-class, white woman in the USA that shapes her resource consumption patterns, and that she has a unique obligation to adopt a sense of environmental responsibility – which she frames through the lens of the GINK ("green inclinations, no kids"). However, she replicates the problematic assumptions that collapse human numbers and environmentally polluting behaviors into one category, specifically by highlighting, then dismissing, the role of environmental values and behaviors. While she does not suggest that all middle-class US residents should avoid having children, she argues that it should be generally easier for women to forgo childbearing, free of shame and blame. Ironically, Hymas' stated goal is to advocate for what she calls "real reproductive freedom" (n.p), including changing cultural norms around childbearing so that women aren't pressured to reproduce, as well as easing restrictions on permanent sterilization procedures such as tubal ligation.

Hymas' argument raises the same question discussed earlier. In addition, a compelling question is: What is the role of agency? While it is tempting to argue that Hymas' approach represents her own agency as an environmentally conscious individual, it is in fact not rooted in individual values, but rather in a context in which women are often expected to be ethical, 'green 'consumers with the transmission of green values to future generations part of women's work

(MacGregor, 2010). Despite the fact that Hymas rejects biological reproduction, she upholds a notion of gendered social reproduction in which women's roles as environmental actors exist primarily at the level of the individual, the household, and the private decision-making sphere. Yet, the issues she describes are in fact public and social embedded within broader contexts, including family and community realities and political and economic conditions. The very notion of agency itself is often constrained by these contexts and what they enable or foreclose for women. Suggesting that achieving GINK status is an individual, free, and ethical decision ignores both the gender socialization governing women's roles as green consumers, as well as the actuality that women's reproductive realities can never be disembedded from their broader social contexts.

As both cases demonstrate, the discourses linking women's reproductive bodies to climate change are grounded in familiar narratives that position women as simultaneously to blame for environmental problems, as well as being responsible for solving them. Whether these discourses are imposed by others or self-adopted, they reproduce a narrow framing of reproduction as always environmentally problematic.

Conclusion: embodying justice

Elizabeth Barajas-Roman and Betsy Hartmann (2009) assert that RJ offers important lessons for CJ movements by "drawing connections between the same powerful forces driving climate change at the global level and environmental injustice and gender discrimination at the local level" (p. 7). This is because RJ draws together biological reproduction with social reproduction, defined as the "intersecting complex of political-economic, sociocultural, and material-environmental processes required to maintain everyday life and to sustain human cultures and communities on a daily basis and intergenerationally" (p. 281). Social reproduction is suggested not as a gendered responsibility for sustaining myriad social conditions that fulfill human needs, but rather as the task of social justice movements and collective action. Environmental movements are vital to helping promote such non-gendered social reproduction across a variety of political, economic, and biophysical contexts, a reality that lays the basis for alliances among and between a range of social groups, across the lines of race, class, gender, geography, citizenship, physical ability, and age. While these alliances are already being forged across scales among movement actors in the EJ, RJ, and CJ movements, there are potential additional alliances to be made in the arenas of environmental public health and disease, disability, and immigration.

Two strategies in particular deserve further attention. First, GCC scholars and activists can help policy-makers shift questions about women's bodies and climate change to a public health perspective emphasizing GCC as an environmental health threat. Solutions would then look primarily to protect population health comprehensively, rather than drawing on a limited reproductive health framework that narrowly focuses on family planning focus. With a broader understanding, we may better respond to climate change's impacts on infertility, infectious diseases, malnutrition and food insecurity, hygiene and sanitation, and natural disasters and heat events, thereby benefitting a range of marginalized women and men, as well as increasing the resources that may be mobilized. It also expands the potential allies with whom activists may develop more comprehensive organizing strategies and frameworks. Second, aligning with local and global movements for just sustainability broadens the framings of both CJ and RJ. Just sustainability sits at the nexus of environmental justice and sustainability as a framing in which "wider questions of social needs and welfare, and economic opportunity, are integrally connected to environmental concerns" (Agyeman et al., 2003, p. 2). Its focus on securing "a better quality of life for all, now and into the future, in a *just* and equitable manner, whilst living within the limits of supporting ecosystems" (ibid., p. 5, emphasis in original) aligns with social reproduction and RJ frameworks that situate environment, climate, and reproduction within the broader social contexts of everyday life. Advocating for just sustainability acknowledges the inextricable links between life forms, economic systems, and political contexts, and pushes advocates to create policies and systems that support, sustain, and affirm life. At the same time, just sustainability can only be truly achieved by recognizing the extent of our dependence on, and entanglement with, all forms of "matter" – and acknowledging that this very dependence is an ongoing challenge in finding more just solutions.

As this chapter has argued, the notion of women's *embodied environmental responsibility* – the idea that, through the careful management of reproduction, women fulfill a social responsibility to help slow global population growth, and associated pressure on the environment – represents a fragmented understanding of bodies, reproduction, climate change, and the environment more broadly. Building cross-movement alliances for social, environmental, climate, and reproductive justice offers a solid path toward necessary reintegrations of body, environment, community, and planet.

References

Agyeman, J., Bullard, R., and Evans, B., eds, 2003. *Just Sustainabilities: Development in an unequal world*. Cambridge, MA: MIT Press.

Alaimo, S., 2010. *Bodily Natures: Science, environment, and the material self*. Bloomington: Indiana University Press.

Asian Communities for Reproductive Justice, 2005. *A New Vision for Advancing our Movement for Reproductive Health, Reproductive Rights, and Reproductive Justice*. Oakland, CA: ACRJ.

Barajas-Roman, E. and Hartmann, B., 2009. Reproductive justice, not population control: Breaking the wrong links and making the right ones in the movement for climate justice. Prepared for the WE ACT for Environmental Justice Conference on *Advancing Climate Justice: Transforming the Economy, Public Health and Our Environment*. New York.

Bennett, J., 2010. *Vibrant Matter: A political ecology of things*. Durham, NC: Duke University Press.

Butler, J., 1993. *Bodies that Matter: On the discursive limits of 'sex'*. New York: Routledge.

Cho, S., Crenshaw, K., and McCall, L., 2013. Toward a field of intersectionality studies: Theory, applications, and praxis. *Signs*, 38(4), pp. 785–810.

Clark, N., 2010. Volatile worlds, vulnerable bodies: Confronting abrupt climate change. *Theory, Culture and Society*, 27(2–3), pp. 31–53.

Connelly, M., 2008. *Fatal Misconception: The struggle to control world population*. Cambridge, MA: Harvard University Press.

Coole, D. and Frost, S., 2010. *New Materialisms: Ontology, agency, and politics*. Durham, NC: Duke University Press.

DiChiro, G., 2008. Living environmentalisms: Coalition politics, social reproduction, and environmental justice. *Environmental Politics*, 17(2), pp. 276–298.

Fairhead, J. and Leach, M., 1996. *Misreading the African Landscape: Society and ecology in a forest-savanna mosaic*. Cambridge: Cambridge University Press.

Gabrielson, T. and Parady, K., 2010. Corporeal citizenship: Rethinking green citizenship through the body. *Environmental Politics* 19(3), pp. 374–391.

Goering, L., 2011. Family planning 'effective' but unpopular climate change solution. Thomson Reuters Foundation. Available at: www.trust.org/item/?map=family-planning-effective-but-unpopular-climate-change-solution (accessed May 20, 2015).

Goldman, M., 2006. *Imperial Nature: The World Bank and struggles for social justice in the age of globalization*. New Haven, CT: Yale University Press.

Haraway, D., 1988. Situated knowledges: The science question in feminism and the privilege of partial perspective. *Feminist Studies*, 14(3), pp. 575–599.

Hartmann, B., 1995. *Reproductive Rights and Wrongs: The global politics of population control*. Cambridge, MA: South End Press.

Hartmann, B., 2006. Liberal ends, illiberal means: National security, 'environmental conflict', and the making of the Cairo Consensus. *Indian Journal of Gender Studies*, 13(2), pp. 195–218.

Hodgson, D. and Cotts Watkins, S., 1997. Feminists and neo-Malthusians: Past and present alliances. *Population and Development Review*, 23(3), pp. 469–523.

Hull, C., 2009. China right to link population to climate. *Canberra Times*, December 26.

Hymas, L., 2011. I am the population problem. RH Reality Check. Available at: http://rhrealitycheck.org/article/2011/08/25/i-population-problem-0/ (accessed May 15, 2015).

Interagency Working Group on Climate Change and Health (IWGCCH), 2009. A human health perspective on climate change: A report outlining the research needs on the human health effects of climate change. *Environmental Health Perspectives*. Available at: www.niehs.nih.gov/climatereport (accessed August 25, 2015).

Klein, N., 2014. *This Changes Everything: Capitalism vs. the climate*. New York: Simon & Schuster.

Kurtz, H.E., 2003. Scale frames and counter-scale frames: Constructing the problem of environmental injustice. *Political Geograph*, 22, pp. 887–916.

Leach, M., 2007. Earth Mother myths and other ecofeminist fables: How a strategic notion rose and fell. *Development and Change* 38(1), pp. 67–85.

Luna, Z. and Luker, K., 2013. Reproductive justice. *Annual Review of Law and Social Science*, 9, pp. 327–352.

MacGregor, S., 2010. A stranger silence still: The need for feminist social research on climate change. *The Sociological Review*, 57(s2), pp. 124–140.

Martin, E., 2001 (1987). *The Woman in the Body*. Boston, MA: Beacon Press.

Mazur, L., ed., 2010. *A Pivotal Moment: Population, environment, and the justice challenge*. Washington, DC: Island Press.

Mohanty, C.T., 1984. Under Western eyes: Feminist scholarship and colonial discourses. *boundary 2*, 12/13, pp. 333–358.

Mosse, D., 2004. *Cultivating Development: An ethnography of aid policy and practice*. London: Pluto Press.

Neimanis, A. and Walker, R., 2014. Weathering: Climate change and the 'thick time' of transcorporeality. *Hypatia*, 29(3), pp. 558–575.

Nelson, J., 2003. *Women of Color and the Reproductive Rights Movement*. New York: New York University Press.

Oppermann, S., 2012. Review of Alaimo, Bodily natures. *Environmental Ethics*, 34, pp. 103–106.

Peet, R., Robbins, P., and Watts, M., eds, 2011. *Global Political Ecology*. New York: Routledge.

Plautz, J., 2014. The climate-change solution no one will talk about. *The Atlantic*, November 1. Available at: www.theatlantic.com/health/archive/2014/11/the-climate-change-solution-no-one-will-talk-about/382197/ (accessed April 1, 2015).

PopOffsets, 2015. What we do and how it works. Available at: www.popoffsets.org/what_we_do.php (accessed September 28, 2015).

Portier, C.J., Thigpen, T., Tart, K., Carter, S.R., Dilworth, C.H., Grambsch, A.E., Gohlke, J., Hess, J., Howard, S.N., Luber, G., Lutz, J.T., Maslak, T., Prudent, N., Radtke, M., Rosenthal, J.P., Rowles, T., Sandifer, P.A., Scheraga, J., Schramm, P.J., Strickman, D., Trtanj, J.M., and Whung, P.-Y., 2010. *A Human Health Perspective On Climate Change: A report outlining the research needs on the human health effects of climate change*. Research Triangle Park, NC: Environmental Health Perspectives/National Institute of Environmental Health Sciences. Available at: www.niehs.nih.gov/climatereport (accessed August 25, 2015).

Pulido, L., 2000. Rethinking environmental racism: White privilege and urban development in southern California. *Annals of the Association of American Geographers*, 90, pp. 12–40.

Resurreccion, B., 2013. Persistent women and environment linkages in climate change and sustainable development agendas. *Women's Studies International Forum*, 40, pp. 33–43.

Roe, E., 1991. Development narratives, or making the best of blueprint development. *World Development*, 19(4), pp. 287–300.

Ross, L.J., 2006. *Understanding Reproductive Justice*. Atlanta, GA: Sistersong Women of Color Reproductive Health Collective.

Sasser, J., 2014. The wave of the future? Youth advocacy at the nexus of population and climate change. *The Geographical Journal*, pp. 102–110.

Silliman, J. and King, Y., eds, 1999. *Dangerous Intersections: Feminist perspectives on population, environment, and development*. Cambridge, MA: South End Press.

Silliman, J., Fried, M., Ross, L., and Gutierrez, E., 2004. *Undivided Rights: Women of color organizing for reproductive justice*. Boston, MA: South End Press.

Solinger, R., 2007. *Pregnancy and Power: A short history of reproductive politics in America*. New York: New York University Press.

Stein, R., ed., 2004. *New Perspectives on Environmental Justice: Gender, sexuality, and activism*. New Brunswick, NH: Rutgers University Press.

Whitmee, S., Haines, A., Beyrer, C., Boltz, F., Capon, A.G., Ferreira de Souza Dias, B.,

Ezeh, A., Frumkin, H., Gong, P., Head, P., Horton, R., Mace, G.M., Marten, R., Myers, S.S., Nishtar, S., Osofsky, S.A., Pattanayak, S.K., Pongsiri, M.J., Romanelli, C., Soucat, A., Vega, J., and Yach, D. 2014. Safeguarding human health in the Anthropocene epoch: Report of The Rockefeller Foundation – *Lancet* Commission on planetary health. *The Lancet Commissions.* Available at: http://press.thelancet.com/ PHCommission.pdf (accessed August 25, 2015).

10 Embracing environmental justice

A brief reflection

Antonia Darder

The Great Mother Wails (2008)

The Earth extends her arms to us;
Revealing through her nature the
changing condition of our existence.

She bends and twists,
Deflecting the swords of
Our foolishness,
Our arrogance,
Our gluttony,
Our deceit.

Unbridled by red alerts or amber warnings,
Her ire gives rise to monsoon winds,
Jarring us from the stupor of
Our academic impunity;
Our disjointed convolutions,
Our empty promises; our
black and white dreams.

Filled with unruly discontent,
we yearn to dominate her mysteries;
reducing her to microscopic dust,
we spit upon her sacredness,
tempting the fury of her seas.

We spill our unholy wars
upon her belly's tender flesh,
blazing dislocated corpses,
ignite her agony and grief.

Still, in love with her creations,
she warns of our complacency
to cataclysmic devastation,
rooted in the alienation of
our disconnection,

our rejection,
our oppression,
our scorn.
And still, we spin ungodly
tantrums of injustice
against her love,
against ourselves,
against one another.

When will we remove blindfolds
from our eyes?
When will we stretch our arms – to her?
When will the cruelty of our
Hatred cease; teaching us to
abandon the impositions of
patriarchy and greed?

Oh! that we might together renew
Our communion with the earth,
She, the cradle of humanity;
She, the nourishment of our seeds;
She, the beauty of the song within;
She, the wailing that precedes.

This poem speaks to critical issues of the environment in ways that hopefully compel us toward environmental justice, in all its manifestations – as collective beings and as individuals in this world we share. Concern and attention to the issue of the environment and climate change then constitute important ecological dimensions of my work as an activist, poet, and radical feminist educator. The treatment and condition of the Earth today is linked in my mind with the same unjust body politics that determine the structures and relationships of inequality and injustice within our everyday conditions. The blatant disregard for life, inherent in the exercise of capitalism, with its accompanying values of patriarchy, racism, homophobia, and incessant push for Western dominion over the Earth, has resulted in an uncertain and precarious future.

In the process, more often than not, we exist in societies that have forgotten that the Earth is a living organism and that climate changes are inextricably tied to the forms of stewardship we provide to both our own lives and that of the planet. Thus, the Earth functions as the great manifestation and essence of life that echoes our collective and individual existence. As such, we must come to recognize that our treatment and relationship with the Earth, upon which humanity depends, must remain a central preoccupation in our political and pedagogical labor as educators, community activists, and cultural workers.

This notion points to both a societal understanding of this relationship, as well as a very personal one; since inherent in the dialectical relationship between nature and humanity, spirit and matter, collectivity and individuality, diversity and universality is found the powerful essence of beingness that unites us, and the delight in our existence that propels our evolution. Hence, the environment, as both planetary and individual phenomena, is both witness and portrayer of our fluidity and the distortions and blockages that result in pain, destruction, and death. The wailing of the Earth is a warning that we must reconsider the decisions by which we live, just as much as the wailing of our souls may signal illness or the need for transformation. Both are reflexive experiences of that collective beingness that requires our participation for the expression of its emancipatory design. Thus, environmental justice reflects an important ethical progression, in the world and within each of us, which moves us toward a greater material manifestation of democratic possibilities, inherently situated in what Paulo Freire often called Unity in Diversity.

Embracing the political

Embracing environmental justice speaks to an emancipatory political expression that contends with the material possibilities of our human survival and our capacities to thrive on Earth. Wherever possible, it is imperative that we link the survival and health of individuals and communities to those larger societal structures and conditions that impact our subsistence. This is even truer today, as we have come to better understand that increasing changes in the Earth's climate are also responsible for dramatic effects in the Earth's potential to sustain life.

Beyond the extremes of focusing on the danger to human subsistence, climate changes caused by the avarice of neoliberal policies and corporate exploitation are on a serious collision course with the biodiversity and well-being of the planet. In response, the Earth's forces are stirred in unexpected ways, leaving us vulnerable to weather calamities and other so-called natural disasters – many of which are of our own making and, thus, could have been prevented. One of the most recent and more obvious examples of this phenomenon in the USA was experienced and witnessed by the residents of New Orleans in 2005, when Hurricane Katrina, considered to be one of the deadliest and most destructive hurricanes in our history, hit the city's coastline. Katrina left nearly 2000 people dead, countless homeless (most of them people of color and poor), and $108 billion in damages, as 80 percent of the city was submerged under water.

There is no question that the magnitude of the tragedy in New Orleans could have been averted had a series of timely, wise, and generous actions been taken and climate change warnings heeded long before the hurricane broke through the city levees. Yet, no sooner had the waters receded than the "shock doctrine" of disaster capitalism did rear its ugly head, and the same class of people who had once failed to heed warnings of impending disaster now positioned themselves to ravenously capitalize on the tragedy suffered by the most disenfranchised. This dynamic has been repeated in a variety of ways each time the Earth's body is further violated, whether it be by military bombings, pharmaceutical testing, agricultural pesticides, factory waste products, or industrial pollution of the air we breathe. Yet, concern, integrity, and accountability seem to be lacking in the ways in which the few who hold power, wealth, and privilege too often flagrantly ignore climate change and its links to all other aspects of life – leaving the majority of the world's population outside to contend with the sins of their greed.

Hence, on the political side of this equation, there is a dire need to study, dialogue, and engage in very real ways what is happening in our neighborhoods, our state, our nation, our world. Understanding our food supply and its movement from production to table is essential. Similarly, considering the many products we consume and the manner in which the carbon footprint of each impacts our future in far-reaching ways. Constructing pedagogical approaches in schools that make questions of the environment commonplace must become fundamental to education in these times. Such a pedagogy also calls upon us to become more conscious about our environment, as well as the consequences of those forms of human production that strip away our humanity and leave the earth vulnerable to its demise. The dialectical relationship, then, between participation in this dance of the environment and our individual lives must be more fully understood, as we engage what it means to be historical subjects of our own destinies.

Embracing the personal

I cannot even begin to embrace the dialectical dance of the political without embracing my personal history, without noting the conditions of my birth and how these, anchored upon the hegemonic imposition of US colonialism and its economic exploitation of Puerto Rico and our people, were to determine dramatically the course of my life. I was born under the shadow of "Operation Bootstrap," a policy wired to solidify US control over the island and the destiny of a people. The consequences were to weigh heavily upon my childhood; my mother, contending with poverty and the impact of her sterilization, social exile from the island, and believing the false promise of a better life in the USA, swept me away from the comfort and familiarity of my country.

The racialized impositions of culture, language, and economic impoverishment that proceeded became major themes and tensions of my existence in this country. In the process, I watched the negative impacts of these demons of migration act upon my mother, my aunts, my grandmother, myself, and others who followed the yellow brick road to the USA. I watched their bodies as they aged prematurely, with illnesses and sorrows that could not be consoled by the whitewashed multiculturalism of a society that perceived, and continues to perceive us, as deficient cultural beings – no matter how much education we possess nor the track record of our accomplishments.

This has been the landscape of my personal environment, upon which these conditions of oppression and invisibility have been both witnessed and reflected within the personal struggles of my life as the daughter of a schizophrenic mother who calmed her demons with booze and child abuse; as a young single mother living in poverty; as a nursing student accused by a professor of plagiarizing my final case study, because she could not believe I could have possibly done such an excellent job; people everywhere who have unscrupulously maligned the intention of my political commitments; men who have accused me of being too emotional to be taken seriously as a partner or as an intellectual. These are but few examples of the many personal assaults I have withstood in my life – assaults that I have come to reconcile with as simply reflections of the larger external environmental forces of poverty, racism, and sexism that have historically plagued populations perceived as anathemas of humanity.

The consequence, here, has been a full-blown personal battle to contend with the ravages of depression, poverty, isolation, and alienation as I have aspired to dig myself out of the societal prison of my colonized subjectivity. My decolonizing battles have been waged at every level of my internal environment – as a daughter, as a parent, as a lover, as a friend, as a comrade, as an artist, as a Puertoriqueña, as a woman, as a physically and psychologically battered human being. It has been a long and bitter struggle over six decades, which has included deep forays into politics, psychotherapy, pedagogy, and spirituality. Even today I am left to contend with the deeply internalized demons of our colonized lineage, which taunt me still into fears of being unworthy of love, respect, or even personal sojourn. This struggle, of course, has been further complicated by the cacophony of those in power who, wittingly or unwittingly, would reinforce the colonizing lies and distortions, were it not for a wild spirit of defiance and justice, who has been my true companion.

However, just as the larger environmental struggle of the planet requires collective action for its transformation, so too is the greatest gift to our personal evolution found in the tenacious forging of personal communities of struggle with those who remind us, when we forget, that what truly matters is that we remain connected to one another, our communities, the Earth, and that spiritual force that enlivens all existence. It is from this deeper knowing *with* others, from whence my identities, my political activism, my theoretical offerings, my pedagogical practices, and my daily spiritual expressions of radical love are born. Hence, just as we must listen to the wailing of our mother, the Earth, so too we must hear her wailing within our own bodies; this guides us to throw off the shackles of our personal and social oppressions and, instead, find that place from which we can labor together with grace, dignity, faith, and hope for the evolution of human consciousness and the making of a truly just world.

Without Borders (2013)

I am a soul without borders,
seeking freedom of life,
moralism squeezes me,
psychology drowns me,
politics represses me,
the academic stifles me.

I want to fly unencumbered
within every sphere
of my humanity,
without fear that
demons will imprison
me or clip my wings.

Tenderly, I seek the
solitude of my desire,
so I may lull my anxieties
with songs in my heart.

Drawing strength
from my courage,
I tumble down the
frescos of tyranny

of manipulation,
of pretension,
of corruption,
of destruction.

Naked and trembling,
I call out for love
to pour hotly over
the frigidness of my fears.

Fighting my shame,
I shatter the ancient
portals that obstruct
pathways to our beauty.

Releasing my passions
to the winds, I run with
the breathe of dreams,
crying,

screaming,
laughing
with joy.

My embrace of environmental justice, then, is vital to a personal and political existence *without borders*. It shapes my relationship between body, politics, and pedagogy, providing the full contours for my personal and political commitment against all forms of imperialism – against all ways of being that defy freedom from militarism, exploitation of labor, cultural domination, linguistic genocide, gender and sexual repression, all matter of violence, and escalating incarceration. And, as such, embracing environmental justice in my daily life, through the political and personal, keeps me ever conscious of the deep and sacred union that prevails within the many expressions of our humanity, as well as the joy that resides in the sanctity and preciousness of all life.

Part II

Air

Figure II.1 Air by Tina Shirshac.

Attitudes   **Philosophy**   **Pedagogy**   **Resilience**   **Theory**

11 The search for authenticity in a climate of denial

Denise Torres and Phoebe Godfrey

Air focuses attention on the connections between Attitudes, Philosophy, Pedagogy, Resilience, and Theory, and the ways in which these abstract concepts invisibly, yet substantially, infiltrate larger discussions and experiences of global climate change (GCC) as well as the social constructs of race, ethnicity, social class, gender, and other identities in very material ways. Just as air provides the necessary lift for birds to fly, Western outlooks and schools of thought create, support, and extend the reality of GCC, as well as the flight of the social narratives surrounding it. Yet, just as each new breath we take offers us the opportunity to become more fully present and therefore to transform our lives, so too can personal and social transformation come from clearly viewing our individual and collective roles in the environmental crises.

Reverberating throughout the works in this section is what climate justice activist Julia Lillian (Butterfly) Hill calls our "disposability consciousness," or modernity's relationship to waste as "one of the ways we *all* have internalized oppression ... those forms of oppression are a disconnect" (n.d.). Ms. Hill is noted for living in and protesting the cutting of redwood trees in California. Our denial of GCC and the disconnect between our attitudes and actions manifests in the fact that we continue to engage in deforestation, despite forests' essential role in creating the very air we breathe. Thus, in the first contribution to Part II, political cartoonist Khalil Bendib's 'Intelligent life' captures the logical disconnect of continued mass and individual consumption as the Earth chokes on our waste and where we see only tree stumps. Similarly, in his poem 'The science proves it (or not),' Lewis Vande Pallen speaks to our collective responsibility for and our blithe response to the progressively real consequences of GCC.

We take a more serious and historical turn with Douglas Herman's 'The canoe, the island, and the world,' which interrogates modernity's 'ego-bound' approach to life with specific attention to the separation of man from nature and science from wisdom. Herman elucidates how traditional knowledge supports wisdom and a view of all in the world as interconnected and related, rather than atomized objects to be used. 'Tlakaelel's view of climate change' demonstrates the resilience of traditional knowledges, with Bert Gunn providing insight into the syncretic indigenous teachings of Tlakaelel, who delivered globally traditional messages of care and connection. Bruno Takahashi and Juliet Pinto fill important gaps in the literature in 'Climate change, commercial news media, and Hispanics: An exploration of cultural processes and mediated environmental information.' They address the complex interaction between and among acculturation, culture, and media messaging, noting that corporate media – whether in English or Spanish – reinforce the dominant culture and ideological discourses, which potentially explains the shift among Latinos from a more traditional consciousness to greater adherence to Western consumerism.

Rebecca Hall's 'A call for climate justice' draws together experiential streams from New Orleans, the Pacific Islands, and the Northeast to demonstrate how our disposability consciousness contributes to the dispossession, dislocation, and disposability of people. Further, given who is stripped of voice and citizenship, she suggests that we reframe our thinking and narratives about the climate crisis, to reconnect it to socio-cultural, economic, and political forces and oppression. Patricia Widener and her student colleagues respond to Hall's call by exploring the transformative potential of combining intersectionality with critical public sociology and critical eco-pedagogy in 'Climate action and literacy through creativity and conversations.' Their reflections lay bare the struggle to more authentically engage others and their investments in their identities to create a coalitional consciousness in their Florida communities and potentially far beyond. Finally, we offer another treeless still from Steve Cutts' 'Man' that depicts the ravages of denial and unbridled consumption.

Reference

Global Oneness Project, n.d., Disposability consciousness. Interview with Julia Butterfly Hill. Available at: www.globalonenessproject.org/library/interviews/disposability-consciousness (accessed September 10, 2015).

12 Intelligent life

Khalil Bendib

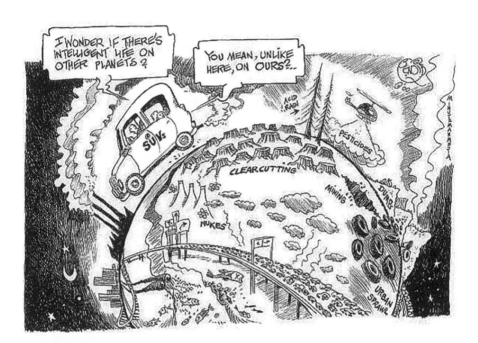

Figure 12.1 Intelligent life by Khalil Bendib.

13 The science proves it (or not)

Lewis Vande Pallen

When you think if what we as a people have done.
It's astounding.
Monumental.
And not in a good way.
The most interesting part about the whole thing is:
We did it together.
We – all of us – worked in unison
For the first time in history
We worked as one,

To kill our planet.

Oh, if we only we could have used our powers for good.

In just over 100 years we have forever changed the earth.
Mind you: it took billions of years to get what we have.
Now,
We have fundamentally changed it.
Forever.

There will definitely be things that we have seen that our children will not
To them, these things will be myth
Legend
Snowcapped mountains.
The North Pole
The South Pole.
Some penguins. Some polar bears

They will see things that we did not
The deserts are growing.
The sea levels are rising.
The earth will still be here but
…
Different

It's not kitschy. It's not science fiction
The earth is getting warmer and we are doing it.
Greenhouse gasses are real.
Some people will tell you the science is inconclusive

Maybe it is.

But the snow caps are still melting
Ocean levels are still rising
It's getting a bit warmer in November

Oh well.
I guess a grizzly bear is just as good as a polar bear.

I mean, it's a bear, right?

14 The canoe, the island, and the world[1]

R.D.K. Herman

During World War II, Ulithi Atoll in Western Micronesia served as the dry dock for the Allied Forces' invasion of Japan. Islanders were moved to and confined on one island of the atoll, while the other three major islands were taken over by different branches of the US military. When the war ended, much of the military material was left behind: bedding, housing, and food supplies, but also jeeps, cranes, and other gas-powered equipment. The atoll's chief, however, determined that none of these vehicles were to be used. The people still used other items left behind, but bit by bit they decayed in the hot, humid tropics. Quonset huts rusted away, fuel ran out, sheets and mattresses rotted, and traditional lifeways resumed. As one islander reflected:

> We were lucky in that, the former chiefs that we had, they didn't allow anyone to use any of those this kinds of [motorized] boat that they had, or any vehicles.... What I respect the old chief for is that, what if he didn't stop us, and we came to depend on what the military had left behind, and suddenly, that runs out – then where are we? So we were lucky that we didn't adopt the use of all those things. Instead, we started making canoes.
>
> (Ramaliol, 2003)

That "wisdom sits in places," as Keith Basso has written, is well understood by Indigenous peoples. As Calistus Legdesog of Ulithi articulated,

> I think we can never get away from our traditions and our culture there, because I think most of them, most of our practices here, have perfected themselves through time to fit the needs of the people here. Given our geographic isolation, geographic size, limitations of natural resources we have, we can move forward in terms of some things, but we cannot abandon what we have at home.
>
> (Legdesog, 2003)

Now, as the culture of modernity seeps into every crack on the globe, it is good to take a step back and realize how and why that culture has worked against place-based wisdom and thereby brought us to this time of global environmental crisis. In order to move forward and address this crisis as a species, we need to evolve beyond the confines of modernity. Modernity was built on the Agricultural Revolution, shaped by the Scientific Revolution, and fueled into high gear by the Industrial Revolution. We have since experienced the Medical Revolution, and a Communications revolution (cf. Albion, 1932; Taylor, 2009). Now we are due – or past due – for the Wisdom Revolution. This requires cutting through the Gordian knot of scientism, racism, sexism, classism, and culturism that characterizes modernity.

Wisdom, as John Kekes defines it, is in part a matter of understanding the significance of what everybody knows (Kekes, 1983, p. 280). It is a kind of knowledge that is theoretically accessible to anyone. It involves the recognition of possibilities and limitations that are the same for all people. But it is the depth of that knowledge, and the discerning of priorities, that turns knowledge into wisdom. That is, wisdom is the ability to construct patterns that are likely to lead to a 'good life' – for individuals or for their groups or communities. It takes a degree of self-control that enables individuals to modify their wants in accordance with their ideals. And it takes a certain steadfastness not to waiver in the face of adversity or temptation, but to exercise constancy in pursuit of one's goals. Most importantly, wisdom must serve the community, or else it has little value or purpose. And here 'community' includes not just the human aspects, but the societies of beings – animal, vegetable, and mineral – that interact and intersect with the people, sharing the place with them. For we are dependent on and nested within the web of life, and to recognize and respect that is part of wisdom.

In this sense, wisdom is at once a clear, content-less insight in the same way that reason is, yet in its embodiment it reflects gender, culture, and other identities in important ways. That is to say, wisdom arises from the conditioned circumstances of the individuals and cultures involved, and rises *above* that conditioning (like compassion, wisdom is not ego-bound), but still reflects, in some degree, the conditions from which it arose. That is, it addresses the milieu that produced it, the issues that prompted it. For example, Lanzetta (2005) offers a specifically feminist approach to wisdom. As such, it is an embodied experience and its articulation will vary according to the identity, experience, and position of its keeper. It will reflect both timeless truth and the specific circumstances of those upon whom it bears. It is important to keep this in mind, lest discourses of wisdom go the way of 'reason' in Western thought, as described below. Reason tends to manifest as an expedient way of thinking, concerned with the immediate moment. Wisdom, conversely, is a mode of thinking about long-term and intersecting consequences, and therefore guiding action in a manner that serves the long-term collective good of the community. An example is the American Indian concept of thinking in terms of 'seven generations' (now a brand name for 'natural' products): a consciousness for decision-making that bears in mind the works and teachings of the three previous generations, then considering the circumstances of your own generation in its time and place, then acting in a way to support your community's descendants three generations hence (Wildcat, 2009). And again, that involves supporting the health of the greater web of life, of which the human community is just one part.

Presumably almost anyone could gain wisdom through persistently applying the all-powerful but informal training of life experience – employing the sense to learn from trial and error, being sensitive to place and its myriad interlinked inhabitants and cycles, and thereby discerning principles of living that we may call wisdom. Others take a more formal approach: the sages of old – the shamans, the yogis, monks, nuns, and other renunciants and spiritual seekers – engaged in practices aimed to strip away that which is distracting and contingent in order to focus their thoughts, words, and deeds on achieving a level of self-knowledge, self-control, connectedness, and humility that would bring true realization. They taught others, especially in those communities living close to the earth, where survival depended on cooperation and an in-depth knowledge of the patterns of the natural world. The evolution of wisdom was rooted in *places* over generations of time. Such wisdom took on a communal presence; it guided the decisions of the communities as a whole. Knowledge and wisdom – and the paths to achieving wisdom – were then passed down to successive generations. Means were developed – rites, rituals, initiations, texts, and instructions (oral or written) – to ensure that wisdom blossomed among members of each generation.

In short, wisdom involves not just knowledge, but the insight to use that knowledge in appropriate ways for the good of oneself and of the community. Wisdom to know which actions would bring success, and which actions would bring disharmony and trouble, even danger. It includes a positive image of the future to lead society on a path that is balanced and harmonious (Ray, 2007). This is true for *any* community in any place at any time.

In Western tradition, however, the path of knowledge took a different trajectory. The growth of Christianity, while not always and everywhere incompatible with wisdom, emphasized faith and church ritual over direct knowledge and experience in the lived world, or transcendence over immanence. For traditional communities, what we now call 'religion' was intertwined with cosmologies, ontologies, and culturally and materially productive practices. The unmanifest or spiritual world existed in conjunction and interaction with the manifest world, with the two being seen as different sides of the same coin. Over time the indigenous ways of European peoples have been largely lost with, for example, the contemporary neo-pagan Reclaiming tradition attempting to reclaim this wisdom and spirituality in the modern context.

Moreover, the Church brought a new (or heightened) patriarchal order, derived from the highly patriarchal Hebrew culture that formed its roots (Fiorenza, 2013). This translated not only into the loss of feminine spiritual aspects of traditional community religions (intimately tied to Nature and the Earth Mother) but also to the actual subjugation of women and feminized Others as social and political (as well as religious) beings (Pillay, 2013; Leonhard, 2006). While the Protestant Reformation, the Enlightenment, and the

Scientific Revolution forcibly separated the pursuit of knowledge from the control of the Church – a good and necessary aim – the new creed was the glorification of a disembodied 'masculine' reason. It is this modern conceptualization that, despite pushbacks and counter-movements, dominates Western thought up until today.

Reason, it was deemed, would free the mind from the capricious and insubstantial, allowing us to realize our genuine humanity. This approach emerged out of seventeenth-century thinkers such as Descartes, Bacon, Galileo, Copernicus, and Newton (Maxwell, 1984; Van Ruler, 2000). Descartes defined wisdom as "prudence in our affairs: a perfect knowledge of all things that mankind is capable of knowing, both for the conduct of life" (moral knowledge), "and for the preservation of health" (medicine), "and the discovery of all manner of skills" (mechanics). In short, for Descartes, wisdom was useful knowledge (Peperzak, 1995, p. 133).

But there were limitations to what would be accepted as knowledge in this new approach that may be called 'science,' and is articulated to its illogical conclusion by the discourses of scientism. As Hutchinson (2011, p. 1) explains, "Scientism is the belief that all valid knowledge is science. Scientism says, or at least implicitly assumes, that rational knowledge is scientific, and everything else that claims the status of knowledge is just superstition, irrationality, emotion, or nonsense." He adds that many leading scientists, and science popularizers, speak and act as if science and scientism are one and the same. This results too easily in confusion between the two, and a rejection of science when scientism is actually the problem (Hutchinson, 2011).

Both science and scientism draw on the Cartesian legacy of 'reason' in seeking a pure form of knowledge separate from the material world. Emotions, feelings, values, art, literature – culture as a whole, in short – played no role in the pursuit of knowledge, because these are all things that cloud reason. Intellectual inquiry is separated from lived experience in the world and the personal, social, political, institutional, and environmental contexts in which it takes place. Science is limited in its focus to only that which can be empirically verified (or proven mathematically). The result is an emphasis on reason that is divorced from the messy vicissitudes of lived reality (Maxwell, 1984). Wisdom exists in the mind, not in places. It takes place in a vacuum, or on the head of a pin, as you prefer. And from it and its antecedents came a whole division of the world across the spectrum of gender, race, class, religion, and culture – as well as the separation of humans from the rest of nature, elevating the former and denigrating the latter.

For Western thought, this created what is often referred to as the 'Cartesian split' between mind and body, therefore between humanity – humans being perceived as the only rational thinking beings – and nature, which is dumb in both senses of the word. This sharp distinction between the human and non-human spheres is mirrored within the self: reason is used to overcome human nature (and to conquer or disempower those Others associated with it, such as women, non-whites, indigenous peoples, the poor, and the many nations of beings that are collectively called 'nature'), just as (white, male) humanity, by exercising reason, overcomes external Nature. 'Man' is thereby defined not as part of the natural, physical, and biological realm but as separate from it and in opposition to it. Nature is posed as being inferior, alien, and oppositional to the authentic, rational human self (Plumwood, 1991). Drawing on Joan Cocks (1989, p. 10), the perceived difference – within this rationalist discourse – may be categorized as shown in Table 14.1.

It is a simple matter to slot Indigenous Peoples and various Others into the same column occupied by 'women.' As Gregory Cajete (2000, pp. 75, 79), for example, describes indigenous epistemology,

> Everything is considered to be 'alive' or animated and imbued with 'spirit' or energy. A stone has its own form of animation and unique energy. Everything is related, that is, connected in dynamic, interactive, and mutually reciprocal relationships. All things, events, and forms of energy unfold and infold themselves in a contextual field of the micro and macro universe.... The ultimate aim is not explaining an objectified universe, but rather learning about and understanding responsibilities and relationships and celebrating those that humans establish with the world.... [It] is also about mutual reciprocity, and which presupposes a responsibility to care for, sustain, and respect the rights of other living things, plants, animals, and the place in which one lives.

If this argument for the reign of 'reason' over Nature were not enough, Francis Bacon urged that science be used to extend man's power over Nature for the benefit of human society and its inhabitants (Reydon, 2012; Bostrom, 2011). White (1967, p. 1203) asserts that this "Baconian creed" for scientific knowledge as a means for technological power over Nature does not manifest much before about 1850 – coincidentally (or not) about the same time that the internal combustion engine is invented, and the petroleum industry begins to take off. The acceptance of technological domination over nature as a normal pattern of action, White (1967) says, "may mark the greatest event in human

Table 14.1 Perceived differences

Men	Women
Egotistical	Relational
Impersonal	Empathetic
Dissection	Organic holism
Definition, logical analysis	Intuitive, receptive, and sensuous
Imposing their own purposes and will on everything around them	Respect the integrity and natural impulse of things in themselves
Master the other	Care for the other
Conflict	Harmony
Civilization, reason, and science	A deep, organic tie to wild, natural forces

history since the invention of agriculture, and perhaps in nonhuman terrestrial history as well" (p. 1203) This "Enlightenment" philosophy turned out to be a successful recipe for industrial capitalism, but as climate change (among many other environmental indicators) shows us, not for human good as is more holistically understood. But it has played a dominant role in the imperial history of the past 200 years.

The alignment of reason and technology with other factors, within this way of thinking, may be categorized as shown in Figure 14.1.

This 'logic' and its evolution from the late eighteenth century may be seen quite clearly in the history of colonialism (cf. Herman, 1995). While Romanticism pushed back against it and created the Noble Savage, this hierarchical, 'rational' worldview held sway as the dominant discourse, and romantics – along with those whom they romanticized – were pushed to the margins as irrational and unscientific. This all the more so with the rise of petro-modernity from the 1860s and the doctrine of progress brought on by rapid expansions in transportation, communication, construction, and warfare technologies.

The wisdom that had derived from millennia of living in this world was eroded over time – if not forcibly excised or discarded – by the Enlightenment and the logic that drew upon it. The result has been called by Weber, Adorno, and others the "disenchantment of the world" (Weber, 1930; Greisman, 1976; Scribner, 1993). While there is some debate over whether there really was such

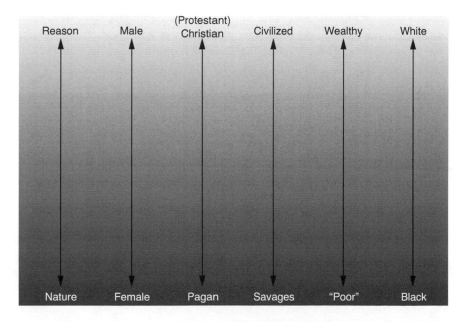

Figure 14.1 The alignment of various factors along a spectrum from the allegedly dark and 'irrational' forces (bottom) to the allegedly light and 'rational' forces (top).

a disenchantment, or whether this was more of an Enlightenment view of the Reformation (Scribner, 1993), the change in attitudes toward Nature is clearly discernible, and has had a profound impact on Western identity constructs as well as intellectual and cultural thought and behavior. Not to mention the Earth itself.

As I have outlined elsewhere (Herman, 2008), there are three main consequences of this shift. The first is the removal of any 'spiritual' aspect to the world – that is, a reduction of the world into pure mechanistic materiality on the one hand and the mental realm of human consciousness on the other. This is a traumatic removal of human existence from its deep, psychic connection with the natural world and itself that provided meaning, guidance, and ultimately wisdom for human communities. In an oft-cited passage, psychoanalyst Carl Jung (1964, p. 95) remarked:

> As scientific understanding has grown, so our world had become dehumanized. Man feels himself isolated in the cosmos, because he is no longer involved in nature and has lost his emotional 'unconscious identity' with natural phenomena.... No river contains a spirit, no tree is the life principle of a man, no snake the embodiment of wisdom, no mountain cave the home of a great demon. No voices now speak to man from stones, plants, and animals, nor does he speak to them believing they can hear. His contact with nature has gone, and with it has gone the profound emotional energy that this symbolic connection supplied.

Second, and related, is that by rendering Nature as mechanistic, it loses any intrinsic values; values come to exist in the mind, not in the world. The natural world, in short, has nothing to teach us. Rather, it is ours to study, to penetrate its mysteries, and to tease apart into a rational construct that we can grasp through reason. It is the passive feminine to the probing masculine of human intellect (see Mathews, 2010; Plumwood, 2009). And with the distorted feminization of Nature comes the feminization of non-white, non-male (and non-white male), and non-human Others, as has been discussed by many (e.g., Enloe, 1989; Minh-Ha, 1989).

Finally, this bifurcation of humanity and Nature poses a conceptual and experiential distance and detachment that allows for the commodification of the material world essential for capitalism. That is, in order to treat Nature as a storehouse of potential resources for industrial production and commodification, it is essential that Nature should *not* be seen as animate or connected directly to the human experience. The emphasis on a disembodied reason made this possible.

From there, as is well known, the relatively new Protestant forms of Christianity rose to ascendance owing to their emphasis on success in the material world as proof of God's grace. Where the Catholic powers of Europe had been dabbling in colonizing foreign lands, Protestant England stepped in with a much more methodical approach to build a commercial and political empire that truly spanned the globe, harvesting resources and raw materials from the colonies, transforming them into manufactured goods back in the home country, then exporting these products all over the world, including back to the colonies. Great Britain's child, the USA, followed suit, though often forgoing the political in favor of the merely commercial.

This has created a 'rational' world order: the global economic system, with its accompanying cultural transformation and exploitation, was called 'modernization' and is now termed 'globalization.' Its pillar of scientific rationalism is distinctly different from the world of experience, interrelatedness, and responsibility found in indigenous traditions. As Artz (1999) states, "The prevailing view of scientific realists is that the world is out there for us to study independent of our perception and understanding. And, our goal is to understand it as it exists." Indigenous traditions understood, however, that reality, as Deloria (1999, pp. 38–39) points out, is

> the experience of the moment coupled with the interpretive scheme that had been woven together over the generations. If there were other dimensions to life – the religious experiences and dreams certainly indicated the presence of other ways of living, even other places – they were regarded as part of an organic whole and not as distinct from other experiences as they are in the approach of Western science.

Then there is 'economic rationalism,' based on the model of *homo economicus*, or Economic Man. Based on the nineteenth-century writings of Adam Smith, but reaching the limits of its fully absurd potential in the twentieth century, this model holds that people will do whatever they can to maximize their own personal benefit (Grampp, 1948). It is the model on which our dominant economic theory is based (Ostrom, 1990), and is the cornerstone of US culture. It is taken as given that all people want to maximize their personal wealth. Peoples who do not ascribe to this approach, but adhere to traditional lifeways (and are therefore close to nature), as well as traditional worldviews and understandings, are deemed backward, heathen, and irrational. This is a powerful discursive force that still dominates the dominant culture and worldview, and has been used to legitimize the disenfranchisement and destruction of such peoples. Indigenous economics presents entirely different logics. With its emphasis on the community rather than the individual, it is rational to ensure that everyone is taken care of. Individual accumulation is to be avoided, except to the extent that accumulation for redistribution gave one 'supernatural capital' (see Voget, 1987). The wealthiest person, in this logic, is the one who has given away the most, and therefore owns the least, materially.

The emphasis on private property in the West (and our rights to it) did assist with the push toward greater human rights, which had been developing along with the growth of reason since the Enlightenment. That has been a very important element in human social evolution over the past few hundred years. But 'human rights' also emphasize the individual as an entity unto itself, an autonomous being free of responsibilities to others (other than to observe their rights), and none to the natural world. It is part of the same rationalist tradition described heretofore, and only in recent years has been stretched to try to accommodate the rights of women, non-whites, and indigenous peoples within its framework.

But efforts to ensure the protection of indigenous communities' knowledges run up against the capitalist forces for individual gain encoded in the Agreement on Trade-related Aspects of Intellectual Property Rights (TRIPS). As Bratspies (2006/2007, p. 326) states, TRIPS "excludes indigenous peoples from the intellectual property rights system much like the way those same groups were systematically denied property rights to their land during the periods of discovery and colonialism." Similarly, the Convention on Biological Diversity, which claims to recognize and protect indigenous knowledge and lifeways, requires only that such actions be "subject to national legislation" and only "as far as possible and as appropriate" (Mugabe, 1999, pp. 23, 25). Indigenous peoples thus find themselves struggling with these imposed Western-style legalities to protect their knowledge, as seen in the case of attempts to patent plants they have cultivated, such as the ayahuasca vine (Fecteau, 2001), basmati rice, and neem (Shiva, 2001).

Finally, rationality bumped up against theology, at first resulting in the notion that knowledge of God can be achieved by reason alone, without the aid of revealed knowledge. Spiritual growth is a matter of the mind, and is not rooted in places and experiences. But the problem with the disenchantment of the world is that not only did it remove any sense of deity or animation in the natural world, it led to the dissolution of any affective notion of deity whatsoever. Van Ruler (2000, p. 389) points out that "If there can be no activity of spirits in the material world, the question soon arises whether God Himself, being of a spiritual nature, could actually influence the course of things." Removing God from the world allowed the flourishing of empirical sciences, and the scientists of the seventeenth and eighteenth centuries held on to their Christian faith. But rationality pushed God out of the picture altogether – the existence of a deity cannot be proven empirically, and seemed increasingly irrelevant to understanding the workings of the world – leaving behind a rationalized and mechanized world empty of divinity and therefore of spiritual values (Surin, 1985).

The result, I suggest, is an increasingly widespread but informal acceptance of the secular humanist stance. As the Humanist Manifesto of 1933 put it,

> Humanism asserts that the nature of the universe depicted by modern science makes unacceptable any supernatural or cosmic guarantees of human values. Obviously humanism does not deny the possibility of realities as yet undiscovered, but it does insist that the way to determine the existence and value of any and all realities is by means of intelligent inquiry and by the assessment of their relations to human needs. Religion must formulate its hopes and plans in the light of the scientific spirit and method.
>
> (Humanist Manifesto I, 1933)

In this secularized, rational, self-seeking, material-only world – in which (certain) humans alone are capable of consciousness and are the sole locus of values – full exploitation of other humans and the environment got underway.

This construction of 'rationality' is a castle made of sand. It is built on a word game in which, by designating one mode as 'rational,' everything else is thereby discredited as 'irrational.' Thus it hides while simultaneously creating the irrationality of the modern world in marching toward ecological suicide on a global scale. In its own logic, it confirms the Chinese proverb that the road to hell is paved with 'good' intentions.

One way to fight back against this is to engage in a new word game. My approach is to call the above mentality 'rationality with a small r.' It is a cohesive system of thought which, like math, makes a great deal of sense internally. But the results externally are fatally flawed. Real 'Rationality' – with a capital R – would be a form of thinking that embraces science and wisdom, recognizing that the former without the latter is a path to destruction. Rationality (with a capital R) might recognize that, while we humans do have the capacity for rapacious individual self-seeking, we also have an innate sense of compassion, charity, and understanding that has helped us survive as long as we have by working together and looking after each other. And it would accept that science and the mechanistic model of the universe cannot address the most important questions about being. As Paul Ray (2007) argues, wisdom not only makes rational decisions based on a longer term context, but uses a positive image of the future to orient society in a balanced direction. Such wisdom would apply the tools of science and technology toward long-term ends that are sustainable and equitable.

To find a model for this, I suggest we look at the traditional peoples of the past, and the Indigenous peoples of the present who still live according to traditional values – not haphazardly or romantically, but realistically and methodically – for the wisdom that Nature has taught our ancestors. I suggest that, except where introduced religion stifled thinking and experimentation, all our ancestors were scientists – or at least, had scientists in their communities in the manner that Cajete and Deloria describe. Those scientists drew not just upon trial and error and experimentation, but also upon lived experience, upon the wisdom of their predecessors, and upon intuition and guidance from visionary experiences. There is really no other way to explain how they came up with, in particular, the herbal concoctions for healing and for vision quests. Out of all the plants in the Amazon jungle, is trial and error enough to determine which plants to use for which ailments? Narby's (1998) discussion on mixing ayahuasca suggests not. There are other paths to knowledge besides reason and the scientific method, if only we allow ourselves to be open to them.

Today, in the Hawaiian Islands, after two centuries of cultural then political colonization – including the near-extinction of the language, the outlawing of traditional medical practices, and the denigration of most other cultural practices and beliefs – there are a growing number of Native Hawaiians who hold advanced academic degrees and at the same time participate in the revitalization of traditional knowledge and practices. Now, not all practices of the past are worth bringing back. Human sacrifice, for example, is best left in the past, along with female genital mutilation. But the detailed, intimate knowledge of the environment, the strategies for managing the land and sea, the values that held the community together, and the understanding of humanity's role in the world, these are very much worth bringing back, and back they are coming – in a new, modern, and scientific way.

The example I find most enthralling is the return of traditional voyaging and navigation. This began in 1976 with the maiden voyage of the Hōkūleʻa, a double-hulled voyaging canoe based on extensive research of traditional designs and explorers' drawings, built of modern materials, and navigated from Hawaiʻi to Tahiti – 2500 miles – using traditional methods. For 40 years now, the Hōkūleʻa has voyaged around the Pacific; at the time of writing, it is undertaking a journey around the globe. The theme of this voyage is Mālama Honua – take care of the Earth (hokulea.org). Why? Because, as navigator Nainoa Thompson says, the sail plan we (humanity) are on is not sustainable. Perhaps this modern version of ancient technology can help raise awareness.

Now, let us bear in mind that the settling of the Island Pacific is one of the greatest human adventures of all time. These voyagers were sailing across an ocean that covers one-third of the planet, finding tiny dots of land, and traveling back and forth to settle these new lands. They were doing this on open, double-hulled canoes built with stone-age tools and materials, and by perhaps AD 1100 had settled the entire Pacific. The voyages of the Vikings, Columbus, or Magellan do not begin to compare with this.

There is a Hawaiian proverb, he waʻa he moku, he moku he waʻa: "the canoe is an island, the island is a canoe." Both are limited vessels surrounded by ocean, and what you have is all you have. The same circumstances and the same values and conditions apply to both situations. This proverb was the basis for a pro-posed exhibition for the National Museum of the American Indian entitled Aloha ʻĀina: Hawaii, the Canoe, and the World. Field research done for that project between 2011 and 2013, and previous fieldwork on my own project, Pacific Worlds (www.pacificworlds.com), enabled me to interview a great many people in Hawaiʻi and Micronesia regarding place-based traditional knowledge and practices. From all that work I derived five values I ascribe to the voyaging canoe as an example of how to live in a finite environment, since the voyaging canoe, like Buckminster Fuller's "Spaceship Earth," tells us a great deal about what it takes to survive when that is all you have:

1   Knowledge (ʻIke): Building and navigating a voyaging canoe required exten-sive knowledge in a range of fields – so many that guilds of experts arose. The most dramatic knowledge is that of navigating across vast ocean dis-tances, finding land, and marking it in a mental map that allowed return voyages. For this, Pacific navigators developed extensive knowledge of the stars at sea and their movements through the heavens at different latitudes. In addition, they read the swells of the ocean, and systematized that

knowledge into a means for guiding the canoe in the absence of stars. They devised several tools for finding land beyond the horizon, including reading the swells and their refractions off land, as well as detailed readings of clouds based on color and movement. They learned which land-based birds (and which stages of life for those birds) could be reliably followed to find land. All of this knowledge is encoded in traditional languages.

I think of knowledge as the science component, but as with all traditional cultures, this science includes intuition and a deep spiritual connection with the world. As master navigator Pius Mau Piailug told his Hawaiian apprentices, there were two kinds of navigators on his home island of Satawal: those who only knew how to navigate, and those who knew both navigation and magic. These latter navigators were considered to be the higher ones. This is where the blinders of the Enlightenment and the focus on reason and empiricism leave modern peoples crippled. We have discarded our sixth sense, and we need to re-learn it. Intuition and connection to the unmanifest is an essential part of the human experience, and informs our transcendent values.

2   *Pursuit of excellence (Po'okela)*: To build a voyaging canoe that can go 2500 miles *and back*, 'good enough' is not good enough. Polynesians did not learn how to build voyaging canoes overnight. It took generations of design and refinement, always pushing the envelope to make something bigger and more seaworthy, using only stone tools and plant products. Here is also another area of knowledge: there were many plants involved in the production of even the simplest canoe, and this required both knowledge and skill to utilize them.

In modern terms, this is about skill and developing technology. In this, humanity currently excels, which is a good thing. At the same time, many things are not made as well as they used to be since the change-over from handmade craftsmen to factory production. This is the culture of consumption: we can afford more because things cost less but are not as good quality. Concerns about production costs can lead to the cutting of corners, even on important projects such as spaceships and nuclear reactors.

Anyone who has crafted something by hand knows the satisfaction of that production – that the process itself was an important experience, not merely the product itself. Thus, the pursuit of excellence is not merely about making a better product, but about developing excellence in one's self and seeking excellence in and for the world.

3   *Rights and responsibilities (Kuleana)*. There are no really accurate translations for *kuleana*. The closest notion is to speak of your 'turf' – the area for which you are solely responsible, but which is also your domain. Rights and responsibilities go together. This is a concept sorely lacking in the parlance of economic rationalism. Even the Universal Declaration of Human Rights uses the word 'responsibility' only once, and then in a most minor way. If we want to save the planet, we need to have not just inalienable rights, but inalienable *responsibilities* – toward each other, toward the myriad nations of

beings with whom we share this planet, and toward the Earth itself. Climate change is teaching us that the Age of the Individual is over. We can no longer do what we want and not concern ourselves with the external impacts on the Earth and on others. If you don't attend to your responsibility, that puts a burden on someone else, and the system starts to break down. *We are all in the same boat.* Do your part.

4   *Pono*: Again, there is no perfect translation for this term, but the concept is easily grasped. It means to act in a way that is balanced, not just socially but cosmically. It is, in short, to *do the right thing* in any given situation, even if that is to your personal disadvantage. *Pono* is to some degree defined in any circumstances by social mores: you should do what is expected of you, given your position. For Hawaiians this meant that inasmuch as the people were responsible to the ruling chiefs, the ruling chiefs were also responsible for the people. We can find this similar attitude in Confucian tradition: children must obey their parents, but parents have the duty of raising their children to be the best possible people. When either party does not fulfill their responsibilities (*kuleana*), they are not acting in a way that is *pono*. Hurting people is not *pono*. Creating imbalance, disharmony is generally not *pono*, unless that is truly what the situation requires.

5   *Mālama*: This refers to nurturance, like a mother looking after a child, or a farmer tending his crops. You nurture what is in your care, feed it, and ensure that it has everything it needs to thrive. You heal or fix things when they are damaged. To *mālama* something also recognizes the integrity or spirit of the thing in itself. This points back to a worldview in which we are not the sole sentient beings on the planet, but live in a rich and vibrant field of sentience in which all things absorb and radiate energy. Treating things with care and kindness not only preserves and protects them, but puts more good energy into the world.

Most importantly, we need to look after the vessel that carries us. On board, this means not only do you *mālama* the canoe, but also each other, and your supplies. What we have is all we have: we need to look after it, make it last, make it endure, make it flourish. And if we think of the Earth as the canoe that carries us all, then we understand why the *Hōkūleʻa*'s World Wide Voyage is themed "*mālama honua*" – to take care of the Earth.

Aloha is the underlying principle that makes all this easier: compassionate, loving kindness. The culture of individualism and the discourses of rationality tend to mitigate in the opposite direction – I, me, mine – but this value does exist in Western culture nonetheless, and can be reclaimed if we accept that we are all in the same boat. What we know now is that the Earth is the canoe; the Earth is the island. It is not just a metaphor; it is reality. Those five values that enabled survival on the canoe and on small islands teach us how to live on this Earth. These are the values of living and working together to succeed as a group. This is a manifestation of wisdom to which we can all relate, regardless of our positions and identities, because we are all already in the canoe – Earth.

Paul Hawken (2007, p. 180) talks about a workshop he conducted with engineers at an agricultural chemical company, based on the "spaceship earth" idea. He asked them to design a spaceship that could leave Earth and return in 100 years with its crew alive, healthy, and happy. The best by far of the solutions produced was the one in which the engineers realized that the people would need a culture: the crew should have artists, musicians, actors, and storytellers. It should have weeds, fungi, bacteria, insects, and small animals. Hawken writes, "Essentially, the winning team created a diverse ecosystem within which a socially just and equitable society practices organic agriculture and designed all objects for disassembly, reuse, and recycling."

This shows that we modern people – even, in this case, chemical engineers – can make the shift to a way of thinking and being that would be more harmonious, more *pono*. This example embraces the principle I call 'Indigeneity.' Indigeneity means being indigenous to a place and means having a depth of knowledge, understanding, and connection to that place. Indigeneity also includes a sense of stewardship and responsibility for managing that place and working respectfully with its non-human inhabitants. This is how all our ancestors lived, and how some Indigenous peoples (and others) live today.

This is not about some romantic notion of returning to nature or to our primal lifestyles. Nineteenth-century Romanticism was indeed about just that, and it was a backlash against the rising dominance of reason and rationality. But there is no going back now, and we do not need romantic images. Instead, we need what Dan Wildcat (2009) calls "Indigenous Realism." That is, the blending of modern science and technology with the holistic worldview and cultural values that encourage sustainability and co-existence – not only among humans, but with the myriad nations of beings with whom we share this planet. The Lakota and other American Indian tribes use the phrase "all my relatives" to talk about the world in which we live. Now science tells us how we are at one with our environment. Earth, air, water, and fire cycle through us (see Suzuki, 2007). We embody, and return to, the animal, vegetable, and mineral kingdoms. Even our DNA tells us that we are related to all other species on the planet (Zimmer, 2013). If we think of the world as "relatives" rather than "resources," we will treat it differently.

Now I look back at the chiefs on Ulithi who had the wisdom to say "we will not use these things." They were able to look far enough ahead to see the consequences, and to nip in the bud what would clearly become a problem for them. Now it is time for all of us to gain, practice, and exercise wisdom: an approach to decision-making that heeds the best teachings of our traditional ancestors, acknowledges the challenges of the present, and acts to provide for generations hence. It is time to insist on wisdom, to unite a culture of wisdom with science into a higher rationality.[2] Without that we are all lost. For we are all in the same boat, and that boat is getting leaky and full of trash. It is up to each of us to understand the places in which we live – from our bodies to our homes to our neighborhoods and communities, then through larger and larger levels up to the Earth itself, and to treat these environments as a part of ourselves, toward which we have the responsibility to manage wisely.

Notes

1 This work draws heavily on my 2015 paper 'Traditional Knowledge in a time of Crisis: Climate Change, Culture and Communication,' *Sustainability Science*. Available at: http://link.springer.com/article/10.1007/s11625–015–0305–9?wt_mc=email. event.1.SEM.ArticleAuthorOnlineFirst.

2 Vine Deloria (1999, p. 39) argued,

> The next generation of American Indians could radically transform scientific knowledge by grounding themselves in traditional knowledge about the world and demonstrating how everything is connected to everything else. Advocacy of this idea would involve showing how personality and a sense of purpose must become part of the knowledge that science confronts and understands.

References

Albion, R.G., 1932. The 'Communication Revolution.' *The American Historical Review*, 37(4), pp. 718–720.

Artz, J.M., 1999. The Decline of Scientific Rationalism. Available at: http://home.gwu. edu/~jartz/alter/next/decrat.html (accessed October 18, 2014).

Basso, K., 1997. Wisdom Sits in Places: Notes on a Western Apache Landscape. In S. Feld and K.H. Basso, eds, *Senses of Place*, Santa Fe, NM: School of American Research Press, pp. 53–90.

Bostrom, N., 2011. A History of Transhumanist Thought. In M. Rectenwald and L. Carl, eds, *Academic Writing Across the Disciplines*. New York: Pearson Longman. Section available at: www.nickbostrom.com/papers/history.pdf (accessed January 10, 2015).

Bratspies, R.M., 2006/2007. The New Discovery Doctrine: Some Thoughts on Property Rights and Traditional Knowledge. *American Indian Law Review*, 31(2), pp. 315–340.

Cajete, G., 2000. *Native Science: Natural laws of interdependence*. Santa Fe, NM: Clear Light Publishers.

Cocks, J., 1989. *The Oppositional Imagination: Feminism, critique, and political theory*. London: Routledge.

Deloria, V., 1999. Relativity, Relatedness, and Reality. In B. Deloria, K. Foehner, and S. Scinta, eds, *Spirit and Reason: The Vine Deloria, Jr. reader*. Golden, CO: Fulcrum Publishing. pp. 32–39.

Enloe, C., 1989. *Bananas Beaches and Bases: Making feminist sense of international politics*. Berkeley: University of California Press.

Fecteau, L.M., 2001. The Ayahuasca Patent Revocation: Raising Questions about Current U.S. Patent Policy. *Boston College Third World Law Journal*, 21(1), pp. 69–104.

Fiorenza, E.S., 2013. *Changing Horizons: Explorations of feminist interpretation*. New York: Fortress Press.

Grampp, W.D., 1948. Adam Smith and the Economic Man. *Journal of Political Economy* 56(4), pp. 315–336.

Greisman, H.C., 1976. 'Disenchantment of the World': Romanticism, Aesthetics and Sociological Theory. *The British Journal of Sociology*, 27(4), pp. 495–507.

Hawken, P., 2007. *Blessed Unrest: How the largest social movement in history is restoring grace, justice, and beauty to the world*. London: Penguin Books.

Herman, R.D.K., 1995. *Kālai'āina – Carving the Land: Geography, desire and possession in the Hawaiian Islands*. PhD dissertation, University of Hawai'i.

Herman, R.D.K., 2008. Reflections on the Importance of Indigenous Geography. *American Indian Culture and Research Journal*, 32(3), pp. 73–88.

Humanist Manifesto I, 1933. *The New Humanist.* Available at: http://Americanhumanist. org/Humanism/Humanist_Manifesto_I (accessed October 18, 2014).

Hutchinson, I., 2011. *Monopolizing Knowledge.* Belmont, MA: Fias Publishing.

Jung, C., 1964. *Man and His Symbols.* New York: Doubleday & Company.

Kekes, J., 1983. Wisdom. *American Philosophical Quarterly,* 20(3), pp. 277–286.

Lanzetta, B., 2005. *Radical Wisdom: A feminist mystical theology.* Minneapolis, MN: Fortress Press.

Legdesog, C., 2003. Farewell to Ulithi. *Pacific Worlds: Yap-Ulithi.* Available at: www. pacificworlds.com/yap/onwards/pau.cfm (accessed September 28, 2015).

Leonhard, B., 2006. St. Paul and Women: A Mixed Record. *St. Anthony Messenger,* July. Available at: www.Americancatholic.org/Messenger/Jul2006/Feature2.asp (accessed January 23, 2015).

Mathews, F., 2010. Environmental Philosophy. In N. Trakakis and G. Oppy, eds, *A Companion to Philosophy in Australia and New Zealand.* Melbourne: Monash University Publishing. Available at: http://books.publishing.monash.edu/apps/bookworm/view/A +Companion+to+Philosophy+in+Australia+and+New+Zealand/56/xhtml/chapter05. html#chapter05sec01 (accessed August 19, 2014).

Maxwell, N., 1984. *From Knowledge to Wisdom: A revolution in the aims and methods of science.* Oxford: Blackwell.

Minh-Ha, T.T., 1989. *Woman, Native, Other: Writing postcoloniality and feminism.* Bloomington, IN: Indiana University Press.

Mugabe, J., 1999. *Intellectual Property Protection, and Traditional Knowledge: An exploration in international policy discourse.* Nairobi, Kenya: ACTS Press, African Centre for Technology.

Narby, J., 1998. *The Cosmic Serpent: DNA and the origins of knowledge.* New York: Jeremy P. Tarcher/Putnam.

Natural Theology, n.d. *The Free Dictionary.* Available at: www.thefreedictionary.com/ Rational+theology (accessed October 8, 2014).

Ostrom, E., 1990. *Governing the Commons: The evolution of institutions for collective action.* Cambridge, MA: Cambridge University Press.

Peperzak, A., 1995. Life, science, and wisdom according to Descartes. *History of Philosophy Quarterly,* 12(2), pp. 133–153.

Pillay, M.N., 2013. Towards Healing and Repenting of the Historic Patriarchy of the Christian Faith. *Continuing Indaba.* Available at: http://continuingindaba.com/2013/ 09/18/towards-healing-and-repenting-of-the-historic-patriarchy-of-the-christian-faith/ (accessed January 23, 2015).

Plumwood, V., 1991. Nature, Self, and Gender: Feminism, Environmental Philosophy, and the Critique of Rationalism. *Hypatia,* 6(1), pp. 3–27.

Plumwood, V., 2009. Nature in the Active Voice. *Australian Humanities Review,* 46.

Ramaliol, P., 2003. Postwar Ulithi. *Pacific Worlds.* Available at: www.pacificworlds.com/ yap/memories/memory3.cfm (accessed September 28, 2015).

Ray, P., 2007. On The Need of Our Time, Draft v.5.2. Available at: www.wisdomuniversity. org/time.pdf (accessed January 24, 2015).

Reydon, T.A.C., 2012. Philosophy of Technology. Internet Encyclopedia of Philosophy. Available at: www.iep.utm.edu/technolo/ (accessed January 8, 2015).

Scribner, R.W., 1993. The Reformation, Popular Magic, and the 'Disenchantment of the World.' *Journal of Interdisciplinary History,* 23, p. 475.

Shiva, V., 2001. Special Report: Golden Rice and Neem: Biopatents and the Appropriation of Women's Environmental Knowledge. *Women's Studies Quarterly,* 29(1–2), pp. 12–23.

Surin, K., 1985. Contemptus Mundi and the Disenchanted World: Bonhoeffer's 'Discipline of the Secret' and Adorno's 'Strategy of Hibernation.' *Journal of the American Academy of Religion*, 53(3), pp. 383–410.

Suzuki, D., 2007. *The Sacred Balance: Rediscovering our place in nature*, 3rd edn. Vancouver: Greystone Books.

Taylor, S.K., 2009. A Communications Revolution. *The Sixteenth Century Journal*, 40(1), pp. 195–197.

Van Ruler, H., 2000. Minds, Forms, and Spirits: The Nature of Cartesian Disenchantment. *Journal of the History of Ideas*, 61(3), pp. 381–395.

Voget, F.W., 1987. The Crow Indian Give-away: A Primary Instrument for Cultural Adaptation and Persistence. *Anthropos* Bd. 82, H. 1(3), pp. 207–214.

Weber, M., 1930. *The Protestant Ethic and the Spirit of Capitalism*. London: Routledge.

White, L., 1967. The Historical Roots of Our Ecologic Crisis. *Science*, 155(3767), pp. 1203–1207.

Wildcat, D.R., 2009. *Red Alert! Saving the planet with Indigenous knowledge*. Golden, CO: Fulcrum.

Zimmer, C., 2013. Genes Are Us. And Them. *National Geographic*. Available at: http://ngm.nationalgeographic.com/2013/07/125-explore/shared-genes (accessed September 28, 2015).

15 Tlakaelel's view of climate change

Bert Gunn, translator

Mexica Tolteca Elder, Tlakaelel Francisco Jimenez Sanchez (1930–2012), teacher, heir and guardian of the oral tradition (or true history), author, and spiritual guide, devoted his life to researching Indigenous traditions. The author of *Nahui Mitl*, he brought a message of peace, cooperation, and the basic unity of all peoples of all colors in one human race to many parts of the world. Tlakaelel was the founder of Kalpulli Koakalko and In Kaltonal, the first Indigenous spiritual organization to be recognized by the Government of Mexico in 500 years. In a ceremony in the chamber of deputies in Mexico City, Tlakaelel was recently named Grandfather of the Grandfathers and spiritual guide of the Mexicanidad by an Indigenous council.

Tlakaelel is of the Tolteca-Chichimeca tradition, closely related to the group of Indigenous peoples of central Mexico. There are 56 Indigenous Nations in Mexico that still exist today, and 25 million Mexicans currently speak Indigenous languages. The term *Toltec*, according to Tlakaelel, refers not to one specific nation, but to the intellectuals or 'wise ones' of the Mexican traditions. For the past 511 years, the Indigenous people of Mexico, as in so many other parts of the world, have been forced into a position as second- or third-class citizens, and most live in poverty. As a champion for the Indigenous peoples of the world, Tlakaelel saw a great commonality among all Indigenous peoples, in their beliefs and also in the way they have so often been decimated by the larger culture. In many ways, the Indigenous peoples are the "canaries in the coalmine" who are the first to show the negative effects of pollution and climate change. Already these changes, as well as cultural changes, are gravely affecting the Indigenous. Tlakaelel has so often said the there is no time to waste – we are getting very close to the point where these effects that we only partly see now may become irreversible, and we will cease to exist as a species on this small planet. Whatever affects any of us, affects all of us.

Tlakaelel has often said that we do not need to save the Earth, but to save our place in it, to leave a livable world for our children and future generations, or else we will go the way of so many other species who are now extinct. I traveled with Tlakaelel for 22 years up until his death in July, 2012, scheduling, organizing, translating, and driving him, helping to deliver Tlakaelel's message around the world. Everywhere we went, especially around the time of December,

2012, so many people would ask him what was going to happen when this auspicious date arrives (December 21, 2012). His answer was always that it is already here. These changes have been going on for many years, and will continue for years to come. The massive changes are happening, and we are seeing the effects of our unsustainable lifestyle. Tlakaelel felt that the causes of these changes are partly natural, but are negatively augmented by our unsustainable lifestyle on this small planet. The following are Tlakaelel's words and responses to questions, some directly translated, and some paraphrased or interpreted from a number of talks he gave in many parts of the world as we traveled together.

When Peta Wilkinson, a good friend of ours in England, was working on a film about the message of the Monarch Butterflies, she asked him for his ideas on the subject. I did not know if Tlakaelel knew anything about the Monarchs, but he had an inspiring response. He replied:

One day we can imagine that we could come together to do a work of peace with a symbol that could represent all of us, of all the colors, with the enthusiastic intention, the great ideal, the grand promise that in future that we can all be united; united in the spirit, united in something that is the patrimony or treasure of all humanity. We have the example of the Monarch Butterfly. It is a message that the Great Spirit sends us – the creator of the universe that we can take as an example that we can overcome all that separates us from a better world. It is necessary to make this reconsideration now because the world is in chaos. We are living enormous changes. We are suffering grand catastrophes; we are seeing the painful birth of a new epoch. For the year 2012, we see a special position of planets in our solar system. Some say that at this time, the Mayan Calendar will end, and some feel the world will end but the world will not end. The Calendar will end when the sun goes out, since the Calendar is based on the movements of the sun and the stars, and that is many years away! [However], this era will end. There will be a change. There will present a possibility for us to understand that all the beings on this planet are our brothers and sisters because the same father – the Great Spirit – created us and the butterflies, the plants, and all that exists, all the elements. We are all one thing. We all are cosmos. We are the universe. All who are born here, all the human beings, we are terrestrials. We are born on this planet. This planet is our homeland. It is our home. What happens in any part of our planet at any time strongly affects us all. Precisely now, the changes are coming, and we know that the butterflies keep flying. It does not matter if the climate is changing. It is not important. Whatever happens, happens. It is an example for the human beings that we can come together. We can act as the animals do by their instincts according to the destiny that Great Spirit has given us all. I give thanks for this opportunity to communicate this message of universality. We are all universal beings. We are connected. We are butterflies, all human beings.

For many years, Tlakaelel and I traveled to the Wichapito Wakan community in Switzerland, guided by Angelique and Kees. Often, Tlakaelel would lead a sweat lodge there and then give a talk after dinner. On one of these occasions, he said:

> We must also know that we are all native of this planet. It's not important the place where we are born. This planet is our home and we are all citizens of this planet. We are citizens of this planet Earth. Universal Citizens. Our future tells us that we will one day all come together as brothers and sisters because we are brothers and sisters. Because we are sharing the same planet and the same destiny. We are universal citizens. We are citizens of this planet as brother and sisters.
>
> I think this is very important because we must learn to live in peace. This might seem a dream, but that day when we are all conscious of what we are, then this will bring forth this human brotherhood. And we will all help one another. This has already happened many times when there is an emergency, like the earthquake in San Francisco. Everybody helped, their personal importance, class, and colors notwithstanding.

We have many dear friends in the UK, Wales, and England. Yeva Gladwin first invited us over to Brighton and Wiltshire, on the suggestion of Colin Andrews, and over perhaps 15 years we were able to visit every year or two, mostly in Brighton and Wales. On one occasion, Sophie Latter arranged a talk at the trendy Inspiral Café in London. Here is an excerpt from that talk:

> We give thanks to Great Spirit. We hope to be useful, to help healing Mother Earth. We have beautiful concepts about what our destiny, what our job is here on this planet. We understand that all of us born here on this planet are brothers and sisters. It doesn't matter where someone is born or what the color of their skin is. Whatever their nationality, or language, this is our home, this is our Mother Earth, the mother of all of us. The destiny that we inherit, we share together.
>
> If we keep on poisoning Mother Earth with chemicals, with gases, with plastic, we ourselves are creating suicide. We all know we are in danger with epidemics, lots of illnesses, and we are using up our planet. We have to think, that all of us are the ancestors of the future and the new generations to come. Those that still aren't born yet may not be able to live normally. They may not be able to breathe the air, or drink the water. They may suffer a lot and speak badly of us. They are going to think that their ancestors poisoned this planet. We are guilty of what's happening. Why are we doing it? Or perhaps they'll bless us. Maybe they'll say, "thank you ancestors that you reacted and thought about us. You cleaned the planet, you've stopped the wars, and everything became creation." That's how we can live in peace.

Tlakaelel had a long friendship with Mary Thunder and her community from West Point, Texas, and we have traveled there for gatherings and ceremonies almost every year. In 2010, she was giving teachings at Big Indian, New York, near Woodstock, and asked Tlakaelel to give a talk there. Here are excerpts from that talk on June 2, 2010:

> You are a slave to the capitalist system. And when the money has ended, then the principal problem is removed. That which is of true value is not the dollar bills; you cannot eat a handful of dollar bills, you cannot chew an ingot of gold or silver, it is not nourishment, just false values.
>
> What is truly valuable is the water, the corn, the wheat, the bean. That has true value. So we understand that what is truly valuable is the land, the plants, the trees, all which surrounds us, the rain, the wind, the Mother Earth. So you are returning to your origin. And when you become very old is when you recognize this and you begin to initiate yourself in your origin. You begin to identify with all that surrounds you. This, after many years, when your faculties are no longer exactly good, then you begin to think of your role in the universe, and you can think of what was your mission. The deer, the rabbit, the buffalo, they are all completing part of the universal project. All sustains life.... You ask yourself; "I, what did I do? Well, I ate the plants and animals. Was my mission just to eat them?" So one may begin to think, "My principal mission that which I worked for, was to create a family, with children. I educated them, fed them, and raised them, and in this manner I preserved life. I preserved the human species. So my main mission is to prepare, to educate the upcoming generations." So we start to identify with the universe. This is the beginning of the universal initiation. What we are looking for is not at the end of life, but that we should have this understanding from when we are children: that we form part of the all, and that we are an important part of the all, and equally important are the deer, the rain, the wind, the fire, or time.... All are equally important. When we have this way of thinking and feeling, then we will be within the universal initiation. These are in simple terms, that which represents for us this universal initiation.

Tlakaelel tells us that we can learn a great deal from the Elders, and from the Indigenous communities. We need to live more in balance with our Mother Earth, so that our children will inhabit an earth that is habitable and inhabited. The future is in the small communities. We can minimize the impact of the Earth's changes by being prepared, and we can each help by being conscious of our impact on the Earth, and living in a way to minimize that impact. Each of us has the responsibility to do our part in the cleaning up and healing of our small planet and improving the health and possible survival of future generations.

16 Climate change, commercial news media, and Hispanics

An exploration of cultural processes and mediated environmental information

Bruno Takahashi and Juliet Pinto

The goal of this chapter is to assess the current state of research on the interfaces among culture, Spanish-language media use, and mediated environmental information aimed at Hispanics (we use the words Hispanic and Latino interchangeably), identify gaps in the literature, and present an agenda for future research. The limited research on global climate change issues (GCC) and Hispanic media is particularly striking because Hispanics are significantly and disproportionately affected by the impacts of climate change, and because Hispanics (derived from political and economic power) can play a key role in determining future policy actions at various levels of government to both mitigate sources of anthropogenic greenhouse emissions (e.g., regulating carbon emissions) and adapt to changing environmental conditions (e.g., severe droughts). This could also be relevant to other racial groups, such as African Americans and Asian Americans.

Hispanics in the USA are gaining importance because of their numbers and purchasing power. The US Census Bureau estimates that the Latino population in the USA accounts for more than 50 million people, 17 percent of the total population, and a growth of 43 percent from the 2000 Census, with estimates that Hispanics will represent 30 percent of the population by 2050.[1] At the same time, Hispanics are becoming more relevant to environmental decision-making within various political arenas. For example, Ard and Mohai (2011) reported that Hispanic members of Congress were more likely than whites, but less likely than African Americans, to vote pro-environmentally throughout the period of study. Nevertheless, Hispanics are not a monolithic group; therefore, we discuss below some differences in ethnicity and other socio-demographic variables.

Hispanics are disproportionately negatively affected by many environmental conditions. One in every two Hispanics in the USA lives in the country's top 25 most ozone-polluted cities (Quintero et al., 2011). GCC is expected to disproportionately affect Hispanic and other minority communities because most live in states that are more vulnerable to the effects of climate change (sea-level rise, droughts, floods) such as Texas, Florida, New York, and California, among others (Martinich et al., 2013; Wilson et al., 2010). In addition, many Hispanics in these areas suffer poverty, unemployment, marginalization, health problems, and socio-economic conditions (Kim et al., 2014; Crowley et al., 2015; Vaughan, 2014).

Closing this research gap could potentially identify best practices and develop evidence-based strategies to communicate threats associated with climate change that take into consideration nuanced differences among Hispanics, their concerns, and a core concern for community. In the following sections we examine current understandings about climate change perceptions, cultural processes, environmental concerns and behaviors, and the media's role in communicating environmental issues to Hispanics from a social scientific perspective – one grounded in media studies.

Climate change perceptions among Hispanics

For Latino audiences, climate change issues should be salient now, according to recent polls. Hispanics' beliefs about climate change (83 percent) are significantly higher than those of African Americans (76 percent) and whites (65 percent) in a March 2015 poll by the University of Texas at Austin (2015). According to a survey by the Pew Research Center, Hispanics report higher levels than whites on beliefs about the role of human activity in global warming (70 percent vs. 44 percent) (Krogstad, 2015). In a study commissioned by the Sierra Club and the National Council of La Raza in 2012, Hispanic voters were also observed to strongly support clean energy and conservation measures, and to believe that climate change is already happening (2012). A 2013 poll for the National Resources Defense Council found that nine out of ten Latinos surveyed believed that the US government should take action on climate change (2014).

Lived experiences particular to various lower income Latino communities have special resonance for these perceptions, as poorer Latino populations are particularly at risk – according to a 2011 Centers for Disease Control and Prevention Report (2011), nearly one in four low-income Hispanic (including Puerto-Rican) children in the USA has been diagnosed with asthma. Indeed, about half of the nation's Latino population lives in urban regions that have heavy traffic, industrial or utility production, or in other areas that often violate clean air rules. Latinos are three times more likely to die from asthma than other racial or ethnic groups, according to the National Hispanic Medical Association (Alvarado, 2011).

Hispanics overall tend to be more concerned about environmental issues than whites (Schultz and Zelezny, 1999; Schultz et al., 2000; Whittaker et al., 2005). However, there is limited academic work that examines perceptions and beliefs about how climate change perceptions among Hispanics may transform along cultural lines, and how media work to influence those processes. Moreover, most of the existing studies do not focus on examining differences among cultural or ethnic groups. The surveys described above point to a significant difference between Hispanics and other racial groups, but they do not explain the reasons for those differences.

Cultural processes and environmentalism among Hispanics

The ways in which we understand environmental and scientific information and news has much to do with cultural and ideological variables, as various scholars have noted. As Corbett (2006) stated, "All environmental messages have ideological roots that are deep and that are influenced by individual experience, geography, history and culture" (p. 26). This has import for understanding news regarding most social issues, including environmental issues, as news stories are "ideological constructions of reality" (Croteau and Hoynes, 2014, p. 153). In the hyper-commercialized, corporately dominated media system in the USA, ratings are more important than public service, and info-tainment is favored over coverage of abstract issues such as climate change (Ungar, 2000; Boykoff and Goodman, 2009). In Spanish-language corporate media, research has found almost a total lack of coverage of climate change, or packaged as something else – for example, a health problem (Takahashi et al., 2015; Villar and Pinto, 2013).

However, from the audience side, public opinion surveys do not match the paucity of environmental coverage for Latino audiences. Indeed, Schultz and colleagues (2000) stated that foreign-born Latinos in the USA show higher levels of environmentalism than do US citizens, something they attribute to the communal nature of environmental issues and the emphasis on tight-knit families in Hispanic societies. Milstein et al. (2011) wrote that, for Hispanics, "nature was largely relational, material, and, even when sublimely spiritual or transcendent, grounded in the social" (p. 502).

Cultural forces are also at work in terms of audiences' understanding or interest in such issues. These differences in environmental concern among cultural groups respond partly to different value orientations and ecological worldviews. Earlier research examining orientations toward Nature found that Latinos saw themselves more as submissive to dominant natural forces, at the mercy of Mother Nature (Kluckhohn and Strodtbeck, 1961). Others have noted deeper cultural ties to the land among Latinos, with a different paradigm of how to view the human–natural relationship as one enmeshing environmental with cultural identity (Lynch, 1993, p. 118). This has clear significance for the design of communication and education efforts to engage Latinos in environmental and Nature-related activities.

There is a robust body of research examining acculturation – defined as a process of interactivity between cultures that involves the maintenance of the original culture and development of relationships with the new culture – as well as the ways in which acculturation can affect various aspects of social interactions among migrants and their descendants. We argue that the process of acculturation can be key in explaining the climate change perceptions described above. Unfortunately, research on this area is scant. Therefore, we extend our discussion on acculturation to broader environmental themes.

Acculturation and environmental concern

Similar differences are found when examining environmental and health risk perceptions. Hispanics and other minorities may dismiss these risks because they are already living with many other highly visible risks (Mays and Cochran, 1988). Similarly, the "white male effect" suggests that whites tend to see the world as less risky than any other group (both other races and women) (Finucane et al., 2000; Flynn et al., 1994). Overall, the evidence suggests that Hispanics perceive higher risks to themselves than to whites (Palmer, 2003; Turner and Kiecolt, 1984; Pipitone et al., 2002; Vaughan and Nordenstam, 1991). These results are even more relevant considering that overall levels of involvement of Hispanics with science, technology, engineering, and math (STEM) fields is low compared to other groups, and the disparities continue to increase (Landivar, 2013). In other words, interest in and concern about environmental and science issues should translate to further involvement with related careers, but systemic factors (e.g., income gaps, racial bias, inadequate K-12 education level, etc.) seem to prevent this from happening. The research described above has not yet examined the more nuanced differences among Hispanics in terms of class, place of residence, language use, education, etc., which are part of the acculturation process. Work in other disciplines such as critical and cultural studies could help inform the social science approach to this area of study.

Overall, the evidence suggests that less acculturated Hispanics tend to score higher in scales measuring environmental concern or attitudes (Johnson et al., 2005; Mukherji, 2005; Noe and Snow, 1990; Schultz et al., 2000). Similarly, some evidence shows that there is no positive relationship between acculturation and self-reported pro-environmental behaviors (Mukherji, 2005; Pearson et al., 2012), including green consumption behaviors (Segev, 2013). The process of acculturation can be modifying these environmental perceptions brought from the immigrants' countries of origin, and now resembling the more dominant environmental discourses of the dominant culture. In summary, what this body of research suggests is that the longer foreign-born Latinos live in the USA, the more acculturated they become; therefore, the levels of environmental concern become similar to those of US residents. Yet, environmental concern of US residents is also a function of other demographic variables, such as income, political ideology, and age, among other factors (Kollmuss and Agyeman, 2002). These dynamics seem to resemble the Hispanic paradox in public health, which suggests that despite lower education and income levels, less acculturated Hispanics have a lower mortality rate and have overall better health than do other racial groups (Markides and Eschbach, 2005; Ruiz et al., 2013). Some factors that explain the paradox relate to cultural variables such as close family ties (Morales et al., 2002). This paradox is another factor that should also drive research into Hispanics and environmentalism.

Environmental concern and activism vary by ethnicity despite similarity for certain socio-economic characteristics (e.g., income) (McCabe et al., 2013) and environmental beliefs (e.g., the belief that we should protect nature over economic gains) (Whittaker et al., 2005). Whittaker et al. (2005) and Schultz et al. (2000) reported important differences in environmental beliefs and behaviors between US-born Latinos and foreign-born Latinos, where the former more closely resembled white environmentalism than the latter. These findings highlight the importance of distinguishing between Latinos by criteria such as immigrant status, acculturation level, language, or country of origin (Johnson et al., 2004). Carter et al. (2012) succinctly described the tension and balance that Latinos, especially Mexican-born immigrants, experience regarding balancing environmental protection and looking after human needs.

Cultural groups labeled as Hispanic should not be considered a composite entity. Mukherji (2005) argued that the design of environmental strategies for ethnic groups must recognize the importance of acculturation and the nuances among the subcultural groups. Johnson et al. (2005) explained that Mexican immigrants and US-born Mexicans offer the most resistance to acculturation when measured by participation in nature-based activities, which is consistent with research in other areas showing resistance to acculturation by Mexican groups. However, Johnson et al. (2005) and Noe and Snow (1990) argued that, although cultural heritage may play a role in determining environmental attitudes, other factors such as exposure to situational opportunities (e.g., park use) and other social factors may also influence those beliefs. Research explaining the role of socio-economic and cultural factors is still inconclusive.

Environmental news consumption and production for Spanish-language audiences

While decades ago there was an explosion of research around Hispanics and media use generally (e.g., Subervi-Velez, 1986; Soruco, 1996), little research has investigated the consumption or production of environmental information by Hispanic audiences and media organizations, with the exception of a growing body of research in comparative 'green consumerism' (e.g., Segev, 2013). If the process of acculturation described above plays such an important role in explaining changes in environmentalism, what is the role of the media in promoting or preventing such changes, and in promoting and preventing acculturation? After all, media are the primary source of information regarding most issues, and as Cohen (1963) famously wrote, they do not tell us what to think, but rather what to think about.

Studies on media coverage of GCC have for the most part focused on the USA (Boykoff and Boykoff, 2007), the UK (Carvalho and Burgess, 2005), and other developed nations (Grundmann and Scott, 2014). Recently, some research studies (Zamith et al., 2012; Takahashi and Meisner, 2013) have started to focus on developing or emergent nations (Billett, 2009; Das et al., 2009; Gordon et al., 2010). Some researchers call for a need for additional multi-nation studies to describe and explain differences in the coverage (Schmidt et al., 2013; Anderson, 2009), which could also help us understand differences in public perceptions. Although we agree that this is an important area of research, we also argue that climate change needs to be explored from local and regional perspectives. A smaller body of research examines local and regional climate change media coverage (Brown et al., 2011; Liu et al., 2008). One of the authors has analyzed media coverage of climate change in the Great Lakes region of the USA and Canada (Takahashi et al., 2014). However, few adopt a cultural perspective to the study of climate change media coverage. With regard to Hispanics, only a study by one of the authors explored the coverage of climate change in Spanish-language media in the USA (Villar and Pinto, 2013).

Here, we discuss media serving Spanish-language audiences. While increasing numbers of Hispanics report getting their news in English, a significant portion still get their news in Spanish, according to a recent survey by the Pew Research Center (Lopez and Gonzalez-Barrera, 2013). Hispanic media cover certain issues, such as labor, voting, political empowerment, healthcare, and immigration, heavily (Abrajano and Singh, 2009; Lozano, 1989; Subervi-Velez, 2008; Vargas, 2000; Vargas and De Pyssler, 1999; Veciana-Suárez, 1990; Vigón, 2010; Villar and Bueno, 2013). Vargas and De Pyssler (1999) reported that health stories dominated the coverage in 17 Latino newspapers (both weeklies and dailies) during a six-month period, followed by crime, immigration, and education stories. Environmental issues, such as climate change, do not receive much attention (Takahashi et al., 2015; Villar and Pinto 2013), and when they are covered they usually focus on direct individual effects (e.g., air pollution affecting respiratory illnesses, or natural disasters) (Takahashi et al., 2015). The limited past research in this area suggests that Spanish-language media have failed at effectively reporting risk issues (Benavides, 2013). However, we contend that environmental realities are another critical area at the top of the Latino agenda and resonant on political social activist agendas across the country, but disconnected to the way in which 'mainstream' media report salient issues of the day to their audiences. As the Latino demographic continues to grow in the USA, the dissonance in terms of the public agenda with that of the media will force changes in media content.

Previous research explains that media catering to Spanish-language audiences function primarily as a mechanism for acculturation in the USA (Subervi-Velez, 2008; Rios and Gaines, 1998), leaving science-based issues relegated to English-language venues. Other research explained the functionalist role of the media from this perspective, where specific traits of the dominant culture are learned and adapted to the individual's or group's culture (Kerevel, 2011).

Recent studies show that media can indeed affect audiences' cognitions about complex scientific issues such as climate change (Brulle et al., 2012; Borick and Rabe, 2010). Scholars have explored the role of political partisanship in relation to media use and perceptions and beliefs about climate change in the general US population (Kim, 2011; Zhao, 2009; Feldman et al., 2012; Hmielowski et al., 2014). However, these studies place limited focus on those cultural differences described above, which is why we argue for a need to integrate these streams of research. A study examining the results of several polls from 2002 to 2010 reported that, although media coverage of climate change was an important factor affecting public concerns about the issue, this influence was a function of elite cues and economic factors (Brulle et al., 2012). In other words, the partisan divide in the political arena, as well as economic crises, are more influential factors. This is particularly relevant to Hispanics, most of whom have traditionally aligned with the Democratic Party (Lopez and Taylor, 2012). However, there is no established literature focusing on minorities in the USA, specifically Hispanics.

Discussion

In this chapter we have reviewed past and current research examining Hispanics' involvement in environmental issues. Using a cultural media studies perspective, we also explored the ways in which media coverage of environmental issues, specifically GCC, can influence perceptions of Hispanics about those issues. Research indicates that Hispanics' views of environmental issues can be impacted by cultural processes, intertwined with ideological and political overtones over time. The role of religious beliefs is also perceived as important, especially considering the large proportion of Catholics among Hispanics and the recent Encyclical on climate change by Pope Francis demanding more political and individual action.[2] But more research is needed to clarify these relationships, similar to research conducted with the overall US population (McCright, 2010; Roser-Renouf and Nisbet 2008), as well as in other countries (Arlt et al., 2011). Much less is known about the influence of gender on Hispanic environmentalism, or media use or production of environmental news for Hispanic audiences.

This chapter suggests that there are considerable gaps in the research about Hispanics and climate change. Although there is a heightened interest in Hispanics' perceptions of the issue, as reflected in some recent polls (see, e.g., the discussion about the NRDC above) exploring Hispanics' perceptions, as well as media coverage focusing on the higher levels of concern of Hispanics compared to other ethnic and racial groups, the scholarly research is clearly lacking. Apart from a single study of climate change media coverage in Spanish-language newspapers (Villar and Pinto, 2013), not much literature is found in this area. Moreover, the literature exploring the ways in which media affect perceptions about the issue is quite limited. While Pinto (2012) utilized feminist theory to deconstruct dualisms in national news regarding a constitutional referendum in Ecuador that would give rights to Nature and natural systems, little is known how gender interacts with culture and Hispanic media content on environmental understandings.

The question remains: In a hyper-commercialized, corporate media system situated within an extremely market-oriented society, how will US Spanish-language media discuss environmental issues with a market segment that has increasing social and political influence and purchasing power? While some research has found that Hispanics are critical of the excessive materialism of

American consumerism and its impacts on the environment and society (Carter et al., 2012), the other literature within acculturation studies suggests that that view changes over time, as Hispanics are also part of the consumerist and individualist society. Unfortunately, the growing literature on Hispanics as consumers (e.g., Dávila, 2012; Korzenny and Korzenny, 2012) has not examined these dynamics in relation to the evolving views about the environment. There is also evidence that Spanish-language media are starting to resemble more and more the corporate model of mainstream English-language media, including similarities in the frequency of coverage across topics (Moran, 2006; Okamoto et al., 2011). Okamoto et al. (2011) explained that mainstream US media are broadening their scope in part due to demographic changes in the country. With billions in spending power and an exponentially growing demographic, these trends will only increase over time.

In a time of increasing effects of climate change impacting almost every sector of society, and particularly the regions where large concentrations of Hispanics live, understanding what is happening and options available to address these changes is of vital importance. For example, the population of Miami-Dade County, Florida is predominantly Hispanic, and at the same time the region is one of the most vulnerable in the world to the effects of sea-level rise. The county is home to many English-, Spanish- and bilingual-media outlets. Yet little is understood about how and why Spanish-only, English-only, and bilingual audiences receive, process, and ultimately understand mediated information on sea-level rise. As information on sea-level rise in English-language general market media increases, how will such diffusion impact or be transformed by Spanish- or bilingual-media content? How will journalism practiced in the region present this type of information to acculturating publics?

Based on the discussion set out in this chapter, we present a set of research questions that could guide the development of the area of study in climate change communication and Hispanics in the USA. First, what are the differences and similarities in the coverage of environmental issues between Spanish-language media and mainstream media, particularly in areas of climate change? Second, how do the news media influence the perceptions and understandings of environmental issues among Hispanics? In this respect, an exhaustive analysis of the sources of information that Hispanic audiences use to learn about, or conversely, to deny, climate change and environmental issues is currently missing. Third, how do Hispanics influence the larger dominant environmental discourse in the country, and how do gender/social class and cultural processes interact in these spheres of influence?

In summary, Hispanics are an increasingly politically and economically powerful sector in the USA; and they show higher levels of environmental concern compared to other racial groups. However, they also receive little, if any news on climate change and associated environmental issues in Spanish-language media, as well as a demographic disproportionately affected by such impacts. In any age of accelerating climate change, scholars and media producers would do well to address this gap by providing more research and content to better inform all sectors of US society. Understanding these dynamics could be useful in the design and implementation of communication efforts that could be based on ideas of community action rather than individual-level behaviors.

Notes

1 See http://quickfacts.census.gov/qfd/states/00000.html.
2 See www.theguardian.com/environment/2015/jun/27/climate-change-hispanic-catholics-message-willing-disciples.

References

Abrajano, M. and Singh, S., 2009. Examining the Link between Issue Attitudes and News Source: The Case of Latinos and Immigration Reform. *Political Behavior*, 31(1), pp. 1–30.

Alvarado, M., 2011. *Report: Hispanics disproportionately affected by air pollution*. The Record.

Anderson, A., 2009. Media, Politics and Climate Change: Towards a New Research Agenda. *Sociology Compass*, 3(2), pp. 166–182.

Anon., 2011. *CDC Health Disparities and Inequalities Report – United States, 2011*. Centers for Disease Control Prevention.

Anon., 2012. *National Latinos and the Environment Survey*. The Sierra Club and the National Council of La Raza.

Anon., 2014. *Overwhelming Majority of Latinos Want Strong Presidential Action to Combat Climate Change, Poll Shows*. National Resources Defense Council.

Anon., 2015. *Energy Poll*. University of Texas at Austin.

Ard, K. and Mohai, P., 2011. Hispanics and Environmental Voting in the US Congress. *Environmental Practice*, 13(4), pp. 302–313.

Arlt, D., Hoppe, I. and Wolling, J., 2011. Climate Change and Media Usage: Effects on Problem Awareness and Behavioural Intentions. *International Communication Gazette*, 73(1–2), pp. 45–63.

Benavides, A.D., 2013. Four Major Disaster Occurrences and the Spanish Language Media: A Lack of Risk Communication. *Disaster Prevention and Management: An International Journal*, 22(1), pp. 29–37.

Billett, S., 2009. Dividing Climate Change: Global Warming in the Indian Mass Media. *Climatic Change*, 99(1–2), pp. 1–16.

Borick, C.P. and Rabe, B.G., 2010. A Reason to Believe: Examining the Factors that Determine Individual Views on Global Warming. *Social Science Quarterly*, 91(3), pp. 777–800.

Boykoff, M.T. and Boykoff, J.M., 2007. Climate Change and Journalistic Norms: A Case-study of US Mass-media Coverage. *Geoforum*, 38(6), pp. 1190–1204.

Boykoff, M.T. and Goodman, M.K., 2009. Conspicuous Redemption? Reflections on the Promises and Perils of the 'Celebritization' of Climate Change. *Geoforum*, 40(3), pp. 395–406.

Brown, T., Budd, L., Bell, M. and Rendell, H. 2011. The Local Impact of Global Climate Change: Reporting on Landscape Transformation and Threatened Identity in the English Regional Newspaper Press. *Public Understanding of Science*, 20(5), pp. 658–673.

Brulle, R., Carmichael, J. and Jenkins, J.C., 2012. Shifting Public Opinion on Climate

Change: An Empirical Assessment of Factors Influencing Concern over Climate Change in the U.S., 2002–2010. *Climatic Change*, 114(2), pp. 169–188.

Carter, E.D., Silva, B. and Guzmán, G., 2012. Migration, Acculturation, and Environmental Values: The Case of Mexican Immigrants in Central Iowa. *Annals of the Association of American Geographers*, 103(1), pp. 129–147.

Carvalho, A. and Burgess, J., 2005. Cultural Circuits of Climate Change in U.K. Broadsheet Newspapers, 1985–2003. *Risk Analysis*, 25(6), pp. 1457–1469.

Cohen, B., 1963. *The Press and Foreign Policy*. New York: Harcourt.

Corbett, J., 2006. *Communicating Nature: How we create and understand environmental messages*. Washington, DC: Island Press.

Croteau, D. and Hoynes, W., 2014. *Media/Society: Industries, images, and audiences*. Thousand Oaks, CA: Sage.

Crowley, M., Lichter, D.T. and Turner, R.N., 2015. Diverging Fortunes? Economic Well-being of Latinos and African Americans in New Rural Destinations. *Social Science Research*, 51, pp. 77–92.

Das, J., Bacon, W. and Zaman, A., 2009. Covering the Environmental Issues and Global Warming in Delta Land: A Study of Three Newspapers. *Pacific Journalism Review*, 15(2), pp. 10–33.

Dávila, A., 2012. *Latinos, Inc: The marketing and making of a people*. Berkeley: University of California Press.

Feldman, L., Maibach, E.W., Roser-Renouf, C. and Leiserowitz, A., 2012. Climate on Cable: The Nature and Impact of Global Warming Coverage on Fox News, CNN, and MSNBC. *The International Journal of Press/Politics*, 17(1), pp. 3–31.

Finucane, M.L., Slovic, P., Mertz, C.K., Flynn, J. and Satterfield, T.A., 2000. Gender, Race, and Perceived Risk: The 'White Male' Effect. *Health, Risk and Society*, 2(2), pp. 159–172.

Flynn, J., Slovic, P. and Mertz, C.K., 1994. Gender, Race, and Perception of Environmental Health Risks. *Risk Analysis*, 14(6), pp. 1101–1108.

Gordon, J.C., Deines, T. and Havice, J., 2010. Global Warming Coverage in the Media: Trends in a Mexico City Newspaper. *Science Communication*, 32(2), pp. 143–170.

Grundmann, R. and Scott, M., 2014, Disputed Climate Science in the Media: Do Countries Matter? *Public Understanding of Science*, 23(2), pp. 220–235.

Hmielowski, J.D., Feldman, L., Myers, T.A., Leiserowitz, A. and Maibach, E., 2014. An Attack on Science? Media Use, Trust in Scientists, and Perceptions of Global Warming. *Public Understanding of Science*, 23(7), pp. 866–883.

Johnson, C.Y., Bowker, J.M. and Cordell, H.K., 2004. Ethnic Variation in Environmental Belief and Behavior: An Examination of the New Ecological Paradigm in a Social Psychological Context. *Environment and Behavior*, 36(2), pp. 157–186.

Johnson, C.Y., Bowker, J.M. and Cordell, H.K., 2005. Acculturation via Nature-based Outdoor Recreation: A Comparison of Mexican and Chinese Ethnic Groups in the United States. *Environmental Practice*, 7(4), pp. 257–272.

Kerevel, Y.P., 2011. The Influence of Spanish-language Media on Latino Public Opinion and Group Consciousness. *Social Science Quarterly*, 92(2), pp. 509–534.

Kim, K.S., 2011. Public Understanding of the Politics of Global Warming in the News Media: The Hostile Media Approach. *Public Understanding of Science*, 20(5), pp. 690–705.

Kim, Y-A., Collins, T.W. and Grineski, S.E., 2014. Neighborhood Context and the Hispanic Health Paradox: Differential Effects of Immigrant Density on Children's Wheezing by Poverty, Nativity and Medical History. *Health and Place*, 27, pp. 1–8.

Kluckhohn, F.R. and Strodtbeck, F.L., 1961. *Variations in Value Orientations*. Evanston, IL: Row Peterson.

Kollmuss, A. and Agyeman, J., 2002. Mind the Gap: Why do People Act Environmentally and What Are the Barriers to Pro-environmental Behavior? *Environmental Education Research*, 8(3), pp. 239–260.

Korzenny, F. and Korzenny, B.A., 2012. *Hispanic Marketing*. Abingdon: Routledge.

Krogstad, J.M., 2015. *Hispanics More Likely Than Whites to Say Global Warming is Caused by Humans*. Pew Research Center.

Landivar, L.C., 2013. *Disparities in STEM Employment by Sex, Race, and Hispanic Origin*. U.S. Census Bureau.

Liu, X., Vedlitz, A. and Alston, L., 2008. Regional News Portrayals of Global Warming and Climate Change. *Environmental Science and Policy*, 11(5), pp. 379–393.

Lopez, M.H. and Gonzalez-Barrera, A., 2013. *A Growing Share of Latinos Get Their News in English*. Pew Research Center. Available from: www.pewhispanic.org/2013/07/23/a-growing-share-of-latinos-get-their-news-in-english/ (accessed December 4, 2014).

Lopez, M.H. and Taylor, P., 2012. Latino Voters in the 2012 Election. *Pew Research Hispanic Center*.

Lozano, J.C., 1989. Issues and Sources in Spanish Language TV: A Comparison of Noticiero Univisión and NBC Evening News. *Frontera Norte*, 1(1), pp. 151–173.

Lynch, B.D., 1993. The Garden and the Sea: U.S. Latino Environmental Discourses and Mainstream Environmentalism. *Social Problems*, 40(1), pp. 108–124.

Markides, K.S. and Eschbach, K., 2005. Aging, Migration, and Mortality: Current Status of Research on the Hispanic Paradox. *The Journals of Gerontology Series B: Psychological Sciences and Social Sciences*, 60(Special Issue 2), pp. S68–S75.

Martinich, J,, Neumann, J., Ludwig, L. and Jantarasami, L., 2013. Risks of Sea Level Rise to Disadvantaged Communities in the United States. *Mitigation and Adaptation Strategies for Global Change*, 18(2), pp. 169–185.

Mays, V.M. and Cochran, S.D., 1988. Issues in the Perception of AIDS Risk and Risk Reduction Activities by Black and Hispanic/Latina Women. *American Psychologist*, 43(11), pp. 949–957.

McCabe, M.B., Corona, R. and Weaver, R., 2013. Sustainability for Hispanics in California: Do They Really Care? *Global Journal of Business Research*, 7(2), pp. 103–112.

McCright, A., 2010. The Effects of Gender on Climate Change Knowledge and Concern in the American Public. *Population and Environment*, 32(1), pp. 66–87.

Milstein, T., Anguiano, C., Sandoval, J., Chen, Y.W. and Dickinson, E., 2011. Communicating a 'New' Environmental Vernacular: A Sense of Relations-in-place. *Communication Monographs*, 78(4), pp. 486–510.

Morales, L.S., Lara, M., Kington, R.S., Valdez, R.O. and Escarce, J.J., 2002. Socioeconomic, Cultural, and Behavioral Factors Affecting Hispanic Health Outcomes. *Journal of Health Care for the Poor and Underserved*, 13(4), pp. 477–503.

Moran, K.C., 2006. Is Changing the Language Enough? The Spanish-language 'Alternative' in the USA. *Journalism*, 7(3), pp. 389–405.

Mukherji, J., 2005. Is Cultural Assimilation Related to Environmental Attitudes and Behaviors? *Advances in Consumer Research*, 32, p. 415.

Noe, F.P. and Snow, R., 1990. Hispanic Cultural Influence on Environmental Concern. *The Journal of Environmental Education*, 21(2), pp. 27–34.

Okamoto, D., Ebert, K. and Violet, C., 2011. ¿El Campeón de Los Hispanos? Comparing the Coverage of Latino/a Collective Action in Spanish- and English-language Newspapers. *Latino Studies*, 9(2), pp. 219–241.

Palmer, C., 2003. Risk Perception: Another Look at the 'White Male' Effect. *Health, Risk and Society*, 5(1), pp. 71–83.

Pearson, H.C., Dawson, L.N. and Breitkopf, C.R., 2012. Recycling Attitudes and Behavior among a Clinic-based Sample of Low-income Hispanic Women in Southeast Texas. *PloS one*, 7(4), e34469.

Pinto, J., 2012. Legislating Rights for Nature: The Mediated Social Construction of Human/Nature Dualisms. In Latta, A. and Wittman, H., eds, *Environment and Citizenship in Latin America: Natures, subjects and struggles*. New York: Berghahn Books, pp. 227–243.

Pipitone, M., Robinson, J.K., Camara, C., Chittineni, B. and Fisher, S.G., 2002. Skin Cancer Awareness in Suburban Employees: A Hispanic Perspective. *Journal of the American Academy of Dermatology*, 47(1), pp. 118–123.

Quintero, A., Jaffee, V., Madrid, J., Ramirez, E. and Delgado, A., 2011. *U.S. Latinos and Air Pollution: A call to action*. NRDC.

Rios, D.I. and Gaines, S.O., Jr., 1998. Latino Media Use for Cultural Maintenance. *Journalism and Mass Communication Quarterly*, 75(4), pp. 746–761.

Roser-Renouf, C. and Nisbet, M.C., 2008. The Measurement of Key Behavioral Science Constructs in Climate Change Research. *International Journal of Sustainability Communication*, 3, pp. 37–95.

Ruiz, J.M., Steffen, P. and Smith, T.B., 2013. Hispanic Mortality Paradox: A Systematic Review and Meta-analysis of the Longitudinal Literature. *American Journal of Public Health*, 103(3), pp. e52–e60.

Schmidt, A., Ivanova, A. and Schäfer, M.S., 2013. Media Attention for Climate Change Around the World: A Comparative Analysis of Newspaper Coverage in 27 Countries. *Global Environmental Change*, 23(5), pp. 1233–1248.

Schultz, P.W. and Zelezny, L., 1999. Values as Predictors of Environmental Attitudes: Evidence for Consistency Across 14 Countries. *Journal of Environmental Psychology*, 19(3), pp. 255–265.

Schultz, P.W., Unipan, J.B. and Gamba, R.J., 2000. Acculturation and Ecological Worldview Among Latino Americans. *The Journal of Environmental Education*, 31(2), pp. 22–27.

Segev, S., 2013. Green Advertising and Behavior. In Kahle, L.R. and Gurel-Atay, E., eds, *Communicating Sustainability for the Green Economy*. Armonk, NY: M.E. Sharpe, p. 104.

Soruco, G., 1996. *Cubans and the Mass Media in South Florida*. Gainesville, FL: University Press of Florida.

Subervi-Velez, F.A., 1986. The Mass Media and Ethnic Assimilation and Pluralism: A Review and Research Proposal with Special Focus on Hispanics. *Communication Research*, 13(1), pp. 71–96.

Subervi-Velez, F.A., 2008. Latinos' Use of Media and the Media's Influence on Political Knowledge and Participation: Findings from the 1989 Latino Political Survey. In Subervi-Velez, F.A., ed., *The Mass Media and Latino Politics*. New York: Routledge, pp. 323–349.

Takahashi, B. and Meisner, M., 2013. Climate Change in Peruvian Newspapers: The Role of Foreign Voices in a Context of Vulnerability. *Public Understanding of Science*, 22(4), pp. 427–442.

Takahashi, B., Huang, K., Fico, F. and Poulson, D., 2014. *Climate Change in Great Lakes Region Newspapers: A Study of the Use of Expert Sources*. Seattle, WA: International Communication Association.

Takahashi, B., Pinto, J., Vigón, M. and Chavez, M., 2015. El Ambiente y Las Noticias: Understanding U.S. Spanish Language Newsrooms. *International Journal of Hispanic Media*, in press.

Turner, R.H. and Kiecolt, K.J., 1984. Responses to Uncertainty and Risk: Mexican American, Black, and Anglo Beliefs About the Manageability of the Future. *Social Science Quarterly* (University of Texas Press), 65(2), pp. 665–679.

Ungar, S., 2000. Knowledge, Ignorance and the Popular Culture: Climate Change Versus the Ozone Hole. *Public Understanding of Science*, 9(3), pp. 297–312.

Vargas, L., 2000. Genderizing Latino News: An Analysis of a Local Newspaper's Coverage of Latino Current Affairs. *Critical Studies in Media Communication*, 17(3), pp. 261–293.

Vargas, L.C. and De Pyssler, B.J., 1999. U.S. Latino Newspapers as Health Communication Resources: A Content Analysis. *Howard Journal of Communications*, 10(3), pp. 189–205.

Vaughan, A.S., 2014. Connecting Race and Place: A County-level Analysis of White, Black, and Hispanic HIV Prevalence, Poverty, and Level of Urbanization. *American Journal of Public Health*, 104(7), pp. e77–e84.

Vaughan, E. and Nordenstam, B., 1991. The Perception of Environmental Risks among Ethnically Diverse Groups. *Journal of Cross-cultural Psychology*, 22(1), pp. 29–60.

Veciana-Suarez, A., 1990. *Hispanic Media. Impact and influence*. Washington, DC: The Media Institute.

Vigón, M., 2010. Covering the news for Spanish-Speaking USA, May 2008. *Journal of Spanish Language Media*, 3, pp. 24–40.

Villar, M.E. and Bueno, Y., 2013. Disparate Health News Frames in English- and Spanish-language Newspapers in Two U.S. Cities. *Howard Journal of Communications*, 24(1), pp. 57–70.

Villar, M.E. and Pinto, J., 2013. Coverage of Climate Change in Leading U.S. Spanish-language Newspapers. *Journal of Spanish Language Media*, 6, pp. 42–60.

Whittaker, M., Segura, G.M. and Bowler, S., 2005. Racial/Ethnic Group Attitudes Toward Environmental Protection in California: Is 'Environmentalism' Still a White Phenomenon? *Political Research Quarterly*, 58(3), pp. 435–447.

Wilson, S.M., Richard, R., Joseph, L. and Williams, E., 2010. Climate Change, Environmental Justice, and Vulnerability: An Exploratory Spatial Analysis. *Environmental Justice*, 3(1), pp. 13–19.

Zamith, R., Pinto, J. and Villar, M.E., 2012. Constructing Climate Change in the Americas: An Analysis of News Coverage in U.S. and South American Newspapers. *Science Communication*, 35(3), pp. 334–357.

Zhao, X.Q., 2009. Media Use and Global Warming Perceptions: A Snapshot of the Reinforcing Spirals. *Communication Research*, 36(5), pp. 698–723.

17 A call for climate justice

Rebecca Hall

> Social justice activists are the hope of the environmental movement because the climate crisis will not be averted without restructuring our economic, social and political systems.

Remember how citizens of the USA became discarded refugees in the aftermath of Hurricane Katrina? African Americans were rounded up and tightly packed into the Superdome, as in some monster slave ship of the infamous Middle Passage on the ocean of flood waters. They were held like captives in slave-pens, without food and water for days, awaiting the slow-to-arrive school buses that would disperse them throughout a land far away from their homes (Hall, 2005, p. 72). Some tried to walk out of the flood-ravaged city rather than wait to drown or die of thirst as our government did nothing. Crossing the Crescent City Connection, the bridge that links the city of New Orleans with the west bank of the Mississippi and the predominantly white town of Gretna, they were met with shotgun fire from white Gretna city police. The group of hundreds of predominantly black citizens stopped at the fire of warning shots. They were instructed that they could not cross. Not knowing where to turn, they began to set up camp in the middle of the bridge. A Gretna police officer yelled at them through a bull horn "Get the f*** off the bridge." Someone asked why they couldn't pass on to safety. They were told, "There will be no Superdome here" (Witt, 2008).

Citizens, who by definition have the constitutional right to travel, who reasonably rely on their own government to help them in a crisis, lost their citizenship rights that night. From the perspective of civil rights and social justice, the entire Hurricane Katrina fiasco was one enormous abrogation of citizenship resulting in death and displacement. A complete betrayal. I remember the shock I felt when I heard news reporters actually refer to the residents of New Orleans as 'refugees.'

As an African American woman and direct descendant of slaves, I am constantly attuned to the fact that people are used and discarded. I was born into a family of activists, and taught that the only way to survive is to fight back. My father was the child of slaves: both his parents were born in 1860, my grandmother Harriet Thorpe was born the property of Squire Sweeney in Howard

County, Missouri and my grandfather Haywood Hall was born the property of Colonel Haywood Hall on his plantation in Tennessee. I never met either of them – they died long before I was born in 1963. My dad was the youngest of the Hall family's children, born in 1898. He never finished eighth grade and worked odd jobs, from shining shoes to waiting tables. Unable to tolerate or comply with racism, he worked outside the system. He was a labor organizer, a communist, and a self-taught worker-intellectual, publishing two books and numerous articles during his life (Hall, 2012). In 1956 my parents were forced to travel to three different states before they could find a judge to marry them – my mom a second-generation White New Orleanian Jew of Russian and German descent, my dad African American and 33 years her senior. My mom had begun her work as an anti-racist activist when she was a teenager in the 1940s and never looked back. She was a teacher and later a professor of history and has written three books on the history of slavery. My parents were black-listed during the McCarthy era and forced to flee the country, which is how I ended up being born in Mexico City and having dual citizenship.

It was Hurricane Katrina that woke me up to the climate crisis, and when I think about it I immediately think of all the disposable people – the people making up half of the planet who are already barely surviving. What will happen when the levees break again?

Pacific Island nations face the destruction and inundation of their lands caused by rising sea levels. There are over seven million Pacific Islanders who live in 22 nations. Tuvalu's Ambassador to the UN explains that his nation was being destroyed by climate change, which represents an unprecedented threat to Tuvalu's "fundamental rights to nationality and statehood, as constituted under the Universal Declaration of Human Rights and other national conventions" (Pita, 2007, n.p.). Following a long fight against slavery and imperialism, after finally gaining their sovereignty in 1978, the First World's consumption of fossil fuels will take their country away again. It is enough to make a grown woman cry.

In 2007 a record was set: human-caused global warming caused the melting of 42 percent of the Arctic ice-cap. That record was broken in 2012, and scientists predict that the Arctic will be ice-free in 20 years (Vidal and Vaughan, 2012). Storms are getting fiercer. Hurricane Sandy appeared soon after that record was broken. Sea levels are rising. I grew up in New York City, and watching the videos of water pouring into the subways and PATH train terrified me. My mom, who grew up in New Orleans and lived in New York for many years, called me crying and said both of her home places were being destroyed.

Ninety-five percent of the world's glaciers are in retreat (Balog, 2009). The snow pack is melting in places where people depend on its water for agriculture and drinking water. Pacific Islanders are already preparing to relocate, to leave their countries underwater and be cast on the mercy of other nations. Who will take them in? Friends of the Earth International and the Australian Refugee Council are urging Australia to step up, to include the category of 'climate refugee' into their asylum program (Lagan, 2013). The current categories that are used worldwide were written in the wake of World War II and were designed to protect those who are in fear of persecution by their government because of their race, religion, or nationality. There is no system in place to handle those forced to leave their nations because of climate change. My fear is that they will continue to be treated as so-called 'economic refugees,' people who, by definition, won't be helped. We in the Global North have no responsibility for people who are in trouble because their economic systems failed them, right? Some scientists estimate that all of the world's mountain snow-pack will be gone by 2030. What happens to the billions of people who need that water to drink and grow food? What will happen when seven million Pacific Islanders, along with the millions of citizens of the low-lying river deltas throughout the world, lose their citizenship along with their nationhood? Will we watch the flags of country after country lowered in front of the United Nations? What will these people be when they are stateless? What will the First World do with all of these stateless people of color from other lands? The US government would not help its own citizens in New Orleans, never mind racialized others who aren't even from here. In 2005 we saw how, within days, African Americans in New Orleans lost their citizenship, as if some Frankenstein-like Chief Justice Taney rose from the dead declaring a new Dred Scott decision (Hall, 2005).

In 2012 the Northeast was hit by the monster Hurricane Sandy, and we saw there that even in the richest part of the world government and humanitarian response systems were overwhelmed, and it mattered what neighborhood you lived in – and, given ongoing segregation, what race and class you belonged to. What happens when we have super-storms every year? It will be a worldwide Dred Scott, as the First World proclaims that the people of the Global South have no rights 'which the white man is bound to respect.'

What can we do? It is not too late to preserve a livable future for humans on this planet, but our window of opportunity is closing. NASA's top climatologist, James Hansen, explains that if we stop the use of coal within the next two decades, phase out the use of conventional petroleum, and ban the use of high-carbon fuels such as shale and tar sands, we have a chance to bring greenhouse gases back down to safe levels (Hansen, 2011). We may not be able to reverse the extensive damage already done, but we can prevent it from getting any worse.

How do we do it? We need more environmentalists to reframe this whole issue of climate crisis as one of social, economic, and racial justice – no small task, but some groups at the more radical end of the environmentalist movement are trying. Here I am calling upon all of us at the other end of this intersection, namely the social and economic justice activists. Over the past three decades the central concept of intersectionality – the ways in which race, class, and gender are mutually constituted and can only be dismantled together – has worked its way into the theory and practice of economic and social justice. Now we need to understand that the climate crisis is the most urgent and deadly issue facing women, poor people, and communities of color today. We must understand that 'the environment' is not something 'out there' that we can think about after we have dealt with poverty, racism, institutionalized male dominance, and heterosexism. The environment is where we live and breathe; it is literally the ground on which we stand and fight. We need a climate justice movement that understands and addresses this intersection.

This movement for climate justice is not merely additive. We don't just join the environmental movement; we redefine it and restructure it. In fact, social justice activists are the hope of the environmental movement because the climate crisis will not be averted without a major restructuring of our economic, social, and political systems. If the standard of success is preserving the Earth, the environmental movement and environmentalism has been a failure. It has been trapped in a world of policy strategies while the democratic system needed to actualize these approaches has collapsed under the weight of corporate capitalism. It has gone along with the prime directive of unbridled growth and all of its sick sequelae that are fundamentally at odds with living on a planet with limits. A climate justice movement will instead demand advancing democracy as well as the health and welfare of the people within systems that will sustain the people and the planet. Environmentalism has attempted change within a system that is designed to 'externalize' the real costs of business as usual – whether that is the burden of subsidizing workers who are paid poverty wages without benefits, or the destruction of local and global ecosystems by international corporations that go wherever they want, and leave behind whatever disasters they make, while creating corporate refugees (people who are being destroyed by this economy). And, unlike corporations, corporate refugees are not allowed to cross borders.

When we demand that our leaders bring carbon emissions back to safe levels, we are requiring them to put the interests of people before the richest corporations in the world. It will require a drastic restructuring of our economy, and provide an opportunity to re-create it in a more just and equitable way. We have moved past environmentalism. The call now is for climate justice, and we must all become climate justice activists.

References

Balog, J., 2009. *Extreme Ice Now*. Washington, DC: National Geographic Society, p. 38.

Hall, G. ed., 2012. *A Black Communist in the Freedom Struggle: The life of Harry Haywood*. Minneapolis: University of Minnesota Press.

Hall, R., 2005. We have lost our citizenship again: Katrina's aftermath; The new Dred Scott. In J.B. Childs, ed. *Hurricane Katrina: Response and responsibilities*. Santa Cruz: New Pacific Press, pp. 70–72.

Hansen, J., 2011. G-8 failure reflects U.S. failure on climate change. *The Huffington Post*. Available at: www.huffingtonpost.com/dr-james-hansen/g-8-failure-reflects-us-f_b_228597. htm (accessed June 18, 2015).

Lagan, B., 2013. Australia urged to prepare for influx of people displaced by climate change. *Guardian*. Available at: www.theguardian.com/environment/2013/apr/16/ australia-climate-change-refugee-status (accessed June 18, 2015).

Pita, A., 2007. *Statement Delivered by His Excellency, Mr. Afelee F. Pita, Ambassador/*

Permanent Representative of Tuvalu to the United Nations at the Special Session of the Security Council on Energy, Climate and Security, April 17, 2007. Available at: www.tuvaluislands.com/un/2007/un_2007-04-17.html (accessed June 15, 2015).

Vidal, J. and Vaughan, A., 2012. Arctic sea ice shrinks to smallest extent ever recorded. *Guardian*. Available at: www.guardian.co.uk/environment/2012/sep/14/arctic-sea-ice-smallest-extent (accessed June 18, 2015).

Witt, H., 2008. Katrina aftermath still roils Gretna. *Chicago Tribune*. Available at: www.chicagotribune.com/news/nationworld/chi-gretna (accessed June 15, 2015).

18 Climate action and literacy through creativity and conversations

Patricia Widener, Carmen Rowe, Ana Marie Estrada, Marcella Ahumada, Martha Eichloff, and Jacquelyn Anderson

How do we initiate conversations on global climate change (GCC), creatively and inclusively, and from the classroom to the community? In Southeast Florida, GCC is or will be impacting the region in several ways, including sea-level rise, salt-water intrusion, higher storm surges, increased flooding, and stronger storms. As residents, we are only just discovering the meaning, relevance, causes, and impacts of GCC, but we are committed to increasing our understanding so as to support each other in becoming conversant and engaged. In this chapter, we explore the power of eco-literacy and public sociology to foster a deeper, local, and more relevant engagement of climate change.

Linking climate change to power and privilege (and their opposites) has proven difficult, and may even require the intimate experience of impact to complete this realization and to push us beyond our ambivalence. As a reflection of our keen sense of responsibility, our work is informed by three approaches: intersectionality (Luft, 2009), public sociology (Burawoy, 2004), and critical eco-pedagogy (Kahn, 2010). The intersectional approach – as a lens, embodiment of, and mechanism for change – incorporates the experience of race, ethnicity, class, gender, age, sexuality, ability, household type, migration story, and nativity (to name a few) in understanding and mobilizing within our locally relevant and globally integrated social world. Intersectionality centers on the idea that multiple axes of identification inform privilege and marginalization, and confronts the tendency to view issues such as discrimination and advantage through a single lens (Crenshaw, 1989, pp. 139–140).

When coupled with critical public sociology, an intersectional approach reveals to a wider audience how social position and participation are contained (when burdensome) and protected (when beneficial). When practiced, public sociology communicates and uses sociological knowledge in a meaningful way to engage and expand public dialogue, while critical sociology pushes us to see and demand alternatives to exclusionary practices and dominant worldviews (Burawoy, 2004). Critical public sociology also dovetails with the intersectional perspective to insert inclusivity into public dialogue and action.

Finally, critical eco-pedagogy (Kahn, 2010) emphasizes an ecological lens to public discussion and effort. Critical eco-pedagogy "is concerned with understanding how political economy and ideology produce the domination of nature," while also promoting the idea that we can transcend these barriers through individual and collective acts of transformation, thereby "transcend[ing] the limited framework of environmental education," by building a more holistic and sustainable experience for students and communities in a globally integrated world (Kellner, 2010, 152–153).

In the classroom

In the fall semester of 2011, 24 undergraduates completed a new course, 'Climate, Disaster and Society,' in which some students not only discovered a sociological perspective, but also started conversations on GCC and sustainability within themselves, among each other, in their households, and with their families, friends, and neighbors. We present how GCC was addressed in the classroom and how students became educators beyond the classroom through personal reflection and creative expression. Indeed, each student was invited by the professor, the first author, to help write this assessment, and the five student contributions analyze how intersectionality is embodied in our understanding and engagement, even though a deep intersectional and critical eco-pedagogical approach at first eluded us as well. We hope our strength in this exercise (collectively linking micro- and macro-level actions) and our limitations (a Northern-centric and intersectional imbalance) serve others in their teaching, scholarship, and/or activism for a richer, more complex understanding and response to GCC.

By way of introduction, of the 24 students only three were men, and approximately 13 were white (54.2 percent), seven black (29.2 percent), and four Latino or white-Latino (16.6 percent); however, this approximation fails to account for diversity in individual or parental ethnicity, place of birth, or nativity. Such a collective offers ethnically and racially diverse views, but a skewed assessment of women's reflections, rather than a more nuanced gender-along-a-continuum understanding. We found that those who participated in this project self-identified gender as a master frame for understanding and responding to GCC and local environmental degradation. We also identified age, and some of the roles or responsibilities associated with age, as an important variable given the diversity in the classroom: students ranged in age from early twenties to early fifties. Students were parents and grandparents, full-time and part-time workers, new arrivals to the country, first-generation Americans, and/or first-generation university students. Of the five student authors who chose to participate, two identify as white-Latina, two as white or Caucasian, and one as Latina. Their own and their families' migration stories were important to them. Three of the student authors had children and one of those three students also had grandchildren.

The class spent seven weeks discussing GCC, including the interconnectedness of impacts (such as social, health, ecological, and economic), climate politics and science, public understanding and confusion, risk and risk perceptions, and mitigation and adaptation strategies. In addition to the standard modes of assessment (in this case a paper and two exams), students were required to complete a creative project, inspired primarily by McNall's (2011) book *Rapid Climate Change* as depicted in this chapter.

Students as educators

As the class progressed, the students began to identify the social value of introducing friends and family to what they were discussing and learning in class: in particular, the local relevance of global warming. Some students engaged their parents, and others their children and grandchildren: "I always recycle and I'm teaching my two year old to practice it too," wrote one. Through their outreach, students sought to align their knowledge and values with others, while also testing new knowledge with those whom they trusted, or to whom they were closest. Or perhaps it was a sense of powerlessness (and worry) that provoked students to return to the safety of their homes, seeking connections in climate discussions with family and friends. In these endeavors, they identified the worth of civic participation (even beyond family and friends) and the importance of public spaces to discuss social problems. They discovered these features as conversational participants and educators, and as early advocates and decision makers.

For many professors, our first, and for some our only, 'public' is found in the classroom (Burawoy, 2004), a venue in which to impart "an invaluable contribution to the construction of a curious, responsible, and informed public" (Pfohl, 2004, p. 114), or to aid in the production of "a sustainable citizenry" (Kahn, 2010, p. 98). Certain subjects may even capture students' attention by stimulating what Pfohl (2004, p. 115) refers to as "their attractions and repulsions, fascinations and fears." Today, GCC may be that trigger. Yet, when asked to rank from 1–5 their priorities for government spending, the majority of the students identified: (1) public education, (2) jobs and economy, (3) healthcare, (4) climate change, and (5) other, including community services, ending war, the military, disaster relief, and programs for inner-city youth. This is not climate ambivalence, but a reflection of the gravity of structural problems pressing in on our students. GCC was compelling enough to elicit the activist educator embodied within many citizens and sociology students. Undeniably, some students were linking their student roles and responsibilities to their social ones, and bridging from the abstract and private space of the classroom to the local and public sphere where they live, work, study, and play; and in this case, educate and create.

Empowered and empowering creativity

Folk, community, and professional artists have a history of capturing through creative expression individual and collective experiences of crisis, fear, pain, or injury. During environmental justice struggles, for example, Hofrichter (2002, p. 91) identified the importance of "cultural activism," whereby communities depict the impact of toxins through the use of videos and photographs in order to "take control of the direction of their communities and build self-esteem." On GCC, artists are also expressing and informing others through dance, photography, video, and music (Wihbey, 2012), endeavors that remind us of art's social importance.

In our class, we reflected, filmed, photographed, painted, created, wrote, and surveyed. We also shared and educated each other to inform ourselves and to influence the opinions and actions of others. In doing so, we found how climate change became more concrete in our relationships with others. When we discovered GCC, we thought of our family, friends, and neighbors, including some of the most vulnerable in our social circles.

As depicted here, tangible expressions of our quest to expand classroom discussion for a wider audience included: (1) a vivid painting linking oil spills and deforestation with consumer youth culture; (2) a memoir on how current worries bridge the past and the future; (3) a witty essay on diapers, landfills, and twins; (4) an essay on coastal landscapes and nuclear power (accompanied by a video); and (5) a survey on the recycling efforts of popular bars (in the absence of clear city policy). Here we let the voices of the five student authors speak for themselves to broaden, enrich, and advance our shared eco-literacy and eco-pedagogy in the practice of public sociology. (Excluding the first author, authorship credit corresponds with order of reflection.)

Consumptive (dis)connectedness

> When science has given us so much, and promises more, why would people resist a finding on which the scientific community has achieved consensus: the Earth is warming.
>
> (McNall, 2011, p. 26)

Science is a part of our society and culture; therefore, the values, beliefs, and ideologies held in a society influence the acceptance or rejection of scientific facts. As McNall noted, some people have rejected the science which states that there are anthropogenic causes to global warming. In an effort to understand inaction and denial from an eco-feminist's perspective, I created this painting (Figure 18.1). My gender identity is woman and my gender matches my biological sex (female). I identify as a feminist; and an eco-feminist perspective informs my understanding of GCC. By linking "the domination of women with the domination of nature," eco-feminism maintains that Western or Northern societies value culture over nature and male over female; this dualistic nature represents a hierarchical mode of thought where ideas of "otherness" and "inferiority" function "to justify superiority and domination" (Kirk and Okazawa-Rey, 2010, G-2, pp. 539–540). Certainly, an intersectional eco-feminist perspective is central to capturing the multiple forces of privilege and oppression as depicted in my art.

Figure 18.1 Consumptive (dis)connectedness artwork by Carmen Rowe.

I based my project on McCright and Dunlap's (2011) article "Cool Dudes," which is about conservative white men whom they identify as being more likely than any other group in the USA to deny GCC. They are also more likely to favor unlimited growth and to disproportionately control the economic industries central to the capitalist system, which produce greenhouse gas emissions. At this time, conservative politics centers on being pro-business and protective of privilege, while rejecting climate science and sustainable practices for being antithetical to capitalist productivity.

The main subject of my painting represents a conservative white man, in his twenties like myself, who rejects climate science while embracing technology that I know and use as well. He consumes the media which reject climate science through his electronic devices – disconnected from Nature, from beliefs other than his own, from people unlike himself, from the environmental costs tied to his consumption, and from a sense of social commitment and obligation. Yet, he can afford his disconnectedness due to his multifaceted privileges. For eyes, he has shields in the form of laptop and television screens, and he has a dollar bill for a mouth, embossed with the words oil, coal, gas, and timber. His black, stained T-shirt signifies oil, representative of the costs of transporting and producing his clothing. Since he is not outdoors, and because he is so disconnected from the world, he is incapable of seeing what is happening (and what he is doing) to the natural and social world around him. This depiction is not meant to dehumanize white men but to humanize all of us who are responding to GCC – hopefully for the sustainability of the planet. Basically, in challenging the status quo, our words and our art are crucial in creating open, inviting, and productive dialogue for change that enhances and equalizes the lives of everyone.

I too know consumption. My parents are professionals, and I grew up in a wealthy Florida suburb. My dad and I were both born in the suburbs of Detroit and my mom was born in Miami. My status afforded me privilege with respect to my early education and my understanding and acceptance of climate science. Yet, I also witnessed unlimited growth and consumption practiced as virtues. I have seen trees torn down to build a shopping center or a luxury residential complex. Given this memory, to the right of the house there is a field of dying grass, a tree chopped down, and an announcement of a new mall. On the left of the house there is water and a tiny strip of land, reflecting sea-level rise. The enlarged sun represents global warming. A small figure, a woman of color, stands near the dirty water. Her size reflects how the conservative white man of my painting views her as less than him in terms of value, importance, and status. From his view, oppressed people are barely noticeable. Although I am white, my mom's family is white and Hispanic – Cuban to be exact – and that gives me some insight into how otherness is experienced. This project has strengthened my commitment to confront privilege and to shift the country toward more sustainable modes of living that are globally and socially just.

Environmental education across the generations/*La educación ambiental a través de las generaciones*

> Climate change is real; it will have disastrous consequences; and we must act.... Acting will protect people's homes, families, children, and grandchildren.
>
> (McNall, 2011, p. 47)

I am a second-generation Cuban born in New York. Now, in my early fifties, I am the only one in my family to earn a Bachelor of Arts degree. I started at a community college in 2004; and it is 2011 as I write. I am the mother of three children and grandmother of two. My parents emigrated from Cuba in 1958 during the Batista revolution before Castro took over, with the clothes on their back and my then-eight-year-old sister. Neither of them completed high school, and they worked hard their entire lives.

In my youth, they kept a garden of herbs, beans, tomatoes, and yucca, and my mom was always picking up garbage (while I helped). I remember asking her why, and she would say, *"Porque nadie sabe comó esta basura daña nuestro mundo."* In other words, "Because no one really knows how this garbage hurts our world." That was in the 1960s. As a child, I don't remember caring, but I followed her steps. My mom is in her eighties now (my dad has passed away), and she is still the recycle queen of the neighborhood. When my children were growing up, we used to do the school and scout recycling projects together, but I was not an adamant or committed recycler. Now, a grandmother of two, I pick up garbage that I find lying around, and I know they are watching me. Sometimes we discuss why I do this, but it is different: we speak in English and I wear gloves to protect myself from infection.

It strikes me as odd that I really learned what my mother was telling me when we discussed GCC in class. She was speaking about the *precautionary principle* – of not releasing anything into the environment unless you know that it will not harm any living being – even before I knew of the precautionary principle. In the past, I would casually recycle, and though I was concerned or cautious about GCC, I did not really think much about it. I believed that it would not harm me or my family. Now I see it as a more urgent threat; and I am alarmed – for my grandchildren's sake.

Our world is going through a global crisis due to world population, poverty, and unsustainable development – but I am not criticizing the world's poor for their poverty or children. I want to know how we can adapt to and survive the conditions that are changing all around us – that we have caused – while ignoring the problems. Or, should we join each other and do something about both GCC and the "toxic discourse" surrounding GCC that is further dividing us (McNall, 2011, p. 27)? It is important for each of us to do our homework and to think about our future. Denying ourselves the truth may lead us to extinction. I know this is a strong word, and yes it could be described as fear-mongering, but I believe that our health, quality of life, and well-being will deteriorate, and we can no longer put bandages on this issue. Free, critical thinkers are needed now, and I intend to share what I know with my grandchildren.

Environmental guilt: 1 + 1 = 4

> Americans use 2.5 million plastic bottles every hour and we throw 70 percent of them away. We also throw away 25 billion Styrofoam cups and make enough plastic film to shrink-wrap Texas, an intriguing thought.
>
> (McNall, 2011, p. 66)

Four years ago our household doubled when my husband and I had twins. Doubling the disposable diapers, the baby bottles, and the amount of every gadget and gizmo I thought I needed added to my environmental guilt. Like any new mom, I was bombarded with information on how many disposable diapers a baby will go through. Depending on the source, numbers varied from 4000 to more than 10,000 for the first two to three years, and I would have to multiply that by two! Worse still was the idea of how long they would sit in a landfill. Those diapers are made with petroleum products; fuel, energy, and resources go into their manufacturing, packaging, and shipping; and then, after a single use, they will sit without degrading for a few hundred years, or some extreme number which implies that future generations will have come and gone and those dirty diapers will still be there. I was a monster … according to most of what I read online.

I am a white-Hispanic woman in my early thirties, born and raised in Miami, with a mostly middle-class upbringing. My Colombian-born Hispanic mother moved to the USA when she was young, and my white father was born in Alabama but was raised in Florida. As for my husband and me, we have both returned to college part-time. We were aware of how companies were buying carbon offsets to minimize or neutralize their carbon footprint. Some of our favorite bands were also touring in buses that ran on recycled cooking oil and recording their albums using only wind and solar power. I started thinking that maybe I could 'offset' my diaper dilemma. I knew I was not going to make a big dent in any of the world's problems, but it would make *me* feel better.

What could I do? There's always the cloth versus disposable argument. Sure, cloth has less of an environmental impact, even after you factor in water and electricity for laundering; it is cheaper in the long run, and it is better for your baby, who will get fewer diaper rashes. Who would not want that? At the time, I didn't! I had twins. Twins are double the work. There was enough laundry without adding cloth diapers. Like people who want their bottled water and Styrofoam® cups, I was going to keep the convenience of disposable diapers. My sanity depended on it. I knew I was paying a price, so I looked elsewhere to make changes; recycling, reusing, and repurposing became a requirement. My guilty conscience depended on it.

I prepared homemade baby food from organic or local food sources whenever possible, and then I started paying attention to what my husband and I ate, since some of what I made was going through a blender for my babies. I tried to make sure our livestock products were fed grass and grains as nature intended. My chickens and eggs had to be free-range, my fish wild but sustainable, my buffalo must have roamed freely on green pastures and not been fed hormones, antibiotics, or 'color-enhanced feed.' I also started reading labels for sourcing – and it didn't make any sense either. I noticed apple juice came from China. That's a long way to go, so I thought I'd buy an organic brand – it came from Turkey. Huh? Don't we have apples in the USA? Shouldn't the orange juice be local since I am in Florida? Nope. In a state full of citrus growers, the juice came from Brazil. WTF?

As my babies grew and became a little bit easier to handle, I revisited the cloth diaper idea. I still depended on the evil disposables when we went out, as convenience wins out over having stinky diapers in my bag, but at home I tried to use cloth. Alas, I got in the game too late, since my girls were able to unsnap their diaper covers and get out of them at will. On the upside I saved 840 disposables from going to the landfill – shrinking my carbon footprint somewhat. The downside? That's only ten weeks' worth of diapers, and they weren't fully potty trained and diaper free until after their third birthday. I stopped counting diapers long before that.

If part of the climate problem stems from not knowing the impact of our behaviors, I decided to see where my family ranked. According to footprintnetwork.org, if everyone wanted to live like us it would take 4.1 Earths, as we take up 18.3 global acres to support our lifestyle. Yet, those 'global acres' seemed too abstract. At coolclimate.berkeley.edu, my results were met with frowning faces and numbers that told me we have the same impact as the average household our size. Given my efforts, I didn't like that either. Of the four different categories, I received a happy face for my shopping habits that were 15 percent better than average, but I still found the frowning faces insulting. I found better news at the government's epa.gov/climatechange/emissions/ind_calculator. If I understood it correctly, my household's footprint is 50 percent less than the average US household of a family of four. I liked those numbers so much I put them on the refrigerator. Yes, I did 'Goldilocks and the Three Bears' those findings: I disregarded the one I didn't understand, ignored the one I didn't like, and picked the one that sounded just right.

While having children can alter the views of parents of any gender, my household experience leads me to believe that my gender, not just being a parent, influences my views. My husband, of a similar race and upbringing, and I have shared values and interests, yet he doesn't view his or our actions as having an impact on the environment in the same way that I do. Our children, who are now aged 4, know what refuse goes in the recycle bin; and when they are unsure they ask. He, on the other hand, would prefer to toss it all in the trash. He begrudgingly recycles the milk cartons and juice bottles because it is easier than hearing me complain. As a parent, I hope that I can raise conscientious people who will understand how their actions affect more lives than their own.

Expanding nuclear power?

> We resist making investments in the planet now, because we do not see an immediate gain.
>
> (McNall, 2011, p. 43)

My particular concerns are extreme weather events, such as floods, hurricanes, droughts, and heatwaves. Due to storms in South Florida, families have lost power for days, and some have lost their most treasured possessions. In 2011, we heard news of flooding across the nation due to hurricanes and tropical storms, and though I understand that GCC and weather are different, I experience weather, and GCC is leading to extreme weather events, which worries me. We watched the destruction of a tsunami that blew through Japan in 2011 and the flooding that contaminated the reactors; we heard about the radiation that went into the atmosphere immediately after the tsunami hit, and the subsequent evacuation that was required for a nine- to 15-mile radius. Having lived for a short time in Okinawa, Japan, I agonized for those affected when I heard or read the news. I worry for my own family and friends: could our nuclear power plant experience the same impacts for the same reasons?

Although there are many competing positions on nuclear energy – from complete opposition and resistance, to slowly retiring old facilities, to expanding existing facilities, to building new ones – proposals to expand a nuclear facility in South Florida are on the table. While nuclear power plants are an important energy source for the country and perhaps a transitional energy from fossil fuels, I am concerned that what happened in Japan could happen here, and I fret that not enough people are doing enough to understand and minimize all the risks. The effects of GCC, especially sea-level rise, and the risks of sea-level rise to the nuclear power plant and to my family and community concern me greatly. Would our families and friends spread out along the Florida coastline stay safe? Would our tourism industry collapse? You see how climate change, rising sea levels, stronger storms, the tourism industry, and proximity to a nuclear plant become quickly intertwined in South Florida.

My family came here from Cuba in 1967 when my mother was a teenager. They built their home, their business, and raised a family in Miami. Who am I? I am a Hispanic white woman, approximately 40 years old, the daughter of Cuban immigrants, and the mother of two daughters. I started college immediately after high school, got married, and had children before completing it. I was actually helping one of my daughters with her recycling project at school while I was working on this climate project for my own class.

In class I was at the edge of my seat, either shocked or exhilarated with the information. I wanted to run out and do something. I wanted to change the world overnight, and then I was overwhelmed because I believed that I couldn't. Talking with people made me depressed or angry because everyone I talked to thought that we had plenty of time – or, worse, that there were no problems at all. By the end of the semester and after visiting the nuclear power plant and touring the coastline, I found myself questioning everything I thought I knew. I found myself wondering whether I would be able to stand in that very same spot in ten years' time and take the same picture. Would parts of Florida – including the nuclear facility – be partially under water? I remember an empty feeling in my stomach: where would I be in 20 years' time and where would my children be in the years to come?

Green efforts, one bottle at a time

> Each of our lives – through the numerous choices we make daily – affects the climate and has a role in creating the future.
>
> (McNall, 2011, p. 48)

After a booze-fueled Saturday night (that did not include touring a nuclear power plant or a diaper landfill), have you ever wondered what happens to the beer bottle left on the bar? One might assume from the abundance of bottles that bars would be required to recycle. But you would be mistaken – at least here in Fort Lauderdale. Those Earth-saving Green Lovelies in Portland (as a white woman, in my thirties, a social worker and an Oregon native, I write this without meaning insult) are glaringly absent on an average night in Fort Liquor-dale.

Thirsty, and curious about the actual number of lost or forgotten bottles, I did some fieldwork. I spoke to four bartenders, and I learned that on an average weekend night a busy, medium-sized bar may go through 30 cases of beer (720 bottles), plus many liquor bottles, and approximately 300 plastic cups. One bartender, who has worked at local bars and restaurants for the past ten years, agreed to an interview:

THIRSTY INTERVIEWER: What are you able to recycle easily?
GREEN BARTENDER: Cardboard and glass.
THIRSTY: Where do you take it?
GREEN: The city of Wilton Manors has a recycling station, a drop-off center – I take it there.

[Wait. What? In the fall of 2011, bartenders were personally – outside of work hours – taking bar recyclables to a recycling center about three miles from where they worked; otherwise the recyclables were thrown out.]

THIRSTY: Are there recycling programs for bars and restaurants in the area?
GREEN: Not unless you do it yourself.

[Another bartender chimes in that independent recycling is only possible if you drive a truck, given the volume.]

THIRSTY: So why do you do it?
GREEN: Because it makes me feel good.

[Unfortunately, the bartender is taken.]

Currently, there is no organized, business-wide initiative to recycle in a neighborhood well known for its nightlife. So, I asked others about the feasibility of recycling and what each felt might motivate bar owners to participate. One expressed that for a smaller establishment, recycling would be easier to implement and more successful. One bar had looked into a green dumpster; however, separating trash would prove challenging and expensive. Most agreed that implementing incentives, as well as penalties, would be most effective, though they struggled with identifying specific incentives and penalties other than monetary ones.

Although the county and the city have recycling and 'going green' as priorities, recycling is clearly not a priority if 2000 beer bottles are thrown into the trash on a weekly basis because recycling is too cumbersome or expensive. Too expensive for a successful Fort Lauderdale bar? Even though the bartenders felt it was wrong not to recycle, they also felt it was too big or overwhelming a problem for them to address. Business owners are the ones who should be concerned. By failing to act now, they may be ensuring the demise of their tourist base, and undermining the future employment and fun for others.

Collective understandings

In this chapter, these student authors have captured their rhythms of consumption, as well as their voices, lifestyles, and behaviors, with flair and early academic rigor. Collectively and collaboratively, we have bridged the micro and the macro in our understanding and depiction of climate change. Embracing Jasanoff's (2010, p. 238) suggestion that "living creatively with climate change will require re-linking larger scales of scientific representation with smaller scales of social meaning," we examined the intersections of micro-level practices (household reuse and recycling), with meso-level experiences (worksites and city- and/or county-wide recycling policy), and macro-level actions (national energy security and a culture of consumption). Similarly, Kahn's (2010, p. 99) eco-pedagogical call "to foster a form of citizenship that is better able to negotiate the complexities of everyday life, labor, and culture" – a citizenship that is "powerfully local and rooted" – was expressed in our endeavors as well. With our identities and histories influencing our climate literacy, consumption became a point of self-awareness and reflection: of household and community activism and education, of personal and group discussions, and as a point to link corporate and political endorsements with individual behaviors. We revealed that we are local, rooted, *and* mobile, with our observations on the intimate and the systemic, navigating along the groups, organizations, and communities to which we belong, have belonged, or are moving toward belonging.

Although our work accentuates individual acts, each act reminds us of the tremendous collective waste produced by industrial and individual activities. For example, Figure 18.1 captures well the tuned-out individual who consumes with abandon, in satiated ignorance, devastating the world's natural resources, while the other authors make clear that individual and community effort is required when we are careless with both the visual and greenhouse gas debris of our consumptive lifestyles. Recycling and repurposing can play an important role in reducing greenhouse gas emissions, though we know that an emphasis on individual or household changes obscures the large-scale industrial waste that is occurring without the institutional guilt, worry, or sense of responsibility that are expressed in our personal essays.

What we produced reflected our lived and embodied intersectionality, weighted by gender, migration stories, race or ethnicity, household and family types, and socio-economic status (our own and our parents'). Only one of us had a parent born and raised in Florida, though we are now Floridians, who understand and express GCC in terms of our social and geographical intimacies. In addition to our mobility, when we expressed our interests and passions, we elevated gender and our gendered roles, thereby *flattening* our other embodiments (Luft, 2009). Mostly, we identified as women and emphasized our role in the education and caretaking of others and the environment.

Our collective worldview of US origins has enabled us to identify and reflect upon gender and ethnicity as told in our migration stories (including the subordination, privilege, and social construction of each), rather than view ourselves based upon our whole being. Yet the gains of women, women of color, and women of migration may have transpired into an indifference, marginalization, or discounting of our complaints in the larger national context. We may have become trained in the awareness of gender and ethnic discrimination and inequality, while simultaneously accepting such disparities and conceding to what appears to be an intractable quality that is based on the hidden machinations of power and privilege. That is, it may be easier to identify and discuss discrimination and inequality in some cases, such as gender, without actually correcting for such obstacles.

This chapter catches us in mid-stride. A comprehensive chain of cause and effect proved exceedingly difficult individually, but possible collectively – which reveals a new starting point: to identify power, privilege, and the political economy of climate change, and then to insert or force our intersectional selves into it. As a group, we are in an early stage of: (1) discovering and illustrating our understanding of GCC; (2) connecting this awareness to ourselves and others; and (3) initiating thoughtful and inclusive discussions so as to act and affect progressive change. Our expressions are initial steps that have encouraged us to look deeper into existing and expected climate inequities and injuries worldwide. We acknowledge that being rooted in our locality we minimized, or struggled to emphasize, the connections between Northern or First World actions and Southern or Third World impacts. Finally, we hope that this chapter serves other lifelong learners and educators, and engaged global citizens, in acting, creating, leading, living, researching, and teaching sustainable societies.

References

Burawoy, M., 2004. Public Sociologies: Contradictions, Dilemmas, and Possibilities. *Social Forces*, 82(4), pp. 1603–1618.

Crenshaw, K., 1989. Demarginalizing the Intersection of Race and Sex: A Black Feminist Critique of Antidiscrimination Doctrine, Feminist Theory, and Antiracist Politics. *University of Chicago Legal Forum*, 140, pp. 139–167.

Hofrichter, R., 2002. Cultural Activism and Environmental Justice. In R. Hofrichter, ed., *Toxic Struggles*. Salt Lake City: University of Utah Press, pp. 85–96.

Jasanoff, S., 2010. A New Climate for Society. *Theory, Culture and Society*, 27(2–3), pp. 233–253.

Kahn, R., 2010. *Critical Pedagogy, Ecoliteracy, and Planetary Crisis*. New York: Peter Lang.

Kellner, D., 2010. Afterward: Mediating Critical Pedagogy and Critical Theory: Richard Kahn's Ecopedagogy. In R. Kahn, ed., *Critical Pedagogy, Ecoliteracy and Planetary Crisis*. New York: Peter Lang, pp. 151–154.

Kirk, G. and Okazawa-Rey, M., 2010. *Women's Lives: Multicultural perspectives*, 5th edn. New York: McGraw Hill.

Luft, R.E., 2009. Intersectionality and the Risk of Flattening Difference: Gender and Race, Logics and the Strategic Use of Antiracist Singularity. In M.T. Berger and K. Guidroz, eds, *Intersectional Approach*. Chapel Hill: University of North Carolina Press, pp. 100–117.

McCright, A.M. and Dunlap, R.E., 2011. Cool Dudes: The Denial of Climate Change among Conservative White Males in the United States. *Global Environmental Change*, 21(4), pp. 1163–1172.

McNall, S.G., 2011. *Rapid Climate Change*. New York: Routledge.

Pfohl, S., 2004. Blessings and Curses in the Sociology Classroom. *Social Problems*, 51(1), pp. 113–115.

Wihbey, J., 2012. Making Climate Media Creative – in the Extreme. *The Yale Forum on Climate Change and The Media*, January 27.

19 MAN still #73

Steve Cutts

Figure 19.1 MAN still #73 by Steve Cutts.

Part III

Earth

Figure III.1 Earth by Tina Shirshac.

Place   Development   Farming   Food   Pollination

20 At the fault lines
Exposing the forces of discontinuity

Phoebe Godfrey

Emerging from Chaos, the Greek Goddess Gaia was the Great Mother of all, giving birth to the entire Universe and subsequently the Earth. Used more specifically now to merely mean the Earth, by referring to Gaia we seek to intentionally situate the Earth as a self-regulating, complex system that creates the conditions for life and is therefore energetically alive. In addition, in naming the Great Mother as Gaia we want to speak her many other names, such as Pachamama for the Incas, Odudua for the Yoruban, Durga for the Hindus, as well as to recognize that cultures around the world have worshiped the Earth as the primal female deity from whom all life has emerged. In returning our social identities to the material reality that all humans are first and foremost "creatures of the Earth" (Suzuki, 2007, p. 49) and by intersectionally traveling the fault lines through the 'gap' between Heaven and Earth, hence through Chaos, we re-situate ourselves firmly upon the earth. And yet intersectionality invites us to recognize that, as social beings, how we collectively interpret our earth-based geographical surroundings – including our bodies – is inseparable from how we interpret ourselves and our experiences. In addition, how we use the Earth to address our physical and social needs is inseparable from GCC, as the pieces in Part III explore in relation to the concepts of Place, Development, Farming, Food, and Pollination. It is not we humans who will 'Save the Earth,' as repeated in the popular dialogue, but the Earth who will save us if we are willing to listen and learn.

Jose Gonzales' powerful image, 'Harvesting poison,' immediately demands a critical view of our current food system wherein what has been seen as a central act for all humans – the gathering and harvesting of food, commemorated with holidays and sacred rituals – has become in our society contaminated by the industrial food system. Chris Williams' 'Contradictions of a sick system: food, climate, and capitalism' takes this theme further with his linking of the profit-driven industrial food system and its direct links to GCC. Parvez Babul then engages with other ways in which structural inequalities intersect by analyzing, from the perspective of a journalist, the intersections of sexism, food insecurity, and GCC in Bangladesh. Recognizing that this same food system contributes about one-third of all CO_2 in the atmosphere, Chantal Bilodeau's excerpt from her play *Sila* situates GCC in the Arctic and explores the intersection of situated knowledges, including that of polar bears. While emblematic of GCC, the precariousness of their continued survival suggests we do not permit polar bears 'voice,' whether mythic or material. Thus, to further accentuate the perspective of non-human animals, a polar bear is featured photographed on Bernard Harbor, Alaska by Arctic photographer and author Subhankar Banerjee. Next, Toban Black, in 'Race, gender, and climate injustice: dimensions of social and environmental inequality,' explores a more theoretical intersectional analysis of climate injustices which weaves together the social and the environmental in order to assert that, without such analyses, there can be no social, let alone climate, justice. Artist Imna Arroyo's image of Mother Earth and her daughter Isis Mattie's companion poem bring us back to the ancient act of anthropomorphizing the Earth. And yet for many such a practice remains vital, as Dylan Harris explores in 'The political ecology of Pachamama: race, class, climate change, and Kallawaya traditions,' which is an intersectional analysis of the Kallawaya – traditional healers of the Bolivian Andes who seek to preserve their ways of living in the face of social and ecological change. Finally, Gabrielle Maughan's poem 'Sandcastle' invites reflection upon the temporal impermanence of human endeavors in the face of more timeless natural forces like the ocean, and ultimately the Earth. Yet from the dissolution of one form comes the pollination and development of the next.

Bibliography

Suzuki, D., 2007. *Sacred Balance: Rediscovering our pace in Nature*. New York: Graystone.

21 Harvesting poison

José G. González

Figure 21.1 Harvesting poison by José González. We depend much on the communities that are often most impacted by environmental inequities such as climate change and pesticide pollution. As they harvest our food, they harvest the impact and consequences of our inaction on climate and environmental justice.

22 Contradictions of a sick system
Food, climate, and capitalism

Chris Williams

A choir of seedlings arching their necks out of rotted tree stumps, sucking life out of death. I am the forest's conscience, but remember, the forest eats itself and lives forever.

(Barbara Kingsolver, *The Poisonwood Bible*)

Capitalist laws of motion

Kingsolver's quote demonstrates the profoundly dichotomous operating principles that underlie nature, versus those that underlie capitalism. The global biosphere of life operates in circular fashion, where the end of one organism becomes the beginning of another. Or, in the words of the late Barry Commoner and his First Law of Ecology: everything is connected to everything else. In contrast, capitalism is a linear system: energy, raw materials, labor, and water are fed in at one end; commodities spew out at the other, along with gigantic quantities of waste, generated at every point along the line. The ultimate purpose of the production system, as Commoner noted, was simply to produce more, faster.

Under capitalism waste signifies success, not failure. The greater the piles of garbage, the deeper the landfills, the more extensive the despoliation of mountains, rivers, and forests, the greater the productive apparatus and thereby profits. The system must constantly produce more so as not to enter into crisis; its maxim: Grow or Die.

Now, however, technology and human labor power applied to constant expansion has become so productive that there are simply not enough people or outlets to consume all of the products churned out. The rate is such that no sensible capitalist can afford to make a product that lasts for long without the need for replacement or upgrades: cultural obsolescence joins technological obsolescence as capitalists vie for market share in ceaseless competition with one another to relentlessly enlarge the sphere of commodities that ripple across the world in a tsunami of short-lived consumer products.

Arguably, no other area of capitalist enterprise personifies the irrational and unsustainable linearity more thoroughly than our industrialized agricultural food system, whereby ten calories of fossil fuel input at one end yield a mere one calorie of food at the other, along with a host of negative side-effects, such as fertilizer run-off, pesticide contamination, soil erosion, aquifer depletion, and carbon emissions – not to mention detrimental health effects on animal and human welfare. Astoundingly, in a world where over one billion go hungry daily, between 40 and 50 percent of all food that is produced is wasted.

With ever-increasing zeal, the time-in-motion studies and strict regimentation so beloved of corporate titans on the factory floor have been successfully exported and implemented in the countryside. The end result is the creation of a wildly unstable and unsustainable mechanized food production system that, through its uniformity and requirement for external inputs, drains the land and agricultural laborers of vitality while forcing an impoverished diet upon those who consume the food, once 'value' has been added through processing. Commenting on the senseless short-termism inherent to capitalist agriculture a century and a half ago, in Volume 1 of *Capital*, Marx (1967, p. 505) wrote:

> Capitalist production ... disturbs the metabolic interaction between man and the earth, i.e. prevents the return to the soil of its constituent elements consumed by man in the form of food and clothing; it therefore violates the conditions necessary to lasting fertility of the soil.... The social combination and organization of the labor processes is turned into an organized mode of crushing out the workman's individual vitality, freedom and independence.... Moreover, all progress in capitalist agriculture is a progress in the art, not only of robbing the worker, but of robbing the soil; all progress in increasing the fertility of the soil for a given time is a progress towards ruining the more long-lasting sources of that fertility. The more a country starts its development on the foundation of modern industry, like the United States, for example, the more rapid is this process of destruction. Capitalist production, therefore, develops technology ... only by sapping the original sources of all wealth – the soil and the worker.

The compulsion to maximize profit at every opportunity, no matter how immoral, extends to predicting the need for increases in certain types of food stocked in supermarkets when natural disasters are impending. Before Hurricane Ivan made landfall in Florida in 2004, Walmart was already planning to stock up on certain products that they predicted would be in high demand:

> We have a remarkable level of real-time visibility into our merchandise planning. So much so that when Hurricane Ivan was heading toward the Florida panhandle, we knew that there would be a rise in demand for Kellogg's Strawberry Pop-Tart toaster pastries.
>
> (Patel, 2007, p. 233)

Writing in *Forbes* after Hurricane Sandy blasted through New York City, Iain Murray (2012, n.p.) used the same logic to argue that big box stores such as Walmart are the most efficient emissaries of disaster relief efforts:

disaster response provides an excellent example of how the invisible hand of the market works to alleviate suffering and bring quick relief to those in need.... Big box stores also don't merely sit and wait. They get out first and help people.... It will be interesting to see whether the lack of big box stores in Manhattan has any effect on the speed of the recovery there. Mom-and-pop stores simply can't do what big stores can in these circumstances.

Capitalism operates best – that is, at its highest rates of profitability – with uniformity, consistency, and a lack of diversity; the better to ensure that machines, laborers, and products are as interchangeable and replaceable as possible.

Capitalism as global simplification project

The fact that much of the developed world now subsists primarily on a radically reduced variegation of diet, consisting in large part of some combination of four grains, four animals, and four species of fish, is testament to the global simplification project that is part and parcel of capitalism's functioning, and is the adjunct to the creation of a globalized and homogenized human culture. Vast monocultures of genetically identical crops and enormous conglomerations of genetically identical animals, reared in Concentrated Animal Feeding Operations (CAFO), formerly known as farms, are the rural counterpart to urban factories, churning out identical products for the mass market in consumer goods.

The simplification and uniformity of product is a corollary of another dynamic central to capitalism: the centralization of capital, as each individual unit becomes larger and entire sectors are dominated by an ever-shrinking number of gigantic, world-girdling corporations. In 1994, three corporations accounted for 80 percent of banana production, five for 77 percent of grain production, and just two companies were responsible for 80 percent of instant coffee (Toussaint, 2005). On the retail end, ten companies supply more than half of all the food and drink sold in the USA. A single company, Wal-Mart, controls 40 percent of Mexico's retail food sector.

Indeed, rather than a human right, food as a commodity is one of the most profitable industries on the planet, alongside mining, energy production, and pharmaceuticals. The concentration of capital in corporations is mirrored by the personal concentration of wealth: in 2002, the world's richest 147 people had more combined wealth than half the planet – or three billion people.

The linearity, waste, and expansion endemic to the capitalist mode of production, coupled with its historical development built on African slave labor, later utilizing fossil fuels for energy, and the incessant drive to replace humans with machines – while turning the remaining humans into machines – is now running into the tolerance limits imposed by the laws of the physical universe that undergird a stable climate. Relative climate equilibrium, with its predictable rainfall, is the basis for successful agriculture. Therefore, a global destabilization of climate, as we are now witnessing, calls into question the foundation of civilization, first established approximately 10,000 years ago. The Final Report from the Commission on Sustainable Agriculture and Climate Change (Beddington et al., 2012, n.p.) highlighted the impact on the poor:

> Extreme weather events and climate change will exacerbate the fragility of food production systems and the natural resource base – particularly in environments prone to degradation and desertification, in areas of widespread or intense water stress, and wherever poverty undermines the capacity of rural people to take the needed preventive steps. Farmers can no longer rely on historical averages of temperature and rainfall, making it harder for them to plan and manage production when planting seasons and weather patterns are shifting. Rainfed agriculture and agropastoral systems are particularly vulnerable to climatic variability.

Such was the escalation in price for food-related material that in 2013 people began stealing bundles of dried grass from remote fields in the Midwest to sell as feed for animals. Along with hay, as desperation and economics cut in over the course of the year, there were increases in the theft of grapes, beehives, and avocados. Deserted rural landscapes were scoured even for broken-down agricultural machinery, which was spirited away for its scrap metal price (Healy, 2013).

That year, grain and other food products which had managed to survive the drought were threatened, as water levels in the Mississippi dropped precipitously. Billions of dollars of food products travel along the USA's largest internal waterway; costs rose as barges could no longer travel fully loaded for fear of grounding (Snyder, 2013). As temperatures continue to warm over the coming years, even if precipitation increases, water-level problems will remain, exacerbated by increased rates of evaporation from higher surface water temperatures and reduced ice cover.

Food robbery: ethanol from corn

The ramifications of drought in the USA go far beyond what happens to food prices here. The USA produces half of all world corn exports; without coordinated governmental action it is all too likely that there will be a replay of the disastrous rise in food prices of 2008, which caused desperate, hungry people to riot in 28 countries. In that instance food was available, but hundreds of millions of people couldn't afford to buy it. Should food prices increase to anywhere near the levels of four years ago, it will be a catastrophe for the two billion people who are forced to scrape by on less than $2 a day.

Supposedly adopted to reduce demand for 'overseas oil' and associated geopolitical concerns after oil topped almost $150 a barrel in 2008, the Obama Administration raised the federal requirement from 13 billion gallons (almost 40 percent of the crop) to 36 billion by 2022, with at least 15 billion coming directly from corn. This is the same as handing a guaranteed market, and hence profits, to corn producers, thus pushing more farmers to plant corn and nothing else. Indeed, despite the claims of industrialized agriculture, 70 percent of all food grown in the world comes from small farms using traditional and low-tech integrated methods of food production. Furthermore, they accomplish the feat of producing 70 percent of the world's food on only 30 percent of the arable land – a statistic reversed for industrialized agriculture. Hence, if we actually want to feed the world, we should find out what peasants in the developing world are doing so effectively with such limited resources, and abolish the petro-agricultural complex (Bittman, 2013).

The poor in developing countries spend 80 percent of their income on food, much of it directly as grain, rather than as manufactured products such as bread or cereal, and so any increase in the price of basic necessities immediately puts them in dire food distress. In the USA, the prices of a loaf of bread or a corn muffin are unlikely to see major increases because – in a nod to capitalist priorities – the cost of those products is largely determined by packaging, advertising, transportation, and storage costs – and ultimately the labor that is embodied in those activities, not the cost of growing the corn or other natural base material (Stecker, 2012). Extensive farm subsidies, the vast bulk of which go to the major producers, mean that US agricultural products can be exported and sold below cost, thus devastating local agriculture in developing countries.

Processed food is cheapest because corporations have spent tens of millions of dollars researching how to make it so; they turn what was once called food and 'process' it by substituting salt, sugar, and fat in the name of convenience and cost. The end result is the poor nutrition and epidemics of obesity and diabetes that can be taken care of by another arm of the capitalist enterprise: the pharmaceutical and healthcare industry. As Michael Moss shows in his 2012 book *Salt, Sugar, Fat*, corporations turn to those three ingredients time and again because they are relentless in their efforts to drive down their own costs in the name of profit and at the expense of the health of their customers.

Rather than any lack of actual food, most analysis indicates that the primary cause of the dramatic escalation in food prices that caused the 2008 food crisis, which was felt most acutely in the developing world, was financial speculation in the food commodity sector. Due to climate-related crop failures, the price of food began to increase over the course of 2007, and at the same time other investments, such as real estate, began to look increasingly risky. Investors began to look for other avenues to make a profit, so speculators began to place bets on food, a process which drove further increases in the cost of food. That is to say, a human tragedy carried out along lines of class, race, and gender, manufactured by the laws of motion of capitalism, rather than the laws of Nature.

In the longer term, measures to raise grain-storage volumes, address infrastructure deficiencies through appropriate investment, re-evaluate inhumane, environmentally destructive, and dangerously unhealthy industrialized livestock feeding practices and examine the location, sustainability, and type of crops and monoculture farming are all issues that need attention. The organization of Western industrial agriculture, which relies so heavily on fossil fuel inputs of artificial fertilizers, pesticides, and herbicides, along with the mechanization that is fueled by petroleum, is, by any definition, completely unsustainable. In one of the more ridiculous, circular irrationalities to emerge from the anarchy of capitalist decision-making in the USA, because ethanol derived from corn – which in turn is derived from oil – is increasing in price because the corn is dying, this ends up raising the cost of ethanol-blended gasoline.

The large-scale effect of such a move on the price of corn is disputed, in part because "It turns out it's really the price of gasoline and the profitability of selling ethanol" (Peterka, 2012) whereby the abstract and impersonal thing called 'the market' determines whether corn will be distilled into ethanol to burn in cars or sold as corn. Oil companies, which are required to blend ethanol into gasoline as part of the utterly inappropriately named 'Renewable Fuel Standard' (RFS), are allowed to carry RFS credits over year to year and thus have 2.4 billion credits available to allow the continued acquisition of corn for ethanol refineries (Peterka, 2012).

It is hard to imagine suddenly freeing up 40 percent of whatever remains of the US corn crop for livestock and human use having a negligible impact on corn prices, even accounting for the activities of the oil companies. As Gawain Kripke, Director of Policy and Research for Oxfam America, has argued, "The federal government can … put an end to the biofuel mandates, which are diverting food into fuel, and work to cut greenhouse gas emissions, which are leading to ever more erratic and extreme weather" (Stecker, 2012, n.p.).

Investment in food infrastructure, such as food storage facilities in the Global South, saw drastic cuts in developing countries throughout the 1980s and 1990s as international lenders demanded cuts to government spending in exchange for loans. In addition, such insurance was seen as unnecessary when 'the market' would automatically adjust for any shortfall; similarly in the USA, grain reserves are low and unable to make up any deficit owing to a reduction in grain storage.

Perhaps more importantly still, if Vilsack, and the Obama Administration in general, had any concern for humanity and the world's poor, and took seriously their democratic mandate on behalf of the people, they could begin an aggressive campaign to re-regulate financial speculation on food prices in international commodity markets. Such an attack on the bankers, stockbrokers, and speculators would no doubt prove wildly popular among the majority of the population. Needless to say, as defenders of the status quo, which rests on economic exploitation and systematic racial and gender oppression, these are not the policies espoused by representatives of the corporate plutocracy.

Rather than downsize the powerful corn-to-ethanol industry, much of it situated in Obama's home state of Illinois, which has the third-largest production capacity (Iowa, a campaign-defining state for Obama in 2008 and a swing state in 2012, produces the most ethanol), Vilsack has instead sacrificed 3.8 million acres of conservation land for grazing and the production of hay in order to circumvent livestock owners' anger directed at ethanol producers.

Thus, while there is a clear and easily achievable short-term solution at hand – reallocation of corn from ethanol distillation to food production – the Agriculture Secretary of the world's biggest corn exporter believes a more useful way of spending his time is in genuflecting to an all-powerful, invisible deity in the sky. The evidence above indicates that Vilsack's object of genuflection seems more aligned with placating corporate ethanol interests than it does a divine entity.

The USA's 2013 $1 trillion farm bill, which will define agricultural policy until 2023, takes things even further in the wrong direction. As Heidi Moore commented in the *Guardian* (2013, n.p.):

> As members of Congress have negotiated over various amendments and riders to the bill, they've set an impressively consistent trend: they mix good ideas and bad ideas and combine them to create the absolutely worst possible policies. Elements of the farm bill, as it stands, will cut food stamps to the poor and the previously incarcerated, thus increasing poverty and possibly crime; add to the growing obesity crisis by encouraging chemical sugar substitutes; push genetically modified food at the expense of public health with the so-called 'Monsanto Protection Act'; and support factory farming at the expense of sustainable food production with abusive crop subsidies.

The collusion between industry and government was highlighted in stark and unabashed form by Dwayne Andreas, former CEO of giant food conglomerate Archer Daniels Midland in response to a question about corporate welfare:

> There isn't one grain of anything in the world that is sold in a free market. Not one! The only place you see a free market is in the speeches of politicians. People who are not in the Midwest do not understand that this is a socialist country.
>
> (Carney, 1995, n.p.)

Climate change and agriculture

At a time when the reality of anthropogenic climate change has become so hard to ignore that even some famous climate skeptics have given up protesting (Muller, 2012), drought is going to be an increasing factor that agricultural planners need to take into account. Therefore, cutting money from programs designed to manage the land more sustainably is a reckless policy, one that can only benefit the profit margins of banks and agricultural concerns while encouraging the further concentration of capital as small farmers are driven to the wall.

As climate blogger Joseph Romm (2012) pointed out in an article in *Nature*, assuming business-as-usual, which is exactly what is going to happen without a class-based mobilization of the people, there will be a cascading series of destabilizing changes that will all negatively impact our ability to grow food:

> Precipitation patterns are expected to shift, expanding the dry subtropics. What precipitation there is will probably come in extreme deluges, resulting

in runoff rather than drought alleviation. Warming causes greater evaporation and, once the ground is dry, the Sun's energy goes into baking the soil, leading to a further increase in air temperature. That is why, for instance, so many temperature records were set for the United States in the 1930s Dust Bowl; and why, in 2011, drought-stricken Texas saw the hottest summer ever recorded for a US state. Finally, many regions are expected to see earlier snowmelt, so less water will be stored on mountain tops for the summer dry season.

(Romm, 2012, n.p.)

Even worse, the recent results of 19 different climate models predict that drought will become a permanent feature of large areas of the North American continent: "If climate change pushes the global average temperature to 2.5 degrees Celsius above pre-industrial era levels, as many experts now expect, [almost all of Mexico, the midwestern United States and most of Central America] will be under severe and permanent drought conditions" (Leahy, 2012, n.p.). In other words, we are only beginning to glimpse the outlines of a situation that will become far worse without drastic ameliorative action in the near-term future. Climate change, caused primarily by the burning of fossil fuels, is creating extended droughts and increasing the frequency and intensity of extreme weather events such as torrential downpours that threaten to undermine agriculture and, thereby, our ability to feed ourselves. Rather than a swift redirection of societal priorities – toward energy conservation, renewable technologies, social justice, equity, and sustainable farming practices – instead there is a continuation and extension of the policies that got us here in the first place.

The process is not limited to the usual Western imperial suspects. Under the guise of 'national development,' governments of the Global South seek to carve out their own piece of economic and military might along capitalist lines. Hence, we see the same irrational process played out in India, which is suffering from a 20 percent shortfall in precipitation, with some states recording 70 percent reductions from historic averages. Sixty percent of India's 1.2 billion people work in agriculture, which accounts for 20 percent of Indian GDP.

Less rain does not just affect farmers directly. Less rain leads to less hydroelectric power, which means farmers have to use their own pumps to obtain water from underground aquifers for crop irrigation to save their harvest. Those pumps run on electricity. Thus, at a time when there was less electricity available because of drought, there was an increased demand for electricity to overcome the drought, a factor contributing to the massive blackout in India. In addition, pumping groundwater has led to aquifers dropping in some areas by 60 to 200 meters, thus requiring bigger, more powerful pumps for deeper wells to continue the unsustainable practice of tapping groundwater supplies at such volumes (Eshelman, 2012). This is despite the fact that, while 90 percent of water use in India is for agriculture, only about 10 to 15 percent ends up reaching the crops, since most of it evaporates on the ground before it gets to them. Rather than investing in sustainable agricultural practices to combat the problem, the Indian government bought heavily into the Western-backed Green Revolution of the 1960s and promoted the planting of water-intensive crops such as rice.

According to Upmanu Lall, Director of the Columbia Water Center at Columbia University's Earth Institute, "the whole water and energy problem [in India] is dire, and it's caused by government policy" (Eshelman, 2012). He gives the example of the Punjab, which has an annual rainfall of 0.40 to 0.80 meters, but now grows rice, which requires 1.8 m of annual rainfall.

The situation has already reached crisis proportions in the USA. USGS research hydrologist Leonard Konikow notes,

> During the 1940s and 1950s, the growth of populations and the expansion of industry meant many more farmers were drilling wells, particularly in the High Plains.... This was land that had never been irrigated before – land that ordinarily wouldn't have supported these kinds of crops.
>
> (Massey, 2013, n.p.)

The international dimension of market priorities

In the realm of energy, representatives of the Ecuadorian government, a left-leaning government that has been much-lauded for its environmental policies, were in Beijing in March 2013 to negotiate a deal with China. The deal was set to allocate rights to three million hectares of undeveloped Amazonian forest – the home to thousands of indigenous people – for oil exploration, along with the construction of a multi-billion-dollar, Chinese-financed oil refinery.

At a protest against the selling of their land, Narcisa Mashienta, a leader of Ecuador's Shuar people, proclaimed defiantly, "What the government's been saying as they have been offering up our territory is not true; they have not consulted us, and we're here to tell the big investors that they don't have our permission to exploit our land" (Kaiman, 2013).

The Brazilian ruling elite, with its imperial aspirations for regional dominance, is similarly opting to scale up exploitation of its natural resources for energy production – specifically in the realm of a truly enormous expansion of hydroelectric power. The gigantic, 11,000-megawatt Belo Monte dam in the Amazon – which will require flooding a vast area of forest, displacing tens of thousands of indigenous people, and cannot be operated at anything like full power without the construction of several more dams – is still being built, despite continual protests and work stoppages by unions and indigenous activists, not to mention a growing international outcry.

Although large environmental groups chose not to publicize this, the Rio+20 Earth Summit itself was part-sponsored and "financed by millions of dollars in financial support from Brazil's largest energy, extraction and petroleum corporations" according to the *New York Times*, including "mining giant Vale, voted the 'worst company of the year' in the 2012 Public Eye Awards, and Electrobras, the state electricity company, a partner with Vale in developing the Belo Monte dam project on the Xingu River" (Barnes, 2012, n.p.).

However, the energy corporations and Brazilian state are facing resistance not only from indigenous groups who are occupying construction sites, but also from workers building the dam. The accelerated dam- and road-building project being pushed by the Brazilian state is leading to labor shortages, which are provoking bitter and widespread labor struggles all across the Amazon:

> In the revolt that broke out in March 2011, apparently after a worker was denied transportation to visit a sick family member in the city, nearly all of the lodgings built to house the 16,000 workers were burnt down, along with other buildings and 60 buses and other vehicles. The uprising gave rise to a lengthy strike demanding wage hikes, better transportation, and more frequent permits allowing workers from distant areas to visit home.
>
> (Osava, 2012, n.p.)

The fact that all of these changes are not only detrimental to indigenous and working people around the world, but are also suicidal when it comes to any hope of maintaining a stable climate, should also indicate our natural allies in the fight for an ecologically sustainable and socially just society. Workers, unions, indigenous groups, and environmental activists need to form a united front against the planet-wrecking priorities of the 1 percent.

Underlying this united front must be a theoretical appreciation for how the laws of motion of capitalism operate inexorably to promote unending growth and international competition over natural resources, factors which will destroy all hope for future generations and lead to the extinction of countless species. In turn, international economic competition threatens constantly to break out into its militarized version, as nation states and antagonistic trading blocs opt for warfare.

The intersection of energy, water, and food with capitalist development and oppression based on gender, race, and class is illustrated in Brazil and India in stark form. But the solution, abstracting the limitations imposed by class society, is once again quite simple, in that crops should be grown where the climate makes most sense, not where they will make the most money, or merely to add to foreign cash reserves or national status. Energy should be produced in the least damaging manner, and with the consent and active and full involvement of the people who will be affected. However, rather than adopting those kinds of measures or addressing climate change, India is building more coal and nuclear plants, and along with Brazil is one of the world's most resistant to taking effective action on climate change.

Real solutions require profound systemic change

Given all of the above, if human civilization is to endure at all, it seems impossible to escape the conclusion that to survive on a planet that looks remotely like the one we were born on, we must confront the system that produces a society at odds both with itself and the natural world for the same reason – class stratification, and inequality resting on gender oppression and structural racism.

That means the building of an organized resistance in every workplace, community, school, and farm across the world fighting for social justice and equality for all. The exploitation and oppression by race and gender that is meted out to the vast majority of the world's population as a consequence of the way the system works is the mirror image of the exploitation of the biosphere, which, ultimately, forms the basis for life – a scientific fact the capitalists seem capable of ignoring.

That is why we have to organize. As just one example, the late Wangari Maathai, who was the first African woman to win the Nobel Peace Prize, described her work in forming the grassroots organization the Green Belt Movement (GBM) in the 1970s to empower rural women by employing over 100,000 of them to plant 15 million trees:

> What we've learned in Kenya – the symbiotic relationship between the sustainable management of natural resources and democratic governance – is also relevant globally.
>
> Indeed, many local and international wars, like those in West and Central Africa and the Middle East, continue to be fought over resources. In the process, human rights, democracy and democratic space are denied.... Unless we properly manage resources like forests, water, land, minerals and oil, we will not win the fight against poverty. And there will not be peace. Old conflicts will rage on and new resource wars will erupt unless we change the path we are on.
>
> (Maathai, 2004, n.p.)

The fact that Maathai saw a clear connection between poverty, the fight for women's rights, political emancipation, and ecological justice – in a country that had lost 98 percent of its forest cover since colonization by the British – is what earned her the enmity of the Kenyan government, not to mention beatings and jail time.

Lack of tree cover from ongoing deforestation and loss of topsoil means that, in Sub-Saharan Africa, women and girls, who are responsible for over 70 percent of water collection, have to travel further and further to obtain it. The UN estimates that women in Sub-Saharan Africa collectively spend 200 million hours per day collecting water for food and farming purposes, or 40 billion hours annually (Deen, 2012).

For the good of the 99 percent of exploited and oppressed humanity and the rest of the biosphere upon which we depend, the capitalist system needs to go. We need to fight to build a world free of a system based on endless growth for profit, inequality, lack of democracy, and the warfare and racial and gender oppression that are its necessary components. We need a world based on cooperation, real democracy, and producing the things we need in harmony

23 Women, climate change, and food security in Bangladesh

Parvez Babul

Food is a basic human right. Yet approximately 1.5 billion people in the world live on less than $1 a day, and every 3.6 seconds someone dies of hunger. Poverty, hunger, and malnutrition silently kill people and take away their ability to work and learn. Thus, adequate food, including access to land to produce food, is needed for survival, ensuring constitutional human rights, and achieving Millennium Development Goals (MDGs) (BBC World Service UK, 2001).

Many rural women come to the cities in Bangladesh to earn a livelihood, but the intersection of gender inequality and poverty limits their opportunities. They live inhuman lives, both in their villages and in cities, without food, shelter, and other basic necessities. Where there is no money, there is no purchasing power, so we need effective and sustainable programs to give such women and their families access to income-generating activities and therefore access to adequate food.

Poor people need a sustainable livelihood as provided by small-scale sustainable farming. Despite the increases in food grain production, around half of the population of Bangladesh remains below the food-based poverty line. Bangladesh will face a major problem in the next 40 years since production of rice will decrease by at least 8 percent, and wheat by 32 percent, while the population will increase by 50 to 75 million (IPCC, 1990). According to the World Bank, approximately 33 million of the 150 million people in Bangladesh cannot afford an average daily intake of more than 1800 calories, which is the minimum standard for nutrition. For the people in most developing countries, the daily calorie average is 2828. In Bangladesh, that average is only 2190 (World Bank).

Poor women are often the first to suffer malnutrition in the family. This has repercussions on their health, productivity, quality of life, and survival. In addition, climate change will affect availability, accessibility, utilization, and systems stability of food. The World Food Programme (WFP) mentioned in its report of 2009 that, in Bangladesh, 46 percent of pregnant women, 39 percent of non-pregnant women, and 40 percent of adolescent girls suffer from anemia because of depleted iron stores during pregnancy and lactation, a consequence of insufficient intake of foods rich in iron and folic acid (WFP, 2015).

Helen Keller International (HKI)-Bangladesh has been working to ensure food security and empowerment of women through different programs. Emily

Hillenbrand, a gender specialist and program manager of HKI-Bangladesh, states: "Gender discrimination is an underlying, structural cause of Bangladesh's alarmingly high rates of food insecurity and malnutrition. Their malnutrition is related to their limited control over economic assets, exclusion from household decisions, and restricted mobility." The International Food Policy Research Institute (IFPRI) agrees that poverty is a major determinant of chronic household food insecurity. The Global Hunger Index 2010 produced by the IFPRI mentioned that malnutrition among children under two years of age is one of the leading challenges in reducing global hunger and can cause lifelong harm to health, productivity, and earning potential (IFPRI, 2010).

The burden of child malnutrition could be cut by 25 to 36 percent by providing universal preventive health services and nutrition interventions for children under two and their mothers during pregnancy and lactation. The health of women, specifically mothers, is crucial to reducing child malnutrition. Mothers who were poorly nourished as girls tend to give birth to underweight babies, perpetuating the cycle of malnutrition. Nutrition interventions should be targeted toward girls and women throughout the life cycle, and especially as adolescents before they become pregnant, the report added (IFPRI, 2010).

Eradicating extreme poverty and hunger has been selected as the target of MDG One. The Millennium Development Goals (MDGs) are eight international development goals established at the Millennium Summit of the United Nations in 2000, following the adoption of the United Nations Millennium Declaration. It aims to:

- Reduce by half the proportion of people living on less than $1 a day.
- Achieve full and productive employment and decent work for all, including women and young people.
- Reduce by half the proportion of people who suffer from hunger.

In fact, MDG One is related to MDG Three: To promote gender equality and women's empowerment; and MDG Five: Women's (maternal) health.

Nobel Laureate of Bangladesh Dr. Muhammad Yunus said: "Economics has a relationship with peace." This means that a poor economy is interlinked with poverty, and familial, social, and political unrest. As a result, poverty hinders our development and economic growth, and creates obstacles on the road to achieving progress, prosperity, sustainable development, food security, adequate nutrition, gender equality, empowerment of women, peace, and Millennium Development Goals (MDGs).

It is wrong to imagine that climate change is taking away attention from domestic causes. You can think in that way only if you believe that climate change is merely about the negotiations, and that the negotiations are only international and not domestic. The climate change challenge is very much domestic. For example, there is currently much greater intensity, frequency, and velocity of cyclones. In May 2015 Cyclone Aila killed at least 125 people and devastated the eastern coast of India. We are also seeing far higher levels of drought and flooding with, for example, flash floods in the deserts of Ladakh and Rajasthan, in areas where there is normally no rain, and the devastating Mumbai floods in 2005. So climate change is very much a domestic threat to the poor, whose very survival is threatened by these events.

The second reason why we must address climate change domestically is that if we do not, we risk adopting policies that make climate change worse – for example, through building an agriculture that makes us more vulnerable, or the rapid expansion of highways without proper impact assessments (www.world-watch.org/node/6189).

Nobel Laureate Dr. Amartya Sen said: "Poverty is caused by the lack of exchange entitlement in a market-economy. Gainful employment is the principal challenge for the poor and the main focus for poverty alleviation." Thus, strategies for poverty alleviation broadly encompass the redistribution and creation of assets in favor of the poor and give guarantee of employment at a reasonable wage and adoption of measures having direct benefit for the poor. This is possible when political leaders express their commitment to implement programs without corruption or discrimination.

Bangladesh is paying a heavy toll for climate change and is facing many big problems, such as increased temperature, salinity, frequent floods, and cyclones. Poverty, food insecurity, and malnutrition are the three silent killers, and climate change has become a severe headache for the nation.

We should take a holistic approach, and the government and development partners should understand the situation, causes, and consequences of food insecurity, hunger, and malnutrition, and emphasize homestead food production.

Addressing the linkages between gender and climate change

Historically, women have been the key to food security. But they are not secure; especially poor women die many times before their death. They receive less, but have no rights to demand more. They say less, but are bound to listen more. They eat less, but must produce more to feed others. If we analyze the lives of poor women, we become amazed at how they survive.

Nazma (not her real name), a 32-year-old woman from Patuakhali whom I interviewed, lost her house and belongings in Cyclone Sidr, which hit in November 2007. Nazma was married at the age of 14 to a 30-year-old day laborer. Nazma had two children, but was abandoned by her husband five years ago due to her ultimate failure to pay extra dowry. He married another woman and now lives in the town of Patuakhali, pulling a rickshaw. Nazma's parents, brother, and sister died in the 1991 severe cyclone; thus, Nazma has no one from whom to seek any support (interview, December, 2007).

Cyclone Sidr took away 3000 lives, destroyed more than one million households, and snatched the livelihoods of millions of people. Before standing up and bracing the loss, the people of the coastal belt faced another cyclone, Aila, in May, 2009. Aila pushed them further back with considerable loss of lives and property. Nazma's two malnourished children survived Sidr and Aila, but then she had nothing to feed them with.

Nazma received some relief material, enough to last her for a few days, but, after that, she found no way to save the lives of her children and herself. So Nazma and her children, like many other victims, begged a launch supervisor to give her free tickets to Dhaka. She had to digest a lot of information from the launch authority. After arriving in Dhaka, Nazma and her children took shelter at Sadarghat launch terminal and passed two days without having any food, and only receiving water from a local restaurant. Nazma looked for a housemaid's job, but people refused to appoint her, as they did not know her. On the third night, two people came to Nazma and proposed to give her a 'good job.'

Nazma agreed and went with one man – the other looked after her children until she returned. Taking her to a room, the man forced her to have sex with him, paid her just Tk.20, and abandoned her. Nazma cried silently and went to reclaim her children, but she lost her way and could not find them. She cried loudly and searched for them everywhere, but she still could not find them. Nazma became very tired searching for the children, passing her time in the city without a job, and begging to survive. After ten days, she somehow returned home to her village in the hope that perhaps her children had returned there, but they had not.

In her village, she could not find work or food, but took shelter on a landlord's verandah for the night. Sidr and Aila had affected many people in her area, so thousands of poor women and children, like Nazma and her lost children, were looking for help. There was profound regret, and wailing among many people, but too little help.

Bangladesh experiences floods, drought, cyclones, and disasters more frequently now, due to climate change. Scientists forecast a further increase in natural calamities in the future. This is why we need to explore the vulnerability of poor women and linkages between gender and climate change issues, particularly in relation to enhancing women's capacities to address climate change to save millions of women like Nazma and her children.

According to the World Health Organization (WHO), public health depends on safe drinking water, sufficient food, secure shelter, and good social conditions. Climate change is likely to affect all of these things. Public health services and high living standards protect some populations from specific impacts; however, the health effects of a rapidly changing climate are likely to be overwhelmingly negative, particularly in the poorest communities. Among these effects are the following:

- Rising temperatures and variable precipitation are likely to decrease the production of staple foods in many of the poorest regions, increasing risks of malnutrition.
- Rising sea levels increase the risk of coastal flooding, and may necessitate population displacement. More than half of the world's population now

lives within 60 km of the sea. One of the most vulnerable regions is the Ganges–Brahmaputra delta of Bangladesh (see different websites).

One of the participants of the Third World Climate Conference (WCC-3), organized by the World Meteorological Organization (WMO) and held from August 31 to September 4, 2009 in Geneva, Switzerland, was our Prime Minister. The theme of the conference was Climate Prediction and Information for Decision-making. Topics covered by this climate conference included the application of climate prediction and information on societal problems to enable better adaptation to climate variability. We need better climate information for a better future.

The sorrows, tragedies, and sufferings of millions of women like Nazma need to be brought before world leaders to help save Bangladesh and its people from the long-term problems of climate change. The United Nations Secretary General said,

> Climate change poses at least as big a threat to the world as war. For that reason, we have a duty to the vulnerable people who contribute least to the problem, but experience its impacts most severely. And we have a responsibility to succeeding generations. We also have a historical obligation to successfully transit to a low-carbon global economy.

In developing countries, owing to poor women's marginalized status and dependence on local natural resources, their domestic burdens are increased; they feel an even greater burden of climate change. They are also underrepresented in decision-making about climate change, greenhouse gas emissions, and, most critically, discussions and decisions about adaptation and mitigation.

So, let us empower our destitute and poor women like Nazma. Their contributions through growing crops, planting trees, and producing homestead food will help the country ensure food and nutritional security, and approach climate change efforts with more than just hope and a begging bowl. The landless, poor women like Nazma are in a desperate plight, having lost all their belongings. Thus, the duty of national and international leaders should be to save their lives, because they, as human beings, have equal rights to stay alive, as have others.

Concerns about women's access to, control over, and ownership of land and resources have been raised over the years at different but interrelated levels. Land and environmental resources are central to the lives of people living in countries whose economic development and subsistence depends on the resources. With regard to environmental resources, women's access to and control over forests, water, and wildlife has come into sharp focus, as it has become clear that the performance of women's day-to-day chores is anchored in these resources. Making access to land and environmental resources equitable is one way to achieve development. The Millennium Development Goals recognize the need to promote gender equality and empower women, the need to alleviate poverty and ensure sustainable environmental management.

References

BBC World Service Millennium Development Goals, 2001. Available at: www.bbc.co.uk/worldservice/specials/1112_mdg/(accessed January 6, 2016).

IFPRI, 2010. Global Hunger Index.

World Food Program, 2015. Available at: www.wfp.org/countries/bangladesh (accessed January 6, 2016).

24 *Sila*

Chantal Bilodeau

Set on Baffin Island in the territory of Nunavut, the play *Sila* examines the competing interests shaping the future of the Canadian Arctic and local Inuit population. The play follows a climate scientist, an Inuit activist and her daughter, two Coast Guard officers, an Inuit Elder, and two polar bears as they see their values challenged and their lives become intricately intertwined. Equal parts Inuit myth and contemporary Arctic policy, *Sila* is a plea for increased collaboration in dealing with the major challenges of our time.

Scene 9

In the following excerpt, Jean, a white male climate scientist from Quebec, approaches Veronica, an Inuit teacher, spoken word poet, and the daughter of a prominent climate change activist, hoping to swiftly fulfill the Nunavut Research Institute's requirement which states that all scientific research on Baffin Island must actively involve the Inuit community. Like most 'Southern' scientists, Jean is showing a remarkable lack of interest for, and sensitivity to, the local culture and sees the Institute's rule more as an annoying impediment to his research than as a real opportunity to build bridges. Veronica, whose cultural baggage includes centuries of racial discrimination, first from European explorers, then from the Canadian federal government, and most recently from corporate interests hoping to cash in on the newly available Arctic resources, is quick to pick up on Jean's narrow-minded agenda. She sets out to teach him a lesson, reminding him that no real collaboration can take place until both parties view each other as equals.

For many years, collaborative relationships between scientists and Inuit were non-existent. (Inuit were used as guides and/or as armed guards to protect scientific crews from polar bears.) Inuit knowledge of the Arctic was unappreciated and unrecognized, most likely a legacy of the colonial concept of indigenous people as 'ignorant savages.' Yet Inuit hunters spend significant time out on the land and have a more intimate and hands-on knowledge of Arctic processes than scientists, who tend to rely mostly on satellite data. Only recently have scientists recognized this phenomenon and begun to design research methods that can combine empirical data with anecdotal observations. The character of Jean is an example of such an effort. Jean is confronted with his own prejudices and becomes more receptive to a different kind of knowledge – something that is acquired, transmitted, and expressed in a completely different way than what he is used to. What this scene – and ultimately the whole play – stresses is that if we are to succeed in addressing climate change, we need to reach across boundaries and cultural divides, and to seek everyone's input in devising successful strategies for action.

Translation of Inuktitut terms

qallunaaq: white man, non-Inuit
nanuq: polar bear
qajaq: kayak
iglu: snowhouse
pitsiaqattautiniq: respect

9

The high school in Iqaluit. A hand-painted banner reads, "Environmental Awareness Day." Veronica stares at Jean, confused.

VERONICA: A presentation?
JEAN: To the students.
VERONICA: What's your name again?
JEAN: Jean. Lefèvre.
VERONICA: And you're –
JEAN: A climate researcher. Well, trying to be.
VERONICA: Were you asked by one of the other teachers?
JEAN: *Non non* it was uh … I'm a friend of Leanna and she –
VERONICA: You know my mother?
JEAN: I helped her with her petition a few years ago.… She said uh *elle m'a dit que vous aviez besoin de quelqu'un pour* – (*gestures toward the banner*). To talk to the students.
VERONICA: My mother asked you to –
JEAN: I'm stuck in Iqaluit for a couple of days and I mentioned that –
VERONICA: What did she say exactly?
JEAN: She uh.… She said you were trying to raise awareness about –
VERONICA: Did she tell you she was the one who was supposed to be here?
[*A beat.*]
She didn't.
JEAN: She was in a hurry. I caught her just as she was leaving for the airport.
VERONICA: Oh. Where's she going this time?
JEAN: *J'me souviens pas trop* … I think it was Vancouver. Or maybe Seattle.
VERONICA: That's just like her. I organized this event around her. I planned this whole day so students would get to know her and her work. And she doesn't even show up.

JEAN: I'm sure she.... Maybe she forgot or −...

VERONICA: Don't make excuses for her.

[*An uncomfortable beat. Veronica examines Jean from head to toes.*]

Jean, you said, right?

JEAN: Uh *oui*. [*extends his hand*] Nice to meet you.

[*They shake hands.*]

I heard nice things about −

VERONICA: Is this your first time on Baffin?

JEAN: *Non pas du tout*. I've been coming here for 15 years.

VERONICA: Impressive.

JEAN: It's my job.

VERONICA: In 15 years, you must have learned some Inuktitut.

JEAN: Not − I mean I'd love to but *la façon dont les choses fonctionnent* most of the time we're in the field and you know ... we have a lot to accomplish in a very short period so there's really no time to −

VERONICA: Typical *qallunaaq*.

JEAN: It's just how science works.

VERONICA: Fifteen years and not a single word?

JEAN: *Ben j'sais pas là* . . .

VERONICA: What about *nanuq*?

JEAN: Sure. *Nanuq* of the North.

VERONICA: *Qajaq*.

JEAN: [*trying to repeat*] Rrrra.... Oh! Kayak right? I didn't realize that was −

VERONICA: *Iglu* . . .

JEAN: OK I guess I know more than I thought.

VERONICA: Colonialism has a sneaky way of leaving its traces. *Qallunaat* got the land but Inuit managed to infiltrate the language.

[*She hands him a flyer.*]

I write spoken word poetry. Come and see me sometime. You might learn something.

JEAN: Sure. I, uh, I'll try.

[*She waits for him to leave. He doesn't.*]

VERONICA: Anything else?

JEAN: Well ... I thought I'd ... [*gestures toward the banner*]

VERONICA: No, thank you.

JEAN: It's really not − I'm happy to. I have to do it anyway.

VERONICA: Have to?

JEAN: *Ça fait partie de* − The Nunavut Research Institute requires that −

VERONICA: Oh. You're here to fulfill your requirement with the Natives so you can get your research license.

JEAN: *Non*. I mean yes in a way but −

VERONICA: You know, Jean, things have changed over the last 15 years. If you want to work in Nunavut, it's not enough to talk AT us anymore. You have to talk WITH us. That's just *pitsiaqattautiniq*.

JEAN: I have no idea what that means.

VERONICA: You're a smart man. Figure it out.
[*She exits.*]

Scene 10

At the top of this scene, Mama, an adult polar bear, and Daughter, her young cub, are celebrating Daughter's first successfully hunted seal. Then as they quiet down and admire the night sky, Mama takes the opportunity to teach her daughter an invaluable life lesson – that each and every one of us is connected to every other living creature. In terms of our survival, this fundamental interdependence is perhaps the single most important value that climate change can teach us.

Polar bears as characters remind us that the animal species is a class unto itself. Like the poor and most vulnerable among us, animals suffer from food scarcity, environmental degradation, and displacement, and many of them are disproportionally affected by the changing climate even though they emit few, if any, greenhouse gases. In addition, within all groups there are interchangeable roles of oppressed and oppressor. Today, it is the seal's turn to give its life force to the bear. But this transaction is temporary; sooner or later, the bear will have to surrender its life force to another animal. This continual flow of energy between species and ecosystems is one of Nature's most complex processes. It is what we humans disrupt when we attempt to control aspects of it. Those disruptions then become the big and small events that constitute climate change.

As Brian Baxter argues in his book, A *Theory of Ecological Justice*, animal species deserve their share of the Earth's resources as much as the human species does. Our Anthropocentric inclinations have created a paradigm where we tend to position ourselves as being more deserving than other species, or, within our own species, we position certain groups as being more deserving than other groups. Throughout history this conception of the world has made it possible to justify far too many injustices and extinctions. If we recognize that the Earth belongs to everyone, then we must shift our paradigm and envision ourselves as equal dwellers.

Translation of Inuktitut terms

anaana: mama
paniapik: my daughter
Nanurjuk: Having the Spirit of a Polar Bear (a star in the Orion constellation)
nanuq: polar bear
Kingulliq: The One Behind (a star in the Orion constellation)
sila: in Inuit cosmology, the primary component of everything that exists

10

Out on the ice. A brilliant and benevolent moon dominates the landscape. Friendly growling and laughter. Mama and her daughter play-fight.

DAUGHTER: I was good, right *Anaana*?
MAMA: You were very good.
DAUGHTER: I stayed low like you said. And I didn't make a sound.
MAMA: You were as silent as a rock.
DAUGHTER: And I waited and waited and waited.... How long did I wait?
MAMA: For as long as it takes the goose to travel to its summer ground.
DAUGHTER: Longer!
MAMA: For as long as it takes the river to grow into an ocean.
DAUGHTER: Longer!
MAMA: For as long as it takes the glacier to crawl down the rocky cliff.
DAUGHTER: Yes! And when I saw the bubbles, I pounced!
MAMA: You're a very good hunter, *paniapik*.
DAUGHTER: I'm a good hunter.
MAMA: I'm proud of you.
[*They look up at the stars as the daughter revels in that thought.*]
DAUGHTER: Look! There's *Nanurjuk*, the *nanuq* who climbed into the sky!
MAMA: Uh-huh. And do you see the hunters chasing after her?
DAUGHTER: Yes. One, two, three.
MAMA: And...?
DAUGHTER: And in the back is *Kingulliq*, the fourth hunter who dropped his mitt and went back to retrieve it.
MAMA: That's right.
DAUGHTER: I'm sad for *Nanurjuk*.... It must be lonely running across the sky like that every night.
MAMA: The only creatures who are lonely are the ones who forget about *sila*.
DAUGHTER: How do you mean, *Anaana*?
MAMA: All life is breath. From the original breath that gave us the miracle of Creation to the world itself, *sila* wraps all around us.
[*The daughter looks around.*]
DAUGHTER: The sky is *sila*?
[*Mama nods.*]
The wind is *sila*?
[*Mama nods.*]
The land, the ice, the ocean?
[*Mama nods.*]
MAMA: And *sila* also moves in and out of our lungs.
[*Mama breathes.*]
See? That's *sila*. And with each breath, *sila* reminds us that we are never alone. Each and every one of us is connected to every other living creature.
[*The daughter breathes.*]

But *sila*'s gift is not ours to keep. We may use our breath while we roam the land
  but we must surrender it once we pass from the land. Creatures who are
  lonely are the ones who hold on to their breath as if it were theirs and
  theirs alone.

[*A beat.*]

One day, you will leave me, *paniapik.*

DAUGHTER: That's not true.

MAMA: One day, it will be time for you to go into the world and meet your own
  destiny.

DAUGHTER: I won't leave you, *Anaanaa.*

MAMA: And I won't be sad and you won't be sad because *sila* will reach across
  the land and bring us the sweet scent of each other's happiness.

[*The sound of a snowmobile fades in, growing louder as it gets closer.*]

DAUGHTER: What's that?

[*Suddenly, silence. They scan the horizon. Gunshot.*]

MAMA: Run!

[*The daughter hesitates. Another gunshot.*]

Quick, run!

[*They run away.*]

25 Polar bear on Bernard Harbor

Subhankar Banerjee

Figure 25.1 Polar bear on Bernard Harbor, Arctic National Wildlife Refuge, Alaska, June 2001 (source: photograph by Subhankar Banerjee). The photograph juxtaposes nature and culture – the bear approaches a whale bone left from previous year's hunt by the Inupiat community of Kaktovik, along the Beaufort Sea coast. On March 19, 2003, Senator Barbara Boxer used a poster-sized version of the image to argue against drilling for oil and gas in the Refuge. To the dismay of the George W. Bush administration, her plea resulted in a victory for the conservationists with a vote of 52 to 48. Since then the image has become one of the most published photographs in the history of photography, while the population of polar bear in the southern Beaufort Sea has declined by 40 percent between 2001 and 2010 due to rapid loss of sea ice caused by Arctic warming.

26 Race, gender, and climate injustice

Dimensions of social and environmental inequality

Toban Black

A climate justice approach is about challenging injustices while mobilizing toward systemic change – for the sake of climate solutions, and much more. I first encountered the concept of climate justice as an activist. This politicization of climate and energy struggles resonated with me immediately. However, I since have found that accounts of the interconnections between climate change and social justice have often been too limited. Attention toward these environmental injustices has frequently only involved highlighting inequality between nations.

These international divisions are certainly relevant and important. Between countries, there are vastly disproportionate greenhouse emissions – particularly when atmospheric emissions are tracked over the course of history to reveal the sum of each nation's carbon emissions. For instance, despite how Caribbean nations have had relatively low emissions per capita, this region is nevertheless facing rising sea levels and intensifying storms. In the meantime, the neighboring USA is a leading source of greenhouse gases, while much of this country is relatively secure from climate impacts. Most of the nearer term impacts are manifesting elsewhere. International climate injustices like these are even more apparent when we take into account what some have called 'outsourced emissions' – that is, greenhouse gases released from areas of the world (such as China) that manufacture products for export to the USA, among other affluent countries.

Other accounts have addressed feminized and racialized causes and consequences of climate change, which are discussed in upcoming pages. The gendered climate injustices noted here include the extended workloads borne by women who gather water amidst droughts, and food from failing crops. Interrelated climate vulnerabilities experienced by many Sub-Saharan Africans and indigenous peoples are pertinent instances of climate injustices along racial lines. Yet, major sources of greenhouse gases have often been bound up more with whiteness, males, and masculinity. Militaries and oil companies are two of several of the institutions that are relevant, given the inequitable control central to their operations. Conversely, marginalized groups have had less input into official climate forums and discourses which claim to be resolving these climate crises. Racialized and gendered patterns such as these have been recognized in declarations from a set of movement networks that have highlighted climate injustices. The founding statement of the international Rising Tide network (2001) pointed to each of these forms of inequality – which were high-lighted by the US Mobilization for Climate Justice (2009) as well.[1] Gendered and racialized inequities have not gone unnoticed, yet these climate injustices have received insufficient attention and recognition, both in movement con-texts and in other relevant writing.

To help fill this gap, this chapter brings together some of the studies which have addressed these dimensions of social and environmental inequality, to reveal some recurring patterns among these disparities, and to indicate the importance of an intersectional approach to climate change. Above all, it will indicate how the causes of climate change are associated more with masculinity and whiteness, even as some females and people of color are offering unrecog-nized contributions to confronting this environmental crisis. This chapter addresses the implications of some previous literature regarding gendered and racialized climate change impacts (as well as related disasters). The core of this chapter consists of two sections regarding climate injustices along lines of gender and race.[2] A separate section will begin to show the relevance of inter-secting economic injustices. Finally, the concluding section suggests that attending to intersecting disparities can help achieve effective responses to climate change, while extending our understanding of social inequality.

I write as a white male residing in Canada, a nation that is at the forefront of driving global climate change. The Canadian government has played a leading role in obstructing international climate negotiations, while aggressively sup-porting the tar sands industry – both domestically and abroad. Although I am not an executive or a statesman, I am nevertheless at least complicit in injus-tices that are important in climate struggles.

This chapter builds on environmental justice analyses which have shown how inequities are bound up with institutions that cause and prolong ecological burdens (see Carmin and Agyeman, 2011; Hofrichter, 1993/2002). Decades of environmental justice analysis have shed light on how racism in industry and government manifests in air pollution, contaminated waterways, and a selective lowering of environmental standards (see, e.g., Bullard, 2005). Eco-feminists (see, e.g., King, 1993), as well as a subset of environmental justice proponents, have stressed how patriarchy bears some responsibility for environmental prob-lems. For instance, these connections are drawn by Sturgeon (2009), who writes: "the biggest environmental polluters and exploiters are governments, militaries, and corporations – all patriarchal institutions in which men have most of the power" (p. 42). As this chapter will show, much the same patterns may be found in the causes and consequences of climate change.

Patriarchy and climate change

Some writings have shown how male privilege and masculine dominance is intertwined with climate disruptions (see, e.g., Dankelman, 2010). In many cases, females have been more vulnerable to flooding, among other disasters. As I will explain, these upheavals involve an intensification of gendered violence

that includes sexual assault. In addition, women can be more at risk of exposure to malaria-bearing mosquitoes that spread due to climate change, and
females will often be expected to take on further care-giving responsibilities as
disease threats mount and family members fall ill. A range of everyday toil
may be expected to magnify for women as their overall living conditions
deteriorate.

At the same time, the major sources of greenhouse gas have often been more
male and masculine. Roberts (2009) concentrates on war and militarism, for
example. There is extensive fossil fuel consumption[3] behind militarized chains
of command, which are largely staffed by men – while the brunt of their organized violence is borne by women and girls (Roberts, 2009). Further insight is
offered by Dankelman (2002), who writes: "it is often male-dominated organisa-
tions that make decisions affecting energy policies and programmes" (p. 25).
Highlighting two of the major institutional bases of greenhouse gas emissions,
Dankelman suggests, "the power and petroleum sectors ... are male-dominated"
(p. 25).

Disasters have been addressed in other literature about gendered threats and
impacts. Women are more vulnerable to a series of disasters that are escalating
amidst climate changes (Dankelman, 2010, pp. 59–62). Given feminized economic disadvantages (addressed below), women are less able to prepare for and
cope with monsoons and tornadoes, among other climate-aggravated disasters.
Roberts (2009) discusses feminized vulnerabilities in the midst of various
upheavals (including wars). She notes: "Of the approximately 50 million people
displaced from their homelands, about 80 percent are women and children"
(p. 39). At the same time, females are also less likely to survive these disasters.
In part, this gendered gap in life expectancy is due to a lack of information
regarding potential hazards, and how to adapt to these. Reduced mobility further
increases these dangers. Regarding natural disasters (rather than human-induced
ones), Ibarrara and colleagues (2009) note, "while women may not be biophysi-
cally more vulnerable to natural disasters ... the difference lies in their social
vulnerability" (p. 560).

Disasters have also entailed increased risks of gendered violence (Dunn,
2009, p. 122). Women and girls may experience further brute force and sexualized violence during these calamities. Disasters are openings for the worst of a
patriarchal culture in which men have been socialized to accept violent aggression, promoted through television, sports, and other such influences (Connell,
1995/2005; Kimmel, 2000). As any day-to-day protections are lost or overturned
amidst disasters and displacement, there is a greater threat that males, in particular, will turn to muscular force. However, social services for women and
other relatively vulnerable groups may become even less accessible and operable
if infrastructure breaks down, and any legal protections would be less enforce-
able under disaster conditions. Refugee camps have often proven to be unsafe as
well. On the whole, everyday violence may be expected to intensify under
disaster conditions as turmoil and chaos overturn any protections that could
help prevent gendered violence. Some other climate threats differ from these

disasters, in that the consequences are more chronic; but uneven risks nevertheless continue to be present.

Certain diseases are transmitted more readily under climate change, and these threats can be gendered as well. For instance, as flooding and other climate disruptions bring additional malaria and diarrhea, females likely will be left to take on more care-giving responsibilities (Dunn, 2009, pp. 117, 122). The same burdens may be expected to come with any additional disease exposure or injuries. Moreover, females who are assigned water-gathering tasks may be exposed to more malaria-bearing mosquitoes (Denton, 2002, p. 15). In various ways, climate impacts may be expected to bring distinctly feminized risks and toil.

As further labor demands are imposed upon them, women and girls may be pressured to make even more sacrifices. In many areas of the world, it is increasingly difficult for women to carry out their underappreciated and undervalued labor. Water-gathering and food cultivation responsibilities are becoming more difficult due to climate impacts. In areas of Sub-Saharan Africa and elsewhere, many women are traveling further to find scarcer water supplies as they also attempt to gather food under less hospitable conditions. Younger females may miss school to take up some of this mounting workload (Denton, 2002, pp. 14–15). Yet, even when women and girls gather food supplies, males in the household may be better nourished (Denton, 2002, p. 14).

The preceding points about gendered injustices are more applicable to 'Third World'[4] contexts, where females may have more direct contact with natural environments, and where climate impacts tend to be more intense. In affluent 'First World' countries, there are fewer straightforward connections between some of these gender inequities and any climate change impacts. Although women in the more affluent areas of the world tend to have relatively more domestic and care-giving responsibilities than their male counterparts, there are additional buffers between these feminized tasks and any climate disruptions, given how mechanized systems have replaced more of the hands-on food and water gathering. This modern equipment and infrastructure may certainly be disrupted by storms however, and gendered vulnerabilities like those described here are to be expected (Khosla and Masaud, 2010). However, when modern 'First World' systems are in operation, some climate impacts can be held at bay (as additional food is shipped from abroad, for instance). Medical programs likewise may provide protections against the spread of diseases from tropical regions.

Wherever females are more vulnerable to climate impacts, it is important to recognize how they are much more than victims. Many women have offered constructive and proactive contributions (see, e.g., Dankelman, 2010, pp. 223–239; Shiva, 2008). Their efforts have included disaster-warning networks, and the cultivation of resilient agriculture. Although self-deprivation is among the forms of feminized coping (which may consist of skipping meals), many female efforts have had noteworthy social benefits in the face of environmental threats (Dankelman, 2010, pp. 63–65). Moreover, there is a great deal of latent potential among females, whose current roles and positions may allow them to make distinct contributions. In Denton's (2002) words: "through close interactions with forest and other ecosystems over many years, women have developed a wealth of indigenous knowledge of plants and their medicinal value" (p. 13). In less modernized settings, females have been closer to the sites of climate disruptions, and such experience and awareness is crucial to our responses to climate change.

Despite how many of the dimensions of climate issues are gendered, the prominent climate policy forums and discourses have failed to recognize the significance of patriarchal conditions and distinctly feminine contributions to addressing climate change (see, e.g. Dankelman, 2010). The organization GenderCC has concentrated on the serious shortcomings of United Nations' climate negotiations, while encouraging further inclusion of gender issues and female voices in these intergovernmental fora. There remains a need for a mainstreaming of feminist perspectives on climate policies, mechanisms, and actions (Dankelman, 2010, p. 2). In the meantime, these viewpoints remain ghettoized, given insufficient support from existing organizations and institutions. Accordingly, feminized perspectives and concerns have also received far too little attention in relevant municipal planning processes (Khosla and Masaud, 2010, p. 89).

Structural racism and climate change

In much of the previous literature about climate change, the attention toward race is rather scattered, and not particularly visible. Yet, there have been noteworthy efforts to show how racialized injustices are linked to climate change (directly or indirectly). In the USA, the National Association for the Advancement of Colored People (e.g. 2011) has focused on the African diaspora as they have called attention to systemic climate racism. Similarly, the National Latino Coalition On Climate Change was formed to convey concerns about unjust climate threats in relation to a particular form of racialization. Moreover, the distinct threats and impacts faced by transnational indigenous peoples have been highlighted in reports, summits, and elsewhere. The Indigenous Environmental Network has played a leading role in addressing climate change and a range of interrelated ecological issues.

Approaching such struggles with a social inequality lens helps link the conditions faced by natives peoples, Latinos, and members of the African diaspora. Ultimately, most of the victims of the earlier and relatively intense climate change impacts are disempowered people of color. In the more affluent nations, those of African descent are among the racialized groups who are more vulnerable to climate impacts – as I will explain below. Air conditioning and disaster assistance are more out of reach for disadvantaged groups, who have also received less attention from officials during hurricanes and other disasters. Nevertheless, climate impacts are more intense for Arctic Inuit peoples and for those who live around the equator. Some climate injustices are a feature of international inequities that are racialized. Levels of affluence and industrialization tend to correspond with degrees of vulnerability, and these uneven conditions usually fall along worldwide color lines as well. This international racialization has a bearing on numerous small island nations, particularly in tropical regions. Comparisons can be drawn with Sub-Saharan Africans who are facing relatively extreme climate impacts, as many residents of small island states have colored skin. Increasingly, these islands are threatened by rising sea levels and extreme storms. By 1989, the disproportionate threats to these islands brought delegates to the Small States Conference on Sea Level Rise.

Desmond Tutu and the United Nations Development Programme (UNDP) (2007) have used the term "adaptation apartheid" to highlight such climate disparities on a global scale. Likewise, there is a racial subtext to statements about how the international distribution of emission sources and climate impacts amounts to a new form of colonialism (see, e.g., Bachram, 2004). There are continuities between various environmental injustices and racialized imperialism in previous centuries (Pellow, 2007, pp. 37–39). Such interconnections with race require further unpacking, but one pattern is evident enough today: relative to 'Third World' countries where climate change tends to be more intense, the 'First World' is much whiter, overall. Yet, much of the racialized injustices of climate change do not line up with the borders of modern nations.

Transnational indigenous peoples are also facing mounting risks of climate disruptions, particularly when tribes try to preserve their traditions. In northern Alaska, and in other inhabited areas of the Arctic, hunting habitats are vanishing as settlements are collapsing (see Trainor et al., 2009). As ice thins out, travel becomes more dangerous. Elsewhere, Pacific natives are among indigenous peoples who can no longer depend on traditional crops (Trask, 2009, p. 37). In many areas, medicinal plants are being lost, alongside many other familiar species that are unable to adapt. From the Tropics to the Arctic, native traditions and subsistence ways of life entail attachments to certain species and natural conditions, which are being lost owing to climate change. Turning to modernized ways of life would entail further colonization of these peoples and traditions, but without providing an escape for racialized native peoples.

Within affluent nations, associated racial injustices have a somewhat different character. The US-based National Latino Coalition on Climate Change (2012) has observed how droughts and extreme weather "undoubtedly affect Latino farm workers and businesses in areas that have traditionally been hit hardest by these natural disasters." African Americans in the USA are also among racially oppressed groups who are more vulnerable to climate change impacts. Hoerner and Robinson (2009) focus on this segment of the US population as they point to threats to "health, housing, economic well-being, culture, and social stability" due to climate change (p. 32). For instance, due to racist discrimination, governments have provided second-rate emergency interventions for black Americans (Ibarrara et al., 2009, p. 561). According to Hoerner and Robinson (2009), "racist stereotypes have been shown to reduce aid donations and impede service delivery to African Americans in the wake of hurricanes, floods, fires and other climate-related disasters as compared to non-Hispanic whites in similar circumstances" (p. 33). To make matters worse, a disproportionate number of black Americans live in zones that are more likely to be hit by hurricanes (Hoerner and Robinson, 2009, p. 32).

Even as racially oppressed groups are facing relatively severe climate vulnerabilities, these peoples are marginalized in official climate forums and discourses. Focusing on African Americans, Hoerner and Robinson (2009) note how "legislation is being drafted, proposed, and considered without any significant input from the communities most affected" (p. 35). Tom Goldtooth of the Indigenous Environmental Network has likewise observed how natives have been "locked out of the process that's happening between the big environmental organizations and big industry" (quoted in Dayaneni, 2009, p. 8). There has been a failure to mainstream anti-racist and feminist priorities and analysis. If unjust climate vulnerabilities continue to be ignored, these are more likely to be prolonged. These patterns of injustice may prove to be difficult to uproot, given how whiteness has involved disproportionate greenhouse gas emissions.

These racialized tiers of responsibility may be observed within the USA. Hoerner and Robinson (2009) note how blacks in the USA "emit nearly 20 percent less greenhouse gases than non-Hispanic whites per capita" (p. 32). Writing about systemic patterns in urban design, land use, and transportation in the USA, Brooks (2009) highlights a history of "white flight" to "sprawling suburban subdivisions," which has involved more car usage relative to inner-city living, where more destinations are closer at hand (p. 66). Unlike walking and cycling, car driving generally entails fossil fuel consumption. Private automobility is also less energy-efficient, relative to mass-transit options which are not as readily available in suburban outskirts of cities. Moreover, larger suburban homes generally require more energy for heating and cooling, which is often powered by fossil fuels (such as 'natural' gas). As with various other systemic causes of climate change, automobility and suburbs are concentrated in whiter 'First World' areas. Such examples demonstrate how whiteness in these locales often comes with a carbon footprint. Exploring these questions further would involve considering the institutional roles of officials, executives, and military commanders who are disproportionately white.

Conversely, people of color have brought distinct contributions to climate struggles. Disadvantaged and disempowered individuals have often played leading roles in climate mobilizations (Moore and Kahn Russell, 2011). In some cases, native peoples have taken this initiative as they have sought to preserve ancestral lands by opposing fossil fuel pipelines and extraction projects that are encroaching onto their territories. Given how these and other indigenous efforts can be additional burdens on people who are already oppressed and marginalized, further recognition is deserved for their opposition toward the sources of greenhouse gas emissions. To give another example: in an effort to preserve Arctic territories, the Inuit Circumpolar Conference filed a lawsuit against the Bush Administration to challenge its support for greenhouse gas-intensive industries. Alongside such interventions, we also find everyday forms of stewardship and awareness which has regularly been underappreciated. Inuit peoples have extensive experience with local ice conditions, for example (Bravo, 2009, e.g., pp. 275–276). Many native peoples and 'Third World' residents have more direct connections to the environments in which climate impacts are occurring; hence, these groups can offer grounded forms of stewardship in and through their traditional ways of life (see Shiva, 2008). Pre-industrial agricultural practices entail less fossil fuel consumption and greenhouse gas emissions, for instance. The same is true of a wide array of native hunting-and-gathering practices. Without idealizing any of these ways of life, we can recognize how certain traditional practices may not involve fossil fueled equipment and industry. Low-carbon ways of life may also be found among pastoralists and small-scale farmers in numerous countries.

Class, economics, and wider injustices

This chapter has indicated how various injustices are default outcomes of climate changes, as these inequities also aggravate climate impacts. Reviewing racialized and gendered forms of inequality in two distinct sections has been oversimplistic and misleading, however. Many individuals are doubly oppressed through gender and race systems, simultaneously (see, e.g., Collins, 2000), and, in many cases, these oppressions intersect with economic injustices. I will now briefly address relevant class factors, to begin to show how race and gender inequities frequently coincide with magnified poverty, and/or financial dependencies.[5] Intersecting class inequities include reduced access to air conditioning and water supplies, as well as a diminished capacity to migrate away from deteriorating environments. On the other hand, those who possess sufficient resources have been able to construct buildings that can rise and fall with the tides (Tokar, 2010, p. 19).

Within and between nations, climate vulnerabilities are more intense for those who have fewer resources. Ibarrara and colleagues (2009) refer to "empirical evidence that the macroeconomic effects [of climate change] will fall largely upon the poor" (p. 562). These impacts include higher mortality rates during and after disasters (Ibarra et al., 2009, p. 556). In part, these risks may be heightened because the social networks of relatively poor individuals "are less likely to contain connections to influential people, such as moneylenders or government officials" (Ibarra et al., 2009, p. 556). In addition, fewer resources will be available for initiatives that could help with adapting to climate change. Dunn (2009) explains how economically impoverished individuals have a more "limited capacity to relocate, repair, rebuild, and replace their houses [and] items" (p. 122). Dunn focuses on economically disadvantaged women (see also Khosla and Masaud, 2010); yet, these observations are relevant to racialized economic inequities. Disproportionate poverty has a bearing on environmental threats to Latinos as well (NLCCC, 2012). The relative vulnerabilities of African-Americans involve a relative lack of medical insurance, along with reduced income, which could otherwise be used to pay for medical costs (Hoerner and Robinson, 2009, p. 33).

Yet, even as economically disadvantaged groups can be more susceptible to climate impacts, they are also less responsible for these conditions, given that they consume and manage fewer resources. Eco-socialist critics such as Foster (e.g., 2009) emphasize the role of class power in driving climate change – sometimes through corporate investments, marketing, and product designs (see also Bond, 2012). Oil and gas industries, car companies, and agribusinesses are particularly important sources of greenhouse gases. Of course, such industrial emissions have very direct ties to capitalist profit-making. As the economy is driven by class power, this inequality involves differing scales of consumption, which tend to correspond with contributions to climate change. As with the aforementioned car driving and home sizes, increased airplane travel dramatically increases fossil fuel consumption. But the gaps in consumption are far more massive than such examples suggest: while some possess mansions and yachts, others live in huts without electricity. Even when frugality is involuntary, personal emissions are generally lowered.

Conclusions

This chapter has been more exploratory than comprehensive. At the risk of oversimplifying, I have concentrated on certain contours of patriarchy and racism as these structural conditions relate to climate change. I have also touched upon class factors to further explore an intersectional approach to climate change. This overview should be sufficient to show that racism and patriarchy are important sides of climate change debates. A fuller look at these inequities would entail considering reproductive rights struggles – to respond to excessive concerns about 'Third World' population growth rates, which have often been presented as a central global problem, but frequently without reckoning with the massive footprints of the economies of affluent nations. The conditions faced by environmental refugees also warrant in-depth attention; in particular, racism toward non-white refugees would be an important topic to address in supplementary analysis. Moreover, forms of disability are among the interlocking injustices that require further attention. The more limited account of climate injustices here has nevertheless shown some recurring patterns which hold implications for understanding social inequality, and for the pursuit of constructive responses to climate change.

The preceding analysis indicates how climate change is adding to pre-existing harm and vulnerabilities for those who are consigned to lower strata of political-economic systems. Declining access to food and fresh water is among the examples of risks that bear down on some much more than on others. In effect, some people are approached as though they are 'redundant' or 'disposable' (Shiva, 2008, p. 2). Steady (2009) likewise observes a "structural expendability" of marginal groups (p. 50). In part, these inequitable risks and impacts are continuing because those who are more victimized are not receiving adequate attention. Relatively severe risks to Inuit peoples are regularly neglected, for example. Even as there is attention toward Arctic warming, the human residents of the far north are frequently ignored (Sturgeon, 2009). Far too often, images of polar bears are shown without acknowledging the existence of Inuit peoples in the region. In Canada, climate impacts are relatively intense in the far north, while climate change is driven by corporations and governments that are based in southern Canada, where climate impacts have been less apparent. Such conditions encourage some to ignore climate injustices, thereby interfering with our capacity to tackle climate change.

The full severity and scope of climate crises has not received enough serious attention from those who have more power to steer societies away from greenhouse gas releases, and toward climate mitigation and adaptation. Disproportionate power and resources allow some to maintain distance from climate threats. Hence, complacency and neglect are widespread in corporate boardrooms, and in various other seats of power. Much the same irresponsibility may be found in the fora that are supposedly dedicated to tackling these problems. In the meantime, the people most affected by climate change have the strongest motivations to support policies and programs designed to resolve its impacts and causes. However, females and people of color are marginalized in official responses to climate change, or they are excluded entirely (as noted above). In UN negotiations and elsewhere, the least vulnerable strata have vastly more input into official processes and discourses that supposedly exist to redress climate problems. Accordingly, such responses have remained deeply compromised, as injustices have largely been ignored. Thus, a cycle continues to be prolonged: even as the power and affluence of some is bolstered with greenhouse gas emissions, these groups remain relatively protected (in part, because of the insurance they can afford). Their insulation from impacts has proven to disconnect them from solutions. These dynamics can be pervasive. My experience in urban Canada is one of deep complacency around climate change: it seems as though only a small minority of the population is very concerned. If the people in other locations who have more of a direct stake in climate change prevention and adaptation continue to have so little capacity to intervene effectively in appropriate political-economic systems, we can expect that climate crises will continue to intensify.

To break out of these cycles it is crucial to support those who live on the frontlines of climate change, to tap into their proactive motivations, as well as their concrete experiences with higher risk conditions where environmental impacts may be the most apparent. Such strategic imperatives are conveyed by Moore and Kahn Russell (2011), who write: "climate solutions that are not populist and serving the needs of disenfranchised peoples will fail. Conversely, climate solutions that serve the needs of those most affected will benefit everyone" (p. 21). Women and people of color are among those who can fill such frontline roles in struggles to address climate change. At times, this frontline campaigning has consisted of supporting those who live in proximity to refineries (see, e.g., Black et al., 2014). A sense of urgency is central to emerging struggles to seek constructive change in collaboration with those who live on the frontlines of such impacts. This strategy may be applied to various sides of the dirty energy systems that are so important in driving climate change. Like many other activists, I have attempted to bring such an approach into campaigning around tar sands, fracking, pipelines, and refineries – while others have applied a frontline strategy to their work on coal-mining and coal plants. Anyone who is not on the frontlines of such climate and energy struggles has a responsibility to take on a supporting role.

In the meantime, social justice considerations are largely being left out of inadequate but prevailing responses to climate change, which merely approach these issues as matters of policy negotiations between nation-states. Inasmuch as injustices are addressed in these negotiations – or in associated discourse from journalists and academics – these issues have usually been reduced to international economic stratification. In the negotiation framework of the UN, the more affluent "Annex I" countries are distinguished from the less industrialized "Non-Annex I" countries. Much the same international binary has been pervasive in climate discourse. Although this approach offers a degree of attention toward disparities, this recognition has been quite limited in practice. Gardiner (2011) suggests that principles of justice have typically not been taken seriously in climate policy fora, as negotiations are reduced to the "political horse trading" of bargaining between state representatives who pledge emissions cuts (p. 312). Thus, international stratification is poorly confronted, even as climate frameworks seem to foreground such issues on paper. Carrying out such negotiations without adequately tackling the massive political-economic divisions in the world amounts to a shallow response to social inequality, and other inequities have been dismissed entirely in the process. Recurring messages to the effect that climate issues are reducible to international relations (see, e.g., Helm and Hepburn, 2009) imply that each nation is a homogeneous mass. Racialized and gendered inequities within nations are simply ignored in such accounts as nations are reified as singular entities. These international emphases are inadequate on their own.

Accounts of the dimensions of racialized and gendered disparities in relation to climate change can extend the more limited forms of climate discourse while deepening our understanding of inequality. Further attention toward these environmental inequities can strengthen social movements, and our analysis. In the meantime, there are major gaps in relevant analysis, and far too little synthesis. To counter this trend, we can continue to contextualize climate issues within a much longer history of patriarchy and racism – along with interlocking injustices and political-economic systems. These understandings should be brought into efforts to pursue social changes and environmental benefits. Feminist and anti-racist insights are among the important contributions toward collective efforts to resolve the climate crisis.

Notes

1 To offer two more examples: a 2008 statement from Climate Justice Now! conveyed concerns about gender (without directly addressing race), and the Environmental Justice and Climate Change Initiative's (2002) "10 principles for just climate Change policies in the U.S." emphasize racialized disparities (without mentioning women or gender).

2 This chapter does not distinguish between cis-gendered and trans-gendered women.
3 "For 2004, military fuel consumption increased 27 percent over the average annual peacetime usage of 100 million barrels. In just three weeks of combat in Iraq, the Army burned 40 million gallons of fuel – or almost two million gallons per day" (Roberts, 2009, p. 39).
4 I am using ironic quotation marks around the phrases 'First World' and 'Third World' to indicate that these labels are crude shorthands for complex disparities between different areas of the globe.
5 Of course, women continue to receive significantly less income, even when they perform the same work.

References

Bachram, H., 2004. Climate fraud and carbon colonialism: The new trade in greenhouse gases. *Capitalism, Nature, Socialism*, 15(4), pp. 5–21.
Black, T., D'arcy, S., Weiss, T., and Kahn Russell, J. (eds), 2014. *A Line in the Tar Sands: Struggles for environmental justice*. Toronto: Between the Lines.
Bond, P., 2012. *Politics of Climate Justice: Paralysis above, movement below*. Scottsville, South Africa: University of KwaZulu-Natal Press.
Bravo, M.T., 2009. Voices from the sea ice: The reception of climate impact narratives. *Journal of Historical Geography*, 35(2), pp. 256–278.
Brooks, K., 2009. SB375 and racism. *Race, Poverty, and the Environment*, 16(2), p. 66.
Bullard, R.D. (ed.), 2005. *The Quest for Environmental Justice*. San Francisco, CA: Sierra Club Books.
Carmin, J. and Agyeman, J. (eds), 2011. *Environmental Inequalities Beyond Borders: Local perspectives on global injustices*. Cambridge, MA: MIT Press.
Climate Justice Now!, 2008. CJN! Poznan Statement. Available at: www.climate-justice-now.org/cjn-poznan-statement-signatures/ (accessed June 30, 2012).
Collins, P.H., 2000. *Black Feminist Thought: Knowledge, consciousness, and the politics of empowerment*. New York: Routledge.
Connell, R.W. 1995/2005. *Masculinities*. Los Angeles: University of California Press.
Dankelman, I., 2002. Climate change: Learning from gender analysis and women's experiences of organising for sustainable development. *Gender and Development*, 10(2), pp. 21–29.
Dankelman, I. (ed.), 2010. *Gender and Climate Change: An introduction*. Washington, DC: Earthscan.
Dayaneni, G., 2009. Carbon fundamentalism vs. climate justice. *Race, Poverty, and the Environment*, 16(2), pp. 7–11.
Denton, F., 2002. Climate change vulnerability, impacts, and adaptation: Why does gender matter? *Gender and Development*, 10(2), pp. 10–20.
Dunn, L.L., 2009. The gendered dimensions of environmental justice: Caribbean perspectives. In F.C. Steady, (ed.), *Environmental Justice in the New Millennium*. New York: Palgrave Macmillan, pp. 115–133.
Environmental Justice and Climate Change Initiative, 2002. 10 principles for just climate change policies in the U.S. Available at: www.ejnet.org/ej/climatejustice.pdf (accessed June 30, 2012).
Foster, J.B., 2009. *The Ecological Revolution*. New York: Monthly Review Press.
Gardiner, S.M., 2011. Climate justice. In J.S. Dryzek, R. Norgaard, and D. Schlosberg,

(eds), *Oxford Handbook of Climate Change and Society*. New York: Oxford University Press, pp. 309–322.

Helm, D. and Hepburn, C. (eds), 2009. *The Economics and Politics of Climate Change*. New York: Oxford University Press.

Hoerner, J.A. and Robinson, N., 2009. Just climate policy – just racial policy. *Race, Poverty, and the Environment*, 16(2), pp. 32–35.

Hofrichter, R. (ed.), 1993/2002. *Toxic Struggles*. Philadelphia, PA: New Society Publishers.

Ibarraran, M.E., Ruth, M., Ahmad, S., and London, M., 2009. Climate change and natural disasters: Macroeconomic performance and distributional impacts. *Environment, Development and Sustainability*, 11, pp. 549–569.

Khosla, P. and Masaud, A., 2010. Cities, climate change and gender: A brief overview. In I. Deckman, (ed.), *Gender and Climate Change: An introduction*. Washington, DC: Earthscan, pp. 78–96.

Kimmel, M.S., 2000. *The Gendered Society*. New York: Oxford University Press.

King, Y., 1993. Feminism and ecology. In R. Hofrichter, (ed.), *Toxic Struggles*. Philadelphia, PA: New Society Publishers, pp. 76–84.

Mobilization for Climate Justice, 2009. Open letter to the grassroots. Available at: www.actforclimatejustice.org/about/open-letter-to-the-grassroots/ (accessed July 30, 2012).

Moore, H. and Kahn Russell, J., 2011. *Organizing Cools the Planet*. Oakland: PM Press.

National Association for the Advancement of Colored People, 2011. From the Bronx to Botswana: Making a climate change connection. Available at: www.naacp.org/blog/entry/from-the-bronx-to-botswana-making-a-climate-change-connection (accessed June 30, 2012).

National Latino Coalition On Climate Change (NLCCC), 2012. What are the impacts of climate change and how will it affect the Latino community? Available at: http://latinocoalitiononclimatechange.org/issues/community_impact/ (accessed June 30, 2012).

Pellow, D.N., 2007. *Resisting Global Toxics: Transnational movements for environmental justice*. Cambridge, MA: MIT Press.

Rising Tide, 2001. Political statement. Available at: http://web.archive.org/web/20010216185944/www.risingtide.nl/statement.html (accessed June 30, 2012).

Roberts, M., 2009. War, climate change, and women. *Race, Poverty, and the Environment*, 16(2), pp. 39–41.

Shiva, V., 2008. *Soil Not Oil: Environmental justice in a time of climate crisis*. Cambridge, MA.: South End Press.

Steady, F.C., (ed.), 2009. *Environmental Justice in the New Millennium*. New York: Palgrave Macmillan.

Sturgeon, N., 2009. *Environmentalism in Popular Culture*. Tucson: University of Arizona Press.

Tokar, B., 2010. *Towards Climate Justice*. Porsgrunn, Norway: Communalism Press.

Trainor, S.F., Godduhn, A., Duffy, L.K., Chapin III, F.S., Natcher, D.C., Kofinas, G., and Huntington, H.P., 2009. Environmental injustice in the Canadian far north: Persistent organic pollutants and Arctic climate impacts. In J. Agyeman, P. Cole, R. Haluza-DeLay, and P. O'Riley, *Speaking for Ourselves: Environmental justice in Canada*. Vancouver: UBC Press, pp. 144–162.

Trask, M., 2009. Indigenous views on climate change. *Race, Poverty, and the Environment*, 16(2), pp. 36–38.

United Nations Development Programme, 2007. Fighting climate change: Human solid-

27 Mother Earth

Isis Mattie and Imna Arroyo

Her peaks pierce the sky
Omniscient sentinels watching over
forest, valley, plain, and shore.

Winged ones circle high in the ether
while land dwellers run, creep or crawl upon her belly
and finned creatures delve deep into her abyss.

Strong and firm,
Bathed in moon light
She is surrounded by her many children.

Displaying her fertile body
Allowing the seeds of life to root and grow within her.
Pregnant with sacred waters
She quenches thirst
and gives us breath.

Figure 27.1 Mother Earth by Imna
Arroyo, linoleum cut 30"×22".
Mother Earth is surrounded by her
children. Spilling from her heart is
the great river goddess Ochun.
The print and poem explore the
relationship between Earth
Mother and Ochun, deity of love,
fertility, beauty, and harmony.
Ochun represents a vast consciousness of divine feminine energies
that sustain humanity and hold
the universal alignment.

28 The political ecology of *Pachamama*

Race, class, gender, climate change, and Kallawaya traditions

Dylan Harris

Introduction

When asked why he thought the river next to his village was dry, Rámon Alvarez, a traditional Kallawaya, lamented that the river was only an omen for what was to come. For the past few years, the potatoes were being harvested earlier. They looked and tasted the same, but Alvarez could feel in his body that he was dying. His bones told him that he would not live as long as his father because the weather is warmer and the river is dryer. For the Kallawaya, an indigenous group of people who depend on their environment for not only physical but also for spiritual sustenance, climate change is doubly detrimental. In addition to changing crops and unprecedented seasonal fluxes, the villages have begun sending their children to work in the cities to find better jobs, leaving the 1000-year-old mountain terraces, which trace the landscape from the river valley to the towering peaks of the Andes, barren and wasted.

This small anecdote taken from my 2012 fieldwork in the *Cordillera Apolobamba* of Northern Bolivia highlights how race, class, gender, and now global climate change (GCC) intermingle, resulting in the potential decline of the centuries-old Kallawaya tradition. The confluence of these relations has become entangled in the local, national, and international climate change discourse, which is especially relevant in the context of the Evo Morales government. Using interviews with Kallawaya healers and experts, information taken from climate change conferences in La Paz, and personal observations from time spent in the *Cordillera Apolobamba* region, this chapter will explore how the intersections of race, class, gender, and now GCC impact the Kallawaya tradition.

A critical introduction to the Kallawaya tradition

As indigenous movements gain more prominence across South America – in Bolivia, Ecuador, and Peru specifically (Cadena, 2010) – increased attention has been paid to what exactly constitutes indigeneity (Dove, 2006). Indigenous groups are often portrayed as having a static and objective identity, meaning that they are studied and discussed as being "frozen in time and space" (Fabricant, 2013, p. 159). However, the Kallawaya, as with other indigenous groups, have a contested identity that is discursively created and mediated throughout time and space.

Despite the presence of considerable research on the Kallawaya tradition, Callahan (2011) contends that previous research is flawed in that it assumes knowledge of Kallawaya expertise. These assumptions create not only representational categories of the Kallawaya tradition but also tension between what researchers consider preservation and the Kallawaya perceive as misrepresentation and exploitation. Previous research offers no account of the interactions between healers or the intramural politics of the communities, such as various divisions among healers or any mention of women. Furthermore, most research with the Kallawaya, with the exceptions of Bastien (1978) and Callahan (2011), has been conducted outside of the communities, meaning that researchers lived elsewhere and traveled back and forth to their research sites. This type of research results in people engaging with individuals, as opposed to communities, and not taking into account the way in which individuals relate to their community or the intra-communal nature of the Kallawaya tradition. Furthermore, previous research has often ignored broader national and global issues (Callahan, 2011, p. 10), such as GCC.

The politics of what constitutes indigeneity have become even more complex as issues like climate change becomes more prominent. The mythos surrounding indigenous cultures often portrays them as 'pure' in the sense that they are untainted by the same modernism that created issues like GCC (Dove, 2006). However, these assumptions tend to create static categories of indigeneity.

As Stengers (2005) points out, engaging with alternative worldviews causes reasoning to slow down and provides space for critical rethinking of complicated problems. It is not the aim of this chapter to prove or discredit Kallawaya indigeneity or indigenous environmental perspectives. Rather, it is meant to reconsider the place of the indigenous in what is considered political and to use the Kallawaya tradition as a backdrop for re-imagining how race, class, gender, and climate change interact and inform one another. It should be noted that the research used for this chapter is more general and not based on years of intra-communal analysis so much as on a few interviews and observations of time spent with the Kallawaya. My role as a white, foreign researcher must be taken into account, as should my lack of time spent with these communities. This chapter is largely speculative and reflective, and serves as a critical entry to further research in these communities. The following section will look more closely at the Kallawaya *cosmovisión* and healing tradition, as a general understanding of the tradition will coordinate the rest of this discussion.

The Kallawaya *cosmovisión* and healing tradition

The Kallawaya, inscribed in 2008 by UNESCO as Representatives of Intangible Cultural Heritage of Humanity (UNESCO.org), trace their tradition back to the era of Tihuanaco cultures of AD 400 to 1145. Their legacy has survived

through the eras of other pre-Inca cultures, the Incan empire, the Spanish conquest, neoliberalization, and into present times. Their reputation as Andean
healers has been tested throughout time, but they have managed to persevere
and continue practicing their healing ceremonies for people across the Andes
who have no access to modern medicine or still prefer traditional medicine
(Krippner, 2006). Although their presence is smaller now than before (Quispe,
2012), they still maintain a sizable population. However, more healers are living
and working as professionals – as healers and other occupations – in cities such
as La Paz, and many of them continue to travel, returning home only at certain
times of the year (Callahan, 2011). Their return home is largely because of their
attachment to the region they call home, which is indelibly influenced by their
particular *cosmovisión*.

Although the word *cosmovisión* is similar to the German concept of a *Weltan-schauung* (i.e., worldview), it differs in that it specifically relates to a view of the
world as understood by Latin American peoples. Similarly, it is akin to the
English idea of cosmology, which is an explanation of how the world came
about, and the values and philosophies maintained in those narratives. A *cos-movisión* is not universal; there are many different cultural narratives in regions
across Latin America.

The Kallawaya *cosmovisión* is based on thousands of years of experiential
knowledge of their environment and is shared among many other communities
across the Andes. As with other well-documented Andean groups, the Kallawaya use a metaphorical understanding of their relationship between individuals, the community, and the environment around them, with specific
reference to the mountains around the area they call home (Bastien, 1985,
p. 608). The mountain and the human body are seen as analogous to one
another, which is the basis of Kallawaya spirituality and healing practice. The
community, or *ayllu*, is crucial to the Kallawaya tradition, as it is for many other
Andean indigenous groups. The *ayllu* is much more than the area in which
people live. According to Oxa (2004, p. 239):

> It is a dynamic space where the whole community of beings that exist in
> the world lives; this includes humans, plants, animals, the mountains, the
> rivers, the rain, etc. All are related like a family. It is important to
> remember that this place [the community] is not where we are from, it is
> who we are.

In short, the *ayllu* defines the Kallawaya as much as the Kallawaya define their
surroundings.

At the center of their *cosmovisión* is the notion that humanity must live in
harmony with the environment. Illness is the result of spiritual dissonance
caused by some disjuncture between a person and his or her environment. One
of the main tenets of the Kallawaya *cosmovisión* is *ayni*, which is an ethic of
respect and reciprocity that is applied equally to people, communities, and the
environment. In a past interview, Walter Álverz Quispe, Executive Director of

the *Instituto Boliviano de Medicino Tradicional Kallawaya*, describes Kallawaya traditions as having a "boomerang law – if you harm others, [a] malevolent act will return to you" (Krippner and Glenney, 1997, p. 217). In order to heal a person, the Kallawaya healer uses various remedies to restore spiritual equilibrium to the person as well as to the community and surrounding environment, ensuring a healthy balance (Chelala, 2009).

Like the human body, the Kallawaya believe that the mountain has an underlying cyclical system that is in a constant state of flux, from death to life and from life to death (Bastien, 1985, p. 597). This metaphor regulates daily life in the communities and also highlights the intrinsic relationship between the land and the communities that live there. One of Bastien's principal informants, Marcelino Yanahuaya, explains:

> I am the same as the mountain, *Pachamama*. *Pachamama* has fluids which flow through her, and I have fluids which flow through me. *Pachamama* takes care of my body, and I must give food and drink to *Pachamama*.
> (Bastien, 1987, p. 68)

Mountains are understood as part of *Pachamama*, an all-encompassing idea/deity similar to Mother Earth. Various ceremonies and medicines are used to make sure that the fluids of the human body are also present in the mountain body. Health is understood as a cycle of fluids that flows uninterrupted between the person, community, and environment. Disease is understood as an accumulation of too much of one fluid or the interruption or cessation of fluid flow (Bastien, 1985, p. 87; 1987, p. 46).

Despite *Bautista Saavedra* – the official name of the Kallawaya – being one of Bolivia's poorest provinces, the home of the Kallawayas hosts a unique ecosystem situated between the glacier-laden peaks of the *Cordillera Apolobamba* and the green semi-tropical climate of the *Yungas*. Although the Kallawaya pantheon centers on *Pachamama*, the distinctive geography of the *Bautista Saavedra* region plays an important role in the Kallawaya tradition. The communities' physical locations correspond to the idea that the body – both human and mountain – is a vertical axis with three levels through which blood and fat flow from the center to the peripheral (Bastien, 1985, p. 87). The mountains are also believed to house spirits that protect those living near the mountains from misfortune. After traveling it is important for the Kallawaya to return to this region in order to maintain a balance between the body, community, and mountain (Chelala, 2009).

According to Quispe, the land now called *Bautista Saavedra* "used to be one land. It was the cradle of the Quechua and Kallawaya culture…. This region – near Charazani [the municipal capital city] – used to be called the lungs of the world" (2012). The sacrality of the Kallawaya homeland ties the healers to the land, but it is also considered to be one of Bolivia's poorest provinces. The following section will discuss the intersections of race, class, and gender that provide insight into why this region, though sacred and rich in resources, is still predominantly poor.

Inequality and indigeneity: race, class, and gender among the Kallawaya

Poverty is a symptom of inequality and is often manifested in relation to other symptoms such as racism, classism, or sexism. The intersection of these issues will be discussed later in this chapter; however, it is important to recognize that it is no coincidence that many of Bolivia's poorest are also its most oppressed. In Bolivia, race and class are so intertwined that discussing them together as reinforcing factors which contribute to the vulnerability of the Kallawaya tradition bears more critical weight. The issue of gender, though less studied than race and class, is important to analyze in the triangulation of the various forms of oppression facing Bolivia's Kallawaya. Before moving on to a discussion of climate change, which only serves to exacerbate inequality, it is necessary to understand the place of race, class, and gender among the Kallawaya.

Race and class

Despite Evo Morales being the first indigenous president, and considering that poverty rates among the indigenous population have improved slightly over the past decade, the indigenous population – especially those living in mountain regions – still remain among the poorest people in Bolivia (Robledo et al., 2004, p. 14). Most of the wealth is concentrated in Bolivia's cities and held by mostly white Spanish descendants, followed by a group of mixed or *mestizo* people who make up a formidable middle class, which leaves the indigenous populations to occupy the lowest rungs of Bolivian society. These wealth-based categories are, however, crude at best. In the same way that indigeneity is a contested topic, there are also contestations over the *mestizo* population (Bruun and Elverdam, 2006, p. 274). For example, Afro-Bolivians, the descendants of escaped African slaves and indigenous Bolivians, make up a sizable population across the country and are often counted among the poorest (Walsh, 2011). This cursory glance at the spread of wealth does, however, highlight how inequality is split along racial lines. In addition, it provides some insight into the ways in which race is often tied to class in terms of European lineage. More specifically, race in Bolivia – as across much of Latin America – is intimately tied to class inasmuch as the two were socially constructed in relation to one another. As such, whiteness is usually linked with prosperity, whereas darker skin is associated with poverty. Ethnicity, unlike race, is usually tied to national or social identity.

Indigeneity in Bolivia is further complicated by the rise of indigenous popular politics, which resulted in the election of Morales. His rise to power is a success story from the nearly two-decades-long march toward indigenous rights (Lupien, 2011, p. 774). As an indigenous president, he has undoubtedly paved the way for better race and class relations for Bolivia's indigenous population (Albro, 2005; Gustafson, 2010; Kohl, 2010; Postero, 2010; Cott, 2003). Still, many of the structural inequalities that pre-date Morales prevail across the country.

A brief look at Bolivia's past shows that the indigenous mountain populations were the most exploited by the Spanish in their conquest for mountain-
based resources, especially silver. In 1545, Potosí's silver deposits attracted the
Spanish, and three years later La Paz was constructed to secure silver transport
through the region to the lowlands of Peru. The extraction and transportation
of silver and other resources were "entirely dependent on the superexploitation
of indigenous labor for its mining and the agricultural production needed to
support it" (Healey, 2009, p. 87). Needless to say, Spanish colonization dramatically altered the indigenous landscape, displacing indigenous communities and
disrupting indigenous ways of life.

This trend manifests in recent history and has only been exacerbated by neoliberal capitalist schemes, and now, arguably, by what Bolivia's Vice President
Álvaro García Linera has coined *Green* or *Amazonian* capitalism. Garcia was
quoted as saying, "Indigenous communities need to be viewed as units of production, not places to do social work" (Postero, 2010, p. 28). Although he operates under the platform of indigenous rights and the abolition of neoliberalism,
Linera has been criticized for essentially reinventing the wheel of neoliberalism
by pushing for a *greener* form of capitalism.

On the other side of the political spectrum, the discourse of Bolivia's right-
wing politicians is wrought with racism. Referencing the Morales government,
former foreign minister Manfredo Kempff argues, "it verges on irresponsibility to
claim that illiterates can legislate." This is a sentiment echoed by the Mayor of
Santa Cruz de la Sierra, Percy Fernández: "from the looks of things, we'll have
to paint our faces and use feathered arrows to be able to exist in this country"
(Tathagatan, 2008, p. 37). Historically, racism has been a catalyst for the
oppression of indigenous people in Bolivia, and this trend is still carried on by
Bolivia's right-wing parties, the same parties that control large portions of the
Bolivia's natural gas industry.

Although this is an oversimplified examination of Bolivia's past and recent
history, one can begin to see how the indigenous population until very recently
has been thrown under the wheels of whatever social, economic, or political
system is in place. Race and class are intimately tied together under the banner
of indigeneity in Bolivia. Bolivia's indigenous populations, especially its indigenous mountain populations, are still among the most oppressed in the country,
and the Kallawaya are no exception. A gendered analysis will add more depth to
this discussion.

Gender

Little research has been done on the role of gender in the Kallawaya tradition
(Callahan, 2011). This lack of research may be due in large part to the previous
nature of the research on the Kallawaya, which, as Callahan (2011) points out,
has been largely based on assumptions about their practice. However, in order
to effectively unpack the complexity of the Kallawaya tradition, a gendered analysis is crucial. Before looking specifically at the available research on gender

and the Kallawaya tradition, a cursory glance at larger issues facing indigenous women will add some clarity to this analysis.

As it stands, much of the world's poor depend on agriculture as a means of survival (Scherr, 2000, p. 479), and there is a long history of the rural poor living on marginal lands in already fragile ecosystems (Barbier, 2010). Among the world's rural poor, women are not only among the poorest of the poor but also work longer and harder hours inside and outside of the household (Sherbinin et al., 2007), which is only exacerbated in marginal ecosystems such as high-mountain regions. However, perhaps because of the reverence given to *Pachamama*, Kallawaya women are still afforded nearly equal status to men in their communities. It is a little-known fact that Kallawaya women can also become healers (Callahan, 2011). Their entry into the healing field, however, is largely mitigated by the politics of the household (Callahan, 2011).

Women, like men, are able to practice medicine, but, owing to a lack of formal apprenticeships and training for women and the sedentary nature of their practice, their trajectories as healers differ from those of the men. Whereas men can travel and gain knowledge through working with other healers, women are often restricted to the land and knowledge surrounding their homes, as they are still expected to take care of anything related to the household. Still, women develop their own healing practice, albeit at a slower pace, and are able to gain just as much attention for their work as men. Women, especially women who do not have children or are widowed, develop healing careers that lead to cities more often than men (Callahan, 2011).

Unfortunately, owing to a lack of research on gender and the Kallawaya tradition, it is not within the scope of this short chapter to explore gender's role more explicitly. However, it is imperative to understand that gender – like race and class – informs the Kallawaya reality. Race, class, and gender are just a few variables in an overall equation of inequality. The previous section pointed out the various ways in which these issues overlap in the context of the Kallawaya tradition. As Callahan (2011) points out, most research on the Kallawaya has isolated them from larger national and global issues. The next section will explore climate change and its impact on the Kallawaya tradition, as it has not only gained prominence globally but especially within Bolivia.

Climate change

Within Bolivia, climate change is understood as something that exists between the Morales discourse, the discourse introduced to communities from foreign researchers, and indigenous perceptions. It is dynamically connected to the issues of race, class, and gender, as reverence for *Pachamama* and the pursuit of indigenous rights are so intertwined in the empowerment schemes of Bolivia's indigenous population that have become mainstream issues under the Morales government. Furthermore, high-mountain populations are among the most vulnerable to climate change globally, and Bolivia's Kallawaya healers are no exception (Espiritu, 2012, p. 2; Robledo et al., 2004, p. 14). The following section will contextualize climate change in terms of race, class, and gender and discuss its impact on the Kallawaya tradition.

At the 2009 United Nations Climate Change Conference in Copenhagen (Copenhagen Summit), President Morales caught policy-makers' attention when he announced, "We come from the culture of life, whereas the Western model represents the culture of death," and called for his own meeting, the 'People's Summit,' to take place the following April in Cochabamba, Bolivia (Aguirre and Cooper, 2010, p. 238). Since Morales rode to the presidency on the shoulders of popular social movements largely composed of indigenous people, what he says not only echoes globally but also resonates especially strong in the ears of his constituency. By taking up the issue of climate change and framing in terms of indigeneity, the Morales' discourse on climate change has filtered across the country, reaching the ears of even the most remote villages. In addition to the Morales' discourse, Bolivia has become a hot spot for studying climate change, which has brought in money and researchers. On the one hand, it is crucial that these communities are aware of the impacts of climate change. On the other hand, it is necessary to not only critically understand the ways in which climate change manifests physically, but also socially.

For the people who live in the *Cordillera Apolobamba* region – home to many of Bolivia's rapidly melting tropical glaciers – the impact of climate change is a lived reality. From risks associated with glacial lake outburst floods (Hoffmann and Weggenmann, 2012) to a decreasing amount of water supply for communities and cities across the Andes (Fabricant, 2013, p. 159), melting glaciers pose a definite threat to the Kallawaya. However, risks from melting glaciers aside, the increase in temperatures due to global warming pose other myriad threats to the Kallawaya tradition. As mentioned earlier, the Kallawaya depend on a complex, experiential knowledge of the land around their home and abroad. As the world's climate continues to shift, there are longer dry seasons and shorter wet seasons that are unprecedented in the Kallawaya legacy. Rivers are in flux, and crops are changing. The Kallawaya healer depends on a diverse pharmacopoeia of plants to perform rituals, and while many plants may actually benefit from a warmer climate (e.g., coca), there is no doubt that a warmer climate will have a negative impact on other plants used by the healers (Janni and Bastien, 2004; Thomas et al., 2008). In addition, the Kallawaya healers use various paths to travel, which may change as a result of warmer temperatures. Because of these issues, many Kallawaya are opting for a more sedentary existence in Bolivia's rapidly growing cities (Gauyagua et al., 2003, p. 288).

The effects of urbanization upon the Kallawaya healing tradition are obvious. The younger generations are moving to the city and are either unable to carry on their traditions or are disinterested in their traditions in lieu of a successful life in the city. While Bolivia's urbanization may be understood as a symptom of uneven neoliberal capitalist development, it has also been discussed as an issue of resource scarcity, which is only exacerbated through climate change (O'Hare and Rivas, 2007, p. 323). However, more so than the resource scarcity or glacial retreat, climate change is taking a more cultural toll, as the shrinking of the glaciers can be tied to the symbolic loss of high-mountain culture (Kaenzig, 2013).

Because the glaciered mountaintops surrounding Kallawaya communities are thought to house spiritual energy crucial to the Kallawaya tradition, the cultural impact of climate change in the area and upon its people is undeniable. The spiritual dissonance caused by climate change disrupts the equilibrium that binds the Kallawaya *ayllus*. The way in which the Kallawaya interact with the environment through the mountain–body metaphor may change. The three levels of the mountain, located in three different ecological zones, house a certain type of agriculture, ensuring that the community has food year-round. It is this mountain–body metaphor that unites the communities. When the ecological zones begin to shift, the metaphor begins to break down, as does the resiliency of the *ayllu* (Bastien, 1978, pp. 190–191). As more people move to the cities and the Kallawaya home continues to shift due to ecological, political, and cultural tumult, the Kallawaya identity may also shift.

Physical climate change and the concurrent political and cultural threats are taking a toll on the Kallawaya tradition. As discussed, the physical realities of climate change are present in the Kallawaya communities, affecting their healing practices and altering the landscape of the region. Because of a lack of resources and an unstable climate, among many other intersecting issues (e.g., uneven neoliberal development schemes), many Kallawaya are opting for life in the city, resulting in more class-based divisions within their communities. However, the cultural and social impacts of climate change pose an even more complex threat to the Kallawaya tradition. In order to effectively address climate change in this region, it is imperative not only to take the complex nature of race, class, and gender into account but also to consider the cultural and symbolic implications of a rapidly changing climate. The next section will attempt to triangulate an understanding of how race, class, gender, and climate change are interconnected through a re-imagining of the intersectionality framework.

Expanding intersectionality: race, class, gender, *and* climate change

The intersectionality framework is meant to accurately describe a situation or circumstance with the mindset that oppression is non-linear, and, as such, must be understood as something more complex and fluid. While this framework is useful for understanding sociological and cultural problems, the discourse could be extended to include physical realities like climate change. As evidenced with the Kallawaya healing tradition, the issue becomes much more complex when climate change is discussed as an actor in relation to the intersecting issues of race, class, and gender. This section will make the case for including climate change in an intersectional analysis with the intention of uncovering new ways to understand the issues at hand, specifically looking at the way in which the natural world is locked into the same ideological discursive terrain out of which other "-isms of oppression" (sexism, classism, racism, etc.) have developed (Fontana, 2013; Warren and Cheney, 1996, p. 251).

One of the best ways to understand the intersections of race, class, gender, *and* climate change is to examine the adaptive strategies of the Kallawaya to explore the ways in which the tradition has changed over time and to understand more completely the potential impact of a rapidly changing climate. The Kallawaya occupy a specific human ecological niche where various ideas about agriculture, health, economy, politics, spirituality, and environment all coalesce to inform their worldview (Hardesty, 1972, p. 8). Throughout history, the Kallawaya tradition has been influenced and altered by external factors, such as neoliberalism and a resurgence in indigenous empowerment. These alterations are redefining the Kallawaya landscape, but as active agents they are able to negotiate these changes (McNeish, 2003, p. 262). Bastien believes that "the mountain and its people change with the seasons, sickness, natural catastrophes, migration, and conquest" (1978, p. 194). But what happens when the people are forced to leave their mountains behind? The addition of the rapidly changing climate coupled with increasingly uneven capitalist development may be the most difficult factor to overcome yet.

The world's poorest populations are the most affected by climate change (Tol et al., 2003). In Bolivia, the poor populations are criticized for their seemingly contradictory use of the land when compared to their indigenous land ethics (e.g., the mining industry). However, the 'degrading' activities of the poor are rarely understood as problems that are exacerbated by inequality (Watts and Peet, 2002), which creates a vicious cycle in which racism, classism, and gender violence not only intersect with but also reinforce one another. Barbier (2010, p. 647) calls this phenomenon a "poverty-environment trap." As the climate continues to change, it could result in the further entrenchment of this cycle, as adaptation will become more difficult.

It is imperative that researchers, activists, and policy-makers understand the interconnected nature of the problems associated with warming temperatures. Understanding the interconnectedness of race, class, gender, and climate change as intersectional, meaning that they exist as a constellation to one another, not only allows for a more in-depth reading of these issues, but it also necessarily adds layers to the discussion about the Kallawaya tradition. When addressing something as complex as climate change, it is imperative to take as many angles into consideration as possible, especially in the context of preserving something as vulnerable at the Kallawaya tradition.

Conclusions

Morales stands on the cusp of effecting the necessary change to address past issues of racism, classism, gender violence, and now climate change in Bolivia, helping preserve the Kallawaya tradition. Still, his rhetoric in many ways has failed to come to fruition in his actions, as evidenced by his contradictory political decisions (such as his approval of a road that will be built from Brazil to

Peru, cutting directly through sacred indigenous land in the Amazon (Aguirre and Cooper, 2010, p. 240)).

Regardless of Morales' regime, an epochal moment has been reached in which indigenous empowerment is enough to disrupt the global status quo (Cadena, 2010). Indigeneity, though contested (Fabricant, 2013), has been tied to environmentally focused social justice movements, for better or worse, which creates a new platform to have discussions about complex ideas like climate change (Dove, 2006). The new-found discursive space afforded to indigenous groups opens doors for new modes of communication and reasoning (Stengers, 2005), and it is in this space where it is crucial to have discussions about how race, class, gender, and other forms of oppression coalesce and inform global problems.

This chapter has examined how race, class, gender, and climate change intersect and inform one another in the context of the Kallawaya tradition. By using the Kallawaya tradition as a lens through which to view these issues, I have made the case for a re-imagination of the ways in which race, class, gender, and climate change are discussed in relation to one another but also in the context of a tradition that is deeply impacted by each issue. While this chapter is in large part speculative, it is meant to encourage more discussions and invite new perspectives on the increasing threat of climate change and to critically engage with the myriad ways it exacerbates inequality. This chapter has teased out the complexities that inform the multi-dimensional situation of the Kallawaya tradition so that these issues – race, class, gender, and climate change – may be addressed by future researchers, policy-makers, and activists more holistically and effectively in future studies and work.

References

Aguirre, J.C. and Cooper, E.S., 2010. Evo Morales, Climate Change, and the Paradoxes of a Social-Movement Presidency. *Latin American Perspectives*, 37(4), pp. 238–244.

Albro, R., 2005. 'The Water is Ours, Cajaro!' Deep Citizenship in Bolivia's Water War. In J. Nash (ed.), *Social Movements: An Anthropological Reader*. Malden, MA: Blackwell, pp. 249–271.

Andean Cosmovision of the Kallawaya, n.d. UNESCO. Available at: www.unesco.org/culture/ich/RL/00048 (accessed January 30, 2013).

Bastien, J.W., 1978. *Mountain of the Condor: Metaphor and Ritual in an Andean Ayllu*. Prospect Heights, IL: Waveland Press.

Bastien, J.W., 1985. Qollahuaya-Andean Body Concepts: A Topographical-Hydraulic Model of Physiology. *American Anthropologist, New Series*, 87(3), pp. 595–611.

Bastien, J.W., 1987. *Healers of the Andes: Kallawaya Herbalists and Their Medicinal Plants*. Salt Lake City: University of Utah Press.

Barbier, E.B., 2010. Poverty, Development, and Environment. *Environment and Development Economics*, 15(6), pp. 635–660.

Bruun, H. and Elverdam, B., 2006. Los Naturistas – Healers Who Integrate Traditional and Biomedical Explanations in Their Treatment in the Bolivian Health Care System. *Anthropology and Medicine*, 13(3), pp. 273–283.

Cadena, M., 2010. Indigenous Cosmopolitics in the Andes: Conceptual Reflections beyond 'Politics'. *Cultural Anthropology*, 25(2), pp. 334–370.

Callahan, M., 2011. *Signs of the Time: Kallawaya Medical Expertise and Social Reproduction in the 21st Century*. Dissertation. Ann Arbor, MI: University of Michigan. Available at: http://deepblue.lib.umich.edu/bitstream/handle/2027.42/84575/molliec_1.pdf (accessed January 3, 2016).

Chelala, C., 2009. Health in the Andes: The Modern Role of Traditional Medicine (Part II). *The Globalist*. Available at: www.theglobalist.com/storyid.aspx?StoryId=7730 (accessed December 6, 2012).

Cott, D.L., 2003. From Exclusion to Inclusion: Bolivia's 2002 Elections. *Journal of Latin American Studies*, 35(4), pp. 751–775.

Dove, M.R., 2006. Indigenous People and Environmental Politics. *Annual Review of Anthropology*, 35, pp. 191–208.

Espiritu, F., 2012. *Identification of Potentialities to Local Adaptation to Climate Change Based Ancient Knowledge in Amarete (Kallawaya Culture)*. Unpublished Master's thesis. Main University of San Andres, Institute of Ecology.

Fabricant, N., 2013. Good Living for Whom? Bolivia's Climate Justice Movement and the Limitations of Indigenous Comsovisions. *Latin American and Caribbean Ethnic Studies*, 8(2), pp. 159–178.

Fontana, B., 2013. The Concept of Nature in Gramsci. In M. Ekers, G. Hart, S. Kipfer, and A. Loftus (eds), *Gramsci: Space, Nature, and Politics*. Sussex: Wiley-Blackwell, pp. 123–141.

Guayagua, G., Máximo, Q. and Angela, R., 2003. The Presence of Aymara Traditions in Urban Youth Culture: Tales About the Multiculture of El Alto, Bolivia. In T. Salman and A. Zoomers (eds), *Imaging the Andes: Shifting Margins of a Marginal World*. Amsterdam: University of Amsterdam Press, pp. 288–299.

Gustafson, B., 2010. When States Act Like Movements: Dismantling Local Power and Seating Sovereignty in Post-Neoliberal Bolivia. *Latin American Perspectives*, 37(4), pp. 48–66.

Hardesty, D.L., 1972. The Human Ecological Niche. *American Anthropologist*, 74, pp. 458–466.

Healey, S., 2009. Ethno-Ecological Identity and the Restructuring of Political Power in Bolivia. *Latin American Perspectives*, 36(4), pp. 83–100.

Hoffmann, D. and Weggenmann, D., 2012. Climate Change Induced Glacier Retreat and Risk Management: Glacial Lake Outburt Floods (GLOFs) in the Apolobamba Mountain Range. In W.L. Filho (ed.), *Climate Change and Disaster Risk Management*. Heidelberg: Springer, pp. 71–87.

Janni, K.D. and Bastien, J.W., 2004. Exotic Botanicals in the Kallawaya Pharmacopoeia. *Economic Botany*, 58, pp. S274–S279.

Kaenzig, R., 2013. A Critical Discussion on the Impact of Glacier Shrinkage upon Population Mobility in the Bolivian Andes. Working Paper 7, Universite de Neuchâtel.

Kohl, B., 2010. Bolivia under Morales: A Work in Progress. *Latin American Perspectives*, 37(3), pp. 107–122.

Krippner, S., 2006. The Future of Ethnomedicine. Presented at the Congress of Ethnomedicine, Munich, Germany, October.

Krippner, S. and Glenney, E.S., 1997. The Kallawaya Healers of the Andes. *The Humanistic Psychologist*, 25(2), pp. 212–229.

Lupien, P., 2011. The Incorporation of Indigenous Concepts of Plurinationality into the New Constitutions of Ecuador and Bolivia. *Democratization*, 18(3), pp. 774–796.

McNeish, J., 2003. Globalization and the Reinvention of Andean Tradition: The Politics of Community and Ethnicity in Highland Bolivia. In T. Brass (ed.), *Latin American Peasants*. London: Frank Cass, pp. 228–269.

O'Hare, G. and Rivas, S., 2007. Changing Poverty Distribution in Bolivia: The Role of Rural–Urban Migration and Urban Services. *GeoJournal*, 68(4), pp. 307–326.

Oxa, J., 2004. Vigencia de la Cultura Andina en la Escuela. In C.M. Pinilla (ed.), *Agruedas y el Perú de Hoy*. Lima: SUR, pp. 235–242.

Postero, N., 2010. Morales's MAS Government: Building Indigenous Popular Hegemony in Bolivia. *Latin American Perspectives*, 37(3), pp. 18–34.

Quispe, W.A., 2012. Personal interview, December 7.

Robledo, C., Fischer, M., and Patiño, A., 2004. Increasing the Resilience of Hillside Communities in Bolivia: Has Vulnerability to Climate Change Been Reduced as a Result of Previous Sustainable Development Cooperation? *Mountain Research and Development*, 24(1), pp. 14–18.

Scherr, S.J., 2000. A Downward Spiral? Research Evidence on the Relationship between Poverty and Natural Resource Degradation. *Food Policy*, 25, pp. 479–489.

Sherbinin, A., Carr, D., Cassels, S., and Jiang, L., 2007. Population and Environment. *Annual Review of Environmental Resources*, 32, pp. 345–373.

Stengers, I., 2005. The Cosmopolitical Proposal. In B. Latour and P. Weibel (eds), *Making Things Public: Atmospheres of Democracy*. Cambridge, MA: MIT Press, pp. 994–1004.

Tathagatan, R., 2008. Bolivia at a Crossroads. *Economic and Political Weekly*, 43, pp. 26–27, 35, 37, 39.

Thomas, E., Vandebroek, I., Geotghebeur, P., Sanca, S., Arrázola, S., and Van Damme, P., 2008. The Relationship between Plant Use and Plant Diversity in the Bolivian Andes, with Special Reference to Medicinal Plant Use. *Human Ecology*, 36(6), pp. 861–879.

Tol, R.S.J., Downing, T.E., Juik, O.J., and Smith, J.B., 2003. Distributional Aspects of Climate Change Impacts. *OECD Workshop on the Benefits of Climate Policy: Improving Information for Policy Makers*. Available at: www.oecd.org/env/cc/2483223.pdf (accessed June 29, 2015).

Walsh, C., 2011. Afro and Indigenous Life – Visions in/and Politics. (De)colonial Perspectives in Bolivia and Ecuador. *Resvista de Estudios Bolivianos/Bolivian Studies Journal*, 18, pp. 49–69.

Warren, K.J. and Cheney, J., 1996. Ecological Feminism and Ecosystem Ecology. In K.J. Warren (ed.), *Ecological Feminist Philosophies*. Bloomington: Indiana University Press, pp. 244–262.

Watts, M. and Peet, R., 2002. *Liberation Ecologies: Environment, development, social movements*. New York: Routledge.

29 Sandcastle

Gabrielle Maughan

on an empty beach
in sunlight
I built my castle
the wind was my architect
together we sculpted
soft curves from the dunes
I found ribbons of seaweed
sprawling like handwriting
in the tideline of debris
washed from the sea of knowledge
with these I garlanded the walls
I made a roof from shells
that giggled stories about crabby hermits
and boring barnacles
someone has spilt black
tar on my castle
ink black sticky stains
that burn where they touch me
that burn

Part IV

Fire

Figure IV.1 Fire by Tina Shirshac.

Conflict   Mobilization   Heat   Resistance   Transformation

30 The struggle for praxis
Forging the uncertainty

Phoebe Godfrey and Denise Torres

As both metaphor and actual material transformer, we use Fire to represent Conflict, Mobilization, Heat, Resistance, and Transformation. From the molten core of the Earth and the fierce eruption of volcanoes to the theft of fire from the Gods, the burning bush on Mt. Sinai, and sacred fireside ceremonies and ritual; to the forging of metals and the secrets of Alchemy; to the fossil fuel revolution; to modern warfare; to the heat of our mammalian bodies; to a world ravaged by wildfires; and to the heating up of the Earth as a result of GCC, the theme of 'uncertainty' and the ways in which it is being forged within the struggle for praxis is explored in this section with particular attention to what exactly is being defined under the term 'praxis,' including questions of how 'we' will live on the Earth and how 'we' is being defined and by whom.

In the human quest for material empowerment through the creation of energy, oil and its sister coal have become the life-blood of the modern world even as their use, abuse, and theft come with an incalculable price. Recognizing oil as created by organic matter being squeezed and heated over millions of years, Julianne Norton's 'Small extinction' starkly suggests what can happen when it is carelessly extracted from the Earth's pulsing veins. Complementing this imagery is Cara Murray's poem 'Crude,' which captures the disastrous real-life consequences of the Valdez oil spill. Exploring aspects of the human price of oil, Defne Sarsilmaz's case study 'Şelmo oil field: A micro-site of global climate change and the global intimate' drills down to surface the complexities of who in Turkey pays – and how – for oil's pleasures.

Priyanka Borpujari takes us to India in 'Singing today, for tomorrow' – a personal narrative of her trials and tribulations as a journalist, set within the literary and visual context of rising temperatures and raised voices amid failing government policies. Such political turmoil links with Albert Fu's 'Global wildfire and urban development: blowback from disaster capitalism,' which looks at the actual experiences of wildfires that are becoming increasingly prevalent.

Meanwhile, Phoebe Godfrey's artwork, 'As the world melts,' captures the failure of the self-indulgent elite classes to intervene or even care about the state of the world, offering a modern interpretation of the anonymous French painting *Gabrielle d'Estrees and One of Her Sisters*, dated 1594.

Soraya Cardenas' 'Personal tale from the environmental wetback: rethinking power, privilege, and poverty in a time of climate change politics' brings us back to North America as she unpacks her personal struggle with the intersecting concepts of identity, power, privilege, poverty, and climate change and seeks to forge a more conscious way to live. In their contribution 'Climate Action Planning (CAP): an intersectional approach to the urban equity dilemma,' Chandra Russo and Andrew Pattison critique US urban climate mitigation strategies, demonstrating how they fail to transform the status quo but persist in privileging growth and resist more equitable approaches. The spoken-word poem by Prince Ea, 'Dear future generation,' invites us to question our current uncertain course with particular attention to the individual and the socially created conflation between having and needing, illuminating the dilemma that we may have all the treasures in the world which ignited fossil fuel has brought, but without a tolerable planet on which to enjoy them they won't ultimately be what we need. Finally, Mr. Fish's cartoon, entitled 'All Yours,' provides a more intimate twist on the question of what we are actually leaving to our future generations. Hence, the struggle burns on.

31 Crude

Cara Murray

Valdez crude coated the coast a deep galactic black. Bligh Reef shimmered,

iridescent in the moonlight, not unlike phosphorescence. Resembling a million wasps' wings woven together, a blanket of portentous error

spread over 11,000 square miles of Prince William Sound, Gulf of Alaska, Pacific Ocean.

It sealed feather against feather, tarred feet and tarnished lungs, froze into shore-shorn

ice and settled: thousands of gallons sit in sand and soil still – until

bilious bacteria break down the slick, hydrocarbon by hydrocarbon – bonds, branched chains, benzene rings – and integrate the insidious sheen – will

the opacity veil an individual forgiveness, or conceal a collective indifference –

32 Small extinction

Julianne Norton

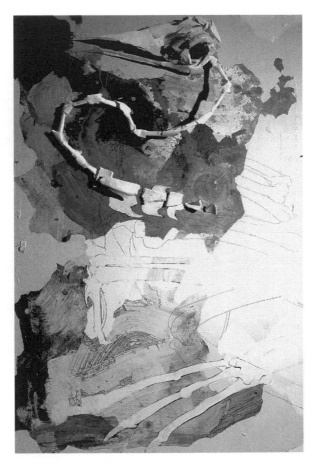

Figure 32.1 Small extinction, acrylic paint, ink, paper, and ceramics by Julianne
Norton. The subject of this piece is a gray heron, native to Nigeria.
Small Extinction is a fantasy narrative that anthropomorphizes the
heron to symbolize environmental, emotional, and psychological
havoc caused by the Delta oil spills in Nigeria. The heron skeleton
is fragile and vulnerable, and the ambiguous background represents
that tragedy based in Nigeria can be felt anywhere.

33 Şelmo oil field

A micro-site of global climate change and the global intimate

Defne Sarsilmaz

You know, these Americans, they come here thinking that they are better than us and they look down on us and call us 'dirty'. You know what happened the other day? This one American worker was urinating behind the trailer! There is a bathroom inside the trailer, why did he have to urinate behind the trailer? Not only is that unhygienic, but also disrespectful! Our land is not a toilet. Tell me, who is the dirty one here? The Americans like to call themselves civilized, but they are dirtier than the people they look down on!

During the summer of 2010, I traveled to the Şelmo oil field in Southeast Turkey as part of my job in human resources at an American oil company. This was the first time I had been to the predominantly Kurdish region. Along with my colleague, who frequented the region, I visited the company's drilling rigs and met with our foreign and local employees. Throughout our trip we held long talks over tea with local workers. I was intrigued by the narratives I had been hearing, such as the one above made by a Kurdish field manager working for an American oil company, causing me to leave the area with more questions than I had before.

These narratives were complex and messy. They were stories of tension: tension between family members, between American and Kurdish employees, between Kurdish guerrillas and foreign investors. I felt there were some deeper meanings that remained to be uncovered. There was something profoundly intriguing about the presence of a multinational corporation in rural Turkey and the friction it created between foreign employees and the local community. In an era with rising concern for global climate change (GCC), I could see how this little oil community was linked to climate change. What I saw was the global bleeding into the local and the embodiment of the global in intimate ways, from unequal pay to fist fights, from exchanging sexual favors for employment to the killing of security guards.

"The global intimate," as Mountz and Hyndman (2006) coin it, is the embodied experience of global trends, observable in the most intimate spaces. Through a focus on embodied experiences, we see that the global and the local are not separate, but continuously inform each other. Grand narratives of globalization tend to overlook the more intricate narratives of the local. The global is constructed as dominant and the local as a passive recipient of globalization. Mountz and Hyndman highlight the importance of an intersectional approach (Crenshaw, 1993; Collins, 1998) for the analysis of the global intimate; "Nationality, gender, race, religion, class, caste, age, nation, ability, and sexuality represent unequal locations within a web of relationships that transcend political borders and scale the global and the intimate simultaneously" (Mountz and Hyndman, 2006, p. 460). Feminist analysis of economic globalization tries to disrupt the grand narrative by breaking the global/local binary, jumping scales, undermining local and global essentialisms, and offering narratives that do not present their subjects as abstract, disembodied, political subjects (Mountz and Hyndman, 2006; Pratt and Rosner, 2006).

The relationship between global climate change and the oil industry is a crucial one: "Oil provides 40 to 43 percent of all energy used by the world" and "Oil and coal each account for 40 percent of global warming emissions from fossil fuels worldwide" (The Price of Oil: Global Warming, n.d., n.p.). Scientific research thus far has focused on environmental consequences, publishing on topics such as climate change-induced hurricanes and droughts, human health impacts, and rising sea levels, but social impacts of GCC have yet to be excavated. Hence, while the environmental impact of climate change has been rightfully analyzed, the human experience lacks articulation (Zehner, 2012). This situation calls for grounded research focusing on how populations unequally located and across different intersections are linked to climate change.

Oil exploration and production (E&P) activities around the world are directly linked to GCC and provide an ideal example for the global intimate. As Oil Change International states, "Any strategy to address climate change must by definition involve the phase out of these fossil fuels, and 'just transition' strategies need to be negotiated for the labor forces and communities that currently produce them" (The Price of Oil: Global Warming, n.d., n.p.). We therefore cannot overlook the immediate sites of oil E&P, and how oil-producing communities are affected by the presence of oil companies. Paying attention to these sites enables us to understand the intricacies of the stage where oil companies act out their neoliberal scripts that directly impact GCC. This chapter shifts the attention from metanarratives to the narratives of micro-sites by moving beyond the local/global binary, thereby addressing an important gap in current literature and articulating the intimate human experience linked to global climate change.

I seek to disentangle some of these narratives by problematizing the local/global binary through the global intimate (Mountz and Hyndman, 2006), taking an intersectional approach (Collins, 1998), and using a postcolonial feminist theoretical lens (Sharp, 2003). These particular lenses provide a much deeper understanding and richer analysis of the oil sector in Southeast Turkey than a potential analysis provided by a traditional political ecology or political geography perspective, because they get into the daily, embodied, and often overlooked experiences. Studies on how oil E&P transform rural communities tend to dichotomize the global and the local, and present rural communities as passive recipients of the global neoliberal project. Hence, one of the main questions I seek to answer in this chapter is: how can the global/local binary be problematized through a feminist analysis? By employing postcolonial feminist theory (Sharp, 2003) and putting it into conversation with intersectionality and the global intimate, I try to complicate the local/global binary and its essentialisms. I surface local agency by explaining how various subjectivities and groups constructed by different nationalities, ethnicities, genders, classes, and political projects intersect and clash while trying to gain access to and control over oil. Finally, in order to see the global intimate at work, I trace the global neoliberal project by looking at how Turkey started partaking in the project through privatization, which consequently opened up its borders to foreign oil companies. This study is important not only because there is a gap in the literature on the Turkish oil industry, but also because it contributes to previous works on the "global intimate" (see Mountz and Hyndman, 2006; Pratt and Rosner, 2006; Wright, 2009) and the socio-environmental effects of global climate change, which is an emerging field of study.

Implications

The comprehensive analysis I provide through in-field experiences and the oil discourse suggests that global climate change is co-produced and co-sustained by transnational companies and coerced local communities. Any action to transition out of the catastrophes of GCC will have to address the local complexities and inequalities, as the case of Şelmo shows their multifaceted entanglement. How can this be done? To begin with, global climate change needs to be contextualized within the region and its history. Second, the various intersections need to be taken into consideration. Finally, any attempts at transitioning out of GCC need to account for the intentionally established dependency of the local communities on the global market.

Methods

This chapter employs qualitative methods, specifically participant observation, content analysis, and critical discourse analysis. Three sets of primary sources were triangulated: (1) field notes from participant observation; (2) a range of websites on petroleum law, exploration, and production; and (3) online news sources. In addition to my observations of the interactions between local and foreign employees during my trip in 2010, I systematically analyzed several Turkish public and government websites in trying to gather historical information on oil E&P in Turkey and understand the oil discourse. The General Directorate of Petroleum Affairs (PIGM) website and the Turkish Petroleum Corporation (TPAO) website were the main sources I relied on for extensive background information. While analyzing these websites, I paid particular attention to (1) language, (2) changing patterns in the oil industry in Turkey, and (3) how the oil industry was portrayed. Finally, I analyzed news articles that covered the attacks on oil companies and their employees in Southeast Turkey.

Setting the scene: history of oil E&P in Turkey

Since the discovery of the first oil field in Raman in 1945 there have been 120 oil and natural gas fields discovered in Turkey, and since 1954 there have been 200 national and foreign companies engaged in E&P operations with 27 continuing operations today, of which five are national companies (Satman, n.d.). Natural gas production is mostly in the Thrace basin, while oil production is mostly found in the Southeast Basin, the latter being the primary focus of this chapter.

Southeast Turkey is heavily populated by Turkey's Kurdish minority, and Şelmo is a prime example of a Kurdish town. For the past three decades there has been ongoing conflict between Turkish military forces and the Kurdistan Workers' Party (PKK), whose goal is to establish an independent Kurdish state in this region. The news articles I analyzed confirm that oil E&P operations put oil companies into direct confrontation with the PKK, who lay claim to Southeast Turkey.

The Kurdish struggle must be situated within the context of postcolonialism in the Middle East. Following the collapse of the Ottoman Empire after World War I, Turkey fought Western colonial powers, and consequently the Republic of Turkey was founded in 1923. The Kurds in the region were unable to obtain backing from European powers to establish their own state; hence they became ethnic minorities in Turkey, Syria, Iraq, and Iran (Ergin, 2012). The Turkish state, founded by the first president Mustafa Kemal Ataturk, took modern Western nation-states as a basis for nation-building. 'Traditional' values were abandoned to a certain extent (the Arabic alphabet was exchanged for the Latin; headscarves were banned in public institutions; women were given the

Map of oil and natural gas basins in Turkey (Transatlantic Petroleum)

Figure 33.1 Map of oil and natural gas basins in Turkey (Transatlantic Petroleum).

right to vote, etc.) and European values were incorporated into the system. Turkish citizenship was primarily constructed around Turkish ethnicity and Sunni belief, marginalizing the various ethnic and religious minorities in the country (Ergin, 2012).

On agency, the global intimate, and intersectionality

Pratt and Rosner define the intimate as "a politicized sphere of overdetermined meaning that feminists can explore and use as a basis for thinking about globalization," and they believe "feminists have the responsibility to approach the world in intimate ways" (2006, p. 21). Feminist analysis unsettles macro-scale narratives of globalization by moving beyond the global/local binary, destabilizing essentialisms and presenting alternative narratives. Both Mountz and Hyndman (2006) and Pratt and Rosner (2006) argue that instead of abandoning scales, they should be used to dismantle hierarchies. Thinking of the body as the finest scale is a constructive way to explore the global as intimate, since it illustrates how global inequalities are reproduced at the level of the individual and the household. This chapter seeks "to show how the intimate and the global intertwine, to try to disrupt grand narratives of global relations by focusing on the specific, the quotidian, and the eccentric" (Pratt and Rosner, 2006, p. 15).

In her analysis of the overlaps between feminist and postcolonial theory, Sharp argues, "In feminist and postcolonial attempts to challenge the binary logic of identity formation, alternative understandings based around fluidity, movement and hybridity have emerged" (2003, p. 61). I draw on this challenge when referring to the local/foreign binary in understanding the tensions in Şelmo. I will not go into the details of identity formation here, but we must keep in mind that subjectivities are not fixed and constantly shift over time and space when looking at how alliances are made and unmade between different groups in Southeast Turkey.

Within the political ecology body of literature, a number of works stand out for their examination of the interplay of geography, history, and identity politics. In *Understories*, Kosek (2006) attends to the complexities of race, ethnicity, and class, showing how resource-rich areas become a crossroad of racism, ethnic tensions, class divisions, cultural identity, and nationhood as a result of extraction processes. I argue that such sentiments, along with the profit associated with oil, add to the messiness of the already historically racialized Southeast Turkey. Kosek problematizes the victim/perpetrator binary, showing that there are combinations of both. Likewise, the concept of 'geographies of terror' found in Escobar's (2008) work in the Colombian Pacific, and Watts et al.'s (2004) work in the Niger Delta, is helpful in understanding the ethnic conflict in Southeast Turkey. In *Territories of Difference*, Escobar's discussion on territory and territorial attachment helps explain Kurdish identity and their historical attachment to Southeast Turkey. Just as Escobar points out how all the different actors in the Colombian Pacific add to the tensions, a similar argument can be made for the Kurdish guerrillas, the Turkish army, the Kurdish local communities, and capitalist multinational corporations in Southeast Turkey. Watts depicts the violence associated with 'black gold,' which he calls 'petro-violence,' while also focusing on social movements. Similar to my claims in this chapter, Watts and colleagues find that in the Niger Delta "the conflicts at the local level emerge from challenges to customary forms of community governance, precipitated by the presence and activities of oil operations, and expressed through struggles over land rights, and access to company rents and resources" (Watts et al., 2004, p. 3).

Although there are examples such as Kosek, Escobar, and Watts' work that are sensitive to local messiness, political geography in general has paid little attention to intersections between race, class, gender, nationalism, etc. (Valentine, 2007). Valentine points out: "the concept of intersectionality that is used to theorize the relationship between different social categories: gender, race, sexuality, and so forth ... this concept of intersectionality has received little attention in geography" (2007, p. 10). This chapter contributes to the literature on intersectionality and local agency within the discourse of neoliberal globalization.

Analysis of the Turkish oil discourse

Representations of the local/global binary

Within a neoliberal understanding, modernization, transformation that signals 'progress,' and incorporation into the global market economy are all desirable outcomes of oil exploration and production. A 2012 IMF survey suggests that Central African countries (Chad, the Central African Republic, Cameroon, Equatorial Guinea, Gabon, and the Republic of Congo) "have been plagued with low levels of growth, widespread unemployment, and high levels of poverty in part due to ineffective management of their oil resources" (IMF, 2012, n.p.). Sharmini Coorey, Director of the IMF's Institute for Capacity Development, says, "If the oil money was used well, it would be transformative. They could, at least, catch up with countries at comparable levels of income" (IMF, 2012, n.p.). He believes that in these countries, "the transformative factor of oil has been limited." Nigeria, which struggles in managing oil effectively, is given as an example of how oil wealth is not always a blessing. The report concludes with familiar neoliberal language:

> The business climate must be improved. Governments should have a more transparent process for setting up businesses; try to eliminate corruption, and improve the investment climate for businesses and for private investment. It is also important to remove trade barriers. The community has very high common external tariffs.... There are also surcharges in addition across borders. Furthermore, there need to be greater efforts to improve labor mobility.

> (IMF, 2012, n.p.)

These are some of the perspectives and desires of international investors and multinational corporations. But what about local communities that are expected to go through these transformations?

In one of its reports, Oil Change International states, "There is an alarming record of human rights abuses by governments and corporations associated with fossil fuel operations, resulting in appropriation of land, forced relocation, and even the brutal and sometimes deadly suppression of critics" (The Price of Oil: Human Rights, n.d., n.p.). The report gives brief descriptions on the negative impacts of oil E&P in different regions of the world, including North America, Africa, and South Asia. The report concludes:

> Around the world minorities, indigenous peoples and local communities struggle, under increasingly difficult odds, to protect their lands from oil and gas extraction. Anger and marginalization, as well as poverty, air, land, and water pollution, and militarization continue in oil producing communities.
> (The Price of Oil: Human Rights, n.d., n.p.)

A report published by a group of American, non-governmental delegates based on their trip to the Niger Delta has found significant environmental, political, and social degradations due to oil E&P. A striking conclusion was how "the presence of multinational oil companies has had additional adverse effects on the local economy and society, including loss of property, price inflation, prostitution, and irresponsible fathering by expatriate oil workers" (Oil for Nothing, 2000, p. 3). Some further findings of the delegation are as follows:

> Organized protest and activism by affected communities regularly meet with military repression, sometimes ending in the loss of life. In some cases military forces have been summoned and assisted by oil companies. Reporting on the situation is extremely difficult, due to the existence of physical and legal constraints to free passage and free circulation of information. Similar constraints discourage grassroots activism.
> (Oil for Nothing, 2000, p. 3)

What we see here is the dichotomization of the global and the local, discussed as separate homogeneous entities. Although assumedly well intended, the above representations of the local and the global essentialize rural communities as the victim and the multinational companies as the vicious exploiter. While there is a plethora of findings confirming these views, alternative narratives remain unnoticed. A more fruitful way of looking at these cases would be to take a feminist and intersectional approach to the specific sites of production.

Privatization: accommodating neoliberalism

Until 2007, Turkish law regarding petroleum E&P was strictly geared toward protecting national interests. Under the 1954 Petroleum Law, Article 12.1 reads:

> No person existing by virtue of law in which a foreign state holds a financial or beneficial interest of such extent or in such form as directly or indirectly to influence his actions, and no person acting for or on behalf of a foreign state may (a) hold a petroleum right or conduct a petroleum operation, or (b) establish or operate installations incidental or to forming part of a petroleum operation.
>
> (PIGM, n.d., n.p.)

This basically restricts petroleum operations to Turkish entities only. An amendment to Article 12.2 in 1973 states that exceptions can be made by the Cabinet as long as requests are in accordance with national interests. Moreover, Article 13.1 of the same law states:

> The petroleum right holders are entitled to export, both in the form of crude and products, 35% of the entire crude and natural gas which they have produced from onshore fields and 45% from offshore fields discovered after 1st January 1980. The remaining part and the entire crude and natural gas produced from fields discovered before 1st January 1980 and the products obtained therefrom shall be allocated for the *needs of the country*.
>
> (PIGM, n.d., n.p., emphasis in original)

This law was meant to protect national interests by allowing only part of the oil and gas production to be exported, thus making Turkey's petroleum need the primary concern.

Interestingly, the Petroleum Law on the PIGM (General Directorate of Petroleum Affairs) website did not include the amended version of the Petroleum Law. Through further research, I found out that Article 5 was officially amended in 2007 to better accommodate foreign interests. Thus, while the 1954 Petroleum Law was geared toward reserving oil for the national interest, the 2007 Law is geared toward expanding commercial interests for foreign oil companies. According to a leftist scholarly website, the new Petroleum Law accepted on January 17, 2007 got rid of Section 5 entitled "Protection of National Interest," and renamed it "National/Foreign Private Company Interests" (Oztaskin, 2007). Moreover, the author writes that Article 12.2 no longer uses the wording "as long as the request is in agreement with national interests," making it easier for foreign companies to apply for petroleum E&P rights and operations while the quota on oil and gas exports has been removed, therefore overlooking national needs.

The most recent amendment to the Petroleum Law was made on June 11, 2013. The final version of the law does not include any restrictions for foreign oil companies; neither does it try to protect national company interests. Under the Amended and Abolished Provisions, Article 27.4 very clearly makes no distinction between "local and foreign petroleum right holders and Turkey branches of foreign companies" (PIGM, n.d., n.p.).

Foreign presence in Southeast Turkey and tensions that arise

The Southeast region is currently the petroleum-richest area in the country and attracts a number of foreign E&P companies. This chapter focuses on the tensions on the field (rural), although, from my personal experience, there are tensions in urban spaces as well. My purpose here is not to victimize the local population or vilify foreign oil companies, but to show the complexity of tensions that arise due to their co-presence. In some cases locals (Kurds) and foreigners collaborate; in some cases they clash. My goal is to simply try to disentangle some of these clashes by paying attention to where and how these interests and tensions intersect, overlap, and even diverge.

Exploitative globalization

There are three types of workers on the field, most of whom are male: foreign employees (mostly American), Turkish employees from other parts of the country, and local Kurdish employees. Drilling rigs are in rural areas, where one is removed from urban life. All fieldworkers work on a rotation basis. This means that foreign employees spend one month on the field and one month off the field. The month off the field is spent in the city (Istanbul or Ankara); in some cases American employees fly back to the USA, although this can become very costly. Turkish and local Kurdish employees work on the rigs for 20 straight days, where they stay in modest trailers, and go home for ten days. Foreign employees have the luxury of staying in local hotels if the rig is close to the city. If they work in remote areas, the company provides them with luxurious trailers.

The difference between an American's salary compared to a local's is dramatic. For example, the typical salary of an American toolpusher is around US$15,000 a month, whereas that of a local toolpusher is US$5000 a month. Due to high unemployment rates in the region, foreign companies offer low wages to the locals and do not raise salaries annually, further exploiting already comparatively cheap labor. Bringing in workers from other parts of Turkey or from outside of the country instead of hiring local workers perpetuates inequality locally.

Americans only: foreign privilege

During my visit to the region in 2010, many of the local employees who worked for the company complained about how they were treated by the American workers. In one incidence, when I was shown around an air-conditioned trailer that had cable television and other perks, the locals told me that an American had chased them out of the trailer, telling them it was for the use of Americans only. Another incident was more violent. Two intoxicated American employees provoked a fight by cursing out a field manager. While the Americans were sent to jail in the latter case, these incidents show a sense of American superiority over the locals, which at times translates into verbal arguments or even physical violence.

Another point of tension is cultural difference. Americans who come to work in the region may be ignorant of local customs, traditional practices, and religious lifestyle. Making fun of the call to prayer, verbally assaulting local women, and consuming alcohol on the streets are some actions displayed by Americans, and which are not welcomed within certain conservative spaces in the host society.

Family against family

One of the most intimate ways the global is embodied is in the household, where it seeps into family relations. Control over resources becomes a major point of conflict, especially among males in an extensive family. Upon getting E&P licenses, companies offer a generous sum to property owners for leases that can run for decades. Given the rural structure of the majority of the region, official land titles may not apply to every property. In some cases, families quarrel over untitled properties, each claiming to be the rightful owner. Such disputes are either resolved informally among family members or end up in legal battle. In more extreme cases physical violence may erupt, resulting in family members murdering each other. Given the patriarchal family structure of the Kurdish household (Hassanpour, 2001; Him and Ho gör, 2011), these family disputes are male-centered. We can imply that quarrels over land titles, threats against family members, and more extreme cases of killings are for the most part carried out by male actors. More research would be required to observe how these tensions are articulated among women.

Local resistance

Drawing on local agency, in this section I would like to elaborate on some cases where local employees resist foreign companies through certain tactics. It is not uncommon for a local to see a foreign company as a 'mint' – a term specifically used in a conversation between a Kurdish and Turkish employee – and to take advantage of the company for financial gain. Indeed, there were 35 local employees who did not show up for work over several months while continuing to receive their monthly salaries from a foreign oil company. The company had to issue a court notice to coerce them into coming to work. I have also heard of instances where local landowners impair oil pipelines that run through their property, causing leakage and demanding legal and financial compensation.

Such instances may be read as resistance tactics, which disrupt the overdetermined narrative of the passive native population as the victim. Local populations form strategic alliances to create spaces of resistance. These alliances may form among the locals or between locals and foreigners. No doubt, there are unequal power structures in place, but it is important to pay attention to interruptions in these power flows and account for resistance tactics, even if they are minuscule.

The masculinized field

During the short period of my visit to the oil fields, I did not encounter another female. The absence of women not only presents the field as a masculinized space, but is also telling of the construction of gender roles. To begin with, this means that women are excluded from high-paying jobs. It also means that men have the freedom and mobility to leave the household, while women are left to attend the household. Thus we can imply that in this scenario, globalization is further inscribing patriarchy and reconstructing normative gender roles.

The global also becomes embodied in the female body through sex work. As in the case of the Niger Delta, the presence of foreign males and the influx of men in the region creates a demand for female sex workers. Yet another way in which women's bodies are commodified is through the construction of their bodies as access points for employment at foreign oil companies. In some instances, local men 'hook up' a local woman with a foreign employee, and in turn the local man secures a position within the company for himself or for a friend/family member. In short, we see a heavily gendered and sexualized space that is the oil field.

Southeast Turkey as a racialized region of conflict

Mollett and Faria describe racialization as "a process whereby human differences are accorded differential treatment based upon hierarchal and stereotypical discourses and imaginings" (2013, p. 3). They explain how "Processes of racialization unveil their historical constructions, which are built on presuppositions infused in racial discourse" and that "these discourses are shaped by the past while simultaneously drawing upon new and contemporary projects without decentering racialized hegemonic orderings" (2013, p. 3). The concept of racialization is crucial in understanding the Kurdish experience and the Turkish–Kurdish conflict, which contributes to the pool of tensions.

Turkish citizenship is built around being ethnically Turkish and religiously Sunni, marginalizing minority groups of various ethnicities and belief systems (see Hassanpour, 2001; Secor, 2007). Recent scholars argue that the Kurds are in fact a racialized group, facing discrimination based on phenotypical, cultural, and political identifiers (Ergin, 2012). Restrictions placed on Kurdish cultural practices by Turkish hegemonic powers has produced a backlash, especially in Southeast Turkey with the Kurdish minority demanding cultural/ethnic recognition, equal citizenship, and freedom to embrace their identity. There are Kurds who have a more ambitious goal, fueled by cultural and political aspirations, which is the creation of a Kurdish state: Kurdistan. In Turkey, the PKK (Kurdish Workers' Party) has taken on the mission of establishing a Kurdish state in Southeast Turkey (Ergin, 2012). The fighting between the PKK and the Turkish Army for the past three decades has cost thousands of lives, displacement, and human rights violations.

More recently, there have been tensions between the PKK and foreign oil companies in Southeast Turkey. The PKK claim Southeast Turkey and parts of

Eastern Turkey as their territory and therefore expect foreign oil companies to obtain permission from them for any E&P operations. A field informant provided a specific example. In 2012, a foreign oil company in Şelmo received a threatening letter from the Kurdish guerrillas which ordered the company to apply for E&P permission from the PKK within the next ten days or else there would be consequences. Seeing this threat as a bluff, the company ignored the request and took no further action. Once the time was up, the company's helicopter was shot at. After this incident, the company closed down all operations and had to release all local employees. A similar incident happened to the oil company I worked at in 2010, according to employee informants. An oilrig was shot at and two transporters and two compressors were burned down, costing the company around US$1 million in damage. Apparently, the company had received a threatening call from the PKK earlier, saying that they were the ones responsible for Southeast Turkey. People are saying that the PKK wants money from foreign companies. For protection, foreign companies hire security personnel, which have proven to be ineffective. These private security companies are seen as 'leeches,' reinforcing the perception of foreign oil companies as 'mints.'

My news analysis confirmed my field notes about the attacks. One Turkish newspaper article reported on an attack on our company's oilrig in 2011 in Southeast Turkey. PKK guerrillas had gunned down three private security officers. The lack of action over the incident caused protests in the town, where victims' families complained that the government had done nothing to bring the murderers to justice eight months after the attack (Ruzgar, 2012). The article also mentions a previous incident back in 1992, when PKK fighters killed three engineers (two Turkish and one foreign) working for Shell, and kidnapped 12 workers who were held for 15 days (Batman'da Petrol, 2011). A BBC article (Explosion in Turkey, 2012) also reports on a pipeline explosion, blamed on the PKK by Turkish officials.

Conclusion

Case studies on micro-sites such as the Şelmo oil field can teach us about the larger issues relating to the global oil industry by showing the interconnectedness of the local/global. Therefore, efforts to minimize the impacts of GCC need to address local politics, ethnic tensions, gender roles, territorial disputes, etc., because these processes are affected by and feed into global practices. Any propositions for a just transition to green energies and community resilience need to focus on building more localized economies.

The oil industry in Turkey remains open to further research. For a more empirical study, I propose going back into the field and exploring Southeast Turkey as a racialized region, while paying closer attention to gender roles and relations. As preliminary research, this work recognizes some of the local tensions that arise due to foreign oil E&P operations in Southeast Turkey, tries to disentangle some of the tensions through a feminist analysis, and shows the micro-sites of GCC. It will be interesting to see how the way the government handles the 'Kurdish Question' may potentially change the picture of the oil industry in Southeast Turkey, especially considering the potential independent state in Northern Iraq that borders the region. For the time being, this study is an important step toward contextualizing global climate change within the region and its history, taking the various intersections into consideration, and taking into account the intentionally established dependency of the local communities on the global market, so that we can begin transitioning out of global climate change.

References

Batman'da Petrol Sahasina Silahli Saldiri: 3 Olu, 2011. *Vatan Gazetesi (Vatan Newspaper)*. Available at: http://haber.gazetevatan.com/batmanda-petrol-sahasina-silahli-saldiri-3-olu/412977/1/Haber (accessed November 24, 2011).

Collins, P.H., 1998. It's all in the family: Intersections of gender, race, and nation. *Hypatia*, 13(3), pp. 62–82.

Crenshaw, K., 1993. Mapping the margins: Intersectionality, identity politics, and violence against women of color. *Stanford Law Review*, 43, pp. 1241–1299.

Ellis, H., 2005. The Baku–Ceyhan pipeline: BP's time bomb. *CorpWatch*. Available at: www.corpwatch.org/article.php?id=12340 (accessed June 2, 2005).

Ergin, M., 2012. The racialization of Kurdish identity in Turkey. *Ethnic and Racial Studies*, pp. 1–20.

Escobar, A., 2008. *Territories of Difference: Place, movements, life, redes*. London: Duke University Press.

Explosion in Turkey hits Iraq Oil Pipeline, 2012. BBC. Available at: www.bbc.co.uk/news/world-europe-18939172 (accessed July 21, 2012).

Fluri, J., 2008. Feminist-nation building in Afghanistan: An examination of the Revolutionary Association of the Women of Afghanistan (RAWA). *Feminist Review*, 89, pp. 34–54.

Hassanpour, A. 2001. The (re)production of patriarchy in the Kurdish language. In S. Mojab (ed.), *Women of a One-state Nation: The Kurds*. San Francisco, CA: Mazda Publishers.

Him, M.S. and Hoşgör, A.G., 2011. Reproductive practices: Kurdish women responding to patriarchy. *Women's Studies International Forum*, 34, pp. 335–344.

International Monetary Fund (IMF), 2012. Oil revenue has power to transform countries of Central Africa. Available at: www.imf.org/external/pubs/ft/survey/so/2012/car100312a.htm (accessed October 3, 2012).

Kosek, J., 2006. *Understories: The political life of forests in Northern New Mexico*. London: Duke University Press.

Mollett, S. and Faria, C., 2013. Messing with geography in feminist political ecology. *Geoforum*, 45, p. 116.

Mountz, A. and Hyndman, J., 2006. Feminist approaches to the global intimate. *Women's Studies Quarterly*, 34(1–2), pp. 446–463.

Oil For Nothing, 2000. Multinational corporations, environmental destruction, death and impunity in the Niger Delta, 2000. *Essential Action*. Available at: www.essentialaction.org/shell/report/ (accessed January 25, 2000).

Oztaskin, M., 2007. Adi Turk, Kendisi Yabanci Petrol Kanunu Yasalasti. *Acik Istihbarat*. Available at: www.acikistihbarat.com/Haberler.asp?haber=6291 (accessed January 29, 2007).

Petroleum Law (PIGM). *General Directorate of Petroleum Affairs*. Available at: www.pigm. gov.tr/english/index.php (accessed November 20, 2012).

Pratt, G. and Rosner, V., 2006. Introduction: The global and the intimate. *Women's Studies Quarterly*, 34(1–2), pp. 13–24.

Profile of Internal Displacement: Turkey, 2004. Global IDP Project, *UNHCR*. Available at: www.unhcr.org/refworld/pdfid/3bd98d600.pdf (accessed April 5, 2004).

Ruzgar, V., 2012. Selmo Olayi Neden Aydinlatilamiyor? *Medya73*. Available at: www. medya73.com/selmo-olayi-neden-aydinlatilmiyor-haberi-997639.html (accessed July 19, 2012).

Saka, Y., 2006. Ozellestirme ve Calisanlar: Tupras Ornegi. *Belgeler*. Available at: www. belgeler.com/blg/pnn/zelletirme-ve-alianlar-tpra-rnei-privatization-and-employees-tpra (accessed November 20, 2012).

Satman, A., n.d. Turkiye'de Petrol. Istanbul Technical University, Petroleum and Natural Gas Engineering Department. Available at: http://web.itu.edu.tr/~pdgmb/ documents/turkiyedepetrol.html (accessed November 20, 2012).

Secor, A., 2007. Between longing and despair: State, space, and subjectivity in Turkey. *Environment and Planning D: Society and Space*, 25(1), pp. 33–52.

Sharp, J., 2003. Feminist and postcolonial engagements. In J.A. Agnew, K. Mitchell, and G. Ó Tuathail, *A Companion to Political Geography*. Hoboken, NJ: Wiley Blackwell, pp. 59–74.

Steward, J., 1959. *The Concept and Method of Cultural Ecology*. Urbana: University of Illinois Press.

The Price of Oil: Global Warming, n.d. *Oil Change International*. Available at: http:// priceofoil.org/thepriceofoil/global-warming/ (accessed November 20, 2012).

The Price of Oil: Human Rights Violations, n.d. *Oil Change International*. Available at: http://priceofoil.org/thepriceofoil/human-rights/ (accessed November 19, 2012).

Turkey Country Profile, n.d. *TransAtlantic Petroleum*. Available at: www.transatlantic-petroleum.com/s/Turkey.asp (accessed November 19, 2012).

Turkiye'de Ozellestirme, n.d. *Word Press*. Available at: http://haticecalkap.wordpress. com/page/2/ (accessed November 20, 2012).

Turkiye'de TPAO, n.d. *Turkish Petroleum Corporation (TPAO)*. Available at: www.tpao. gov.tr/tp2/sub_tr/sub_icerik.aspx?id=29 (accessed November 20, 2012).

Valentine, G., 2007. Theorizing and researching intersectionality. *The Professional Geographer*, 59(1), pp. 10–21.

Watts, M., Okonta, I., and Von Kemedi, D., 2004. *Economies of Violence: Petroleum, politics and community conflict in the Niger Delta, Nigeria*. Working Paper No. 1. Berkeley, CA: Institute of International Studies, University of California.

West, P., 2012. *From Modern Production to Imagined Primitive: The social world of coffee from Papua New Guinea*. London: Duke University Press.

Wright, M., 2009. Gender and geography: Knowledge and activism across the intimately global. *Progress in Human Geography*, 33(3), pp. 379–386.

Yeni Petrol Yasasi Milli Menfaatlere Aykiri, n.d. *Enerji2023 Organization*. Available at: http://enerji2023.org/index.php?option=com_content&view=article&id=100:yen-petrol-yasasi-mll-menfaatlere-aykiri&catid=10:doalgaz&Itemid=24 (accessed November 21, 2012).

Yucel, O., n.d. 5574 Sayili Turk Petrol Kanunu'nun Incelenmesi. *Turk Hukuk Sitesi*. Available at: www.turkhukuksitesi.com/makale_526.htm (accessed November 21, 2012).

Zehner, O., 2012. *Green Illusions: The dirty secrets of clean energy and the future of environmentalism (our sustainable future)*. Lincoln, NE: University of Nebraska Press.

34 Singing today, for tomorrow

Priyanka Borpujari

I call myself an Indian. I call myself a Mumbaikar. I call myself an Assamese. I consider myself homeless. I consider myself of-this-Earth. What defines an identity? Is it merely the address on the passport, on the telephone bill, on the birth certificate – if there is one? Or is it about the people whose love and patient ears and mirthful hearts provide us with that sacrosanct space one can call home? Is it the work we undertake, the weather we choose to endure, the relationships we choose to maintain, or the place where we park our toothbrushes that makes a place 'home'?

This has been a quest even before I knew that was what it could be called. And to get to this understanding, as with everything else, I ought to return to remembering that....

I always wanted to be a painter as a child. I got crayons as birthday gifts. I wrote poems as a teenager in the leftover pages of the previous academic years' notebooks. I wanted to save the world with my journalism when I was a 20-something. Cameras were handed down to me from a cousin, and then better ones by two consecutive ex-boyfriends.

But I always bought my own railway tickets. The Indian railway website threw open the possibility to explore my country. Through the long days and nights on a rail route, with eyes fixated on the window as the trees flew by, I entered and exited new territories. India, I learned, was not merely a diverse country; it was a continent, and with every 150 km the landscape transformed into something entirely different and beautiful. The water, soil, air, trees, flowers, birds, food, people, human habits, words, dialects, handicrafts, beliefs, customs....

Notebook, camera, and enthusiasm in hand, I began to seek to understand how this diversity is being torn down to create a homogeneous idea of a place that is unrecognizable, and at worst unsustainable. One need not go too far to realize this painting of a singular color on a textured canvas. Take the weather, for instance. Every change of season brings initial cheer, and then the stench of death. Scores of Indians perish in the summer heat. Scores of Indians get washed away in the furious monsoon floods. Scores of Indians numb and freeze in the piercing winter. Scores of women, men, children, cattle, dogs are perishing in the heatwave of 2015. The mercury has been swaying wildly in the past few years, and so have the wind sails.

In recent times, each of the seasons has dawned upon us more acutely. Temperatures are soaring and dipping abnormally; rains tease farmers until the earth is scorched, or lash out until all that has ripened goes under a white bed of flood.

This wild dance of nature, worsened by humanity's inconsiderate lifestyles, is in stark visibility in the Indian countryside, and among India's poor. With an increasing consciousness of my place of immense privilege, I grew up learning that it was morally not acceptable that our world is so unjust. As an adult, and as a journalist, the moral denial was strengthened by mere facts: that a quarter of the world's children-under-five deaths are from India, and the rate of obesity in India is increasing too. About 400 million continue to survive on less than a dollar a day, while the third highest number of billionaires in the world live in India. But we know which statistics make it to the headlines more often.

My train sojourns have shown me the irreconcilability of these two Indias, which is evident by infrastructure, mining, dam, and energy projects that have been allowed their cancerous growth, with little or no concern for the ecology or the communities. Companies that are hailed internationally have been found to be violators of human rights, forcing peoples out of their homes and off their land. The government's procedures to assess the cost to such 'development' efforts are easily forged. For every one arrest or murder of an activist protesting a mine, a dam, a deforestation, a nuclear plant, there are at least 30 others whose whereabouts are unknown. Our police stations are filled with undocumented detainees.

Where, then, do I stand, as I see the beautiful and bitter in my country changing rapidly into a homogeneous grey space where malls grow taller, and the girth of slums around the malls grows wider? Could my stories of women, children, and men – of their struggles and triumphs, their dignity and loss – convey the immense importance of realizing oneself in the other, and the other in one's own self, and thereby a consciousness in living?

Through the decade of relentlessly pursuing my journalism, in trying to choose dignity before provocative headlines, even as some days and nights are determined to prove the pointlessness of it all, I have thus learned: people survive. They find the means; they find novel ways. They plan, they gerrymander, they coax, they protest, they go on hunger strikes, they pay petty bribes of a chicken or two, they write letters. Indigenous peoples raise their tattooed chins and puff out their decorated chests to protect their mountains from being razed, the rivers from being dammed or polluted.

The struggle for their survival – and for the survival of their future generations through the survival of the ecology – is as diverse as my country, my home. Somewhere, where city folk may not care to hear them sing, people are singing the song of the dark nights and the optimism of bright mornings. They sing so that home continues to be home. They know better to protect it whatever little they have, for they have seen the ungodly summers and the ferocious rains and dark winters that were never a part of their old tales. They look up to the sky, to the soil, to the rivers, and they know that 'climate change' – the buzzword that is new to them – has begun to change their lives, without their consent or doing.

In their struggles I have found my answer to my own identity and that luxury idea called 'purpose of life.' That borders are not defined in a human spirit is something I have soaked in as I embarked upon journeys to try to give a voice to those whom I assumed needed a voice. How foolish, and futile! The voices have always been there, but were only silenced. The idea of an identity and privilege

Figure 34.1 In the summer of 2011, about 20,000 women and men of indigenous peoples from across the state of Maharashtra walked for 15 days, to eventually culminate their walk in the state capital of Mumbai – also India's financial capital. The rally was christened 'Jungle Haq Sangharsh Yatra' (Forest Rights Movement Rally). This photograph (and the one overleaf) demonstrate a massive show of protest was born out of the state government's callous attitude toward the implementation of the Forest Rights Act of 2006. The Act had promised to rectify decades of injustice, and validate the right of indigenous peoples over the land and forest in which they have lived for generations. However, negligible justice has been delivered since. Of the 288,000 land claims, 170,000 had been rejected. Further, the approved claims did not promise sufficient land for an economic holding. Is this rejection partly to not grant those land holdings at all, and thus ensure that the government gets a free hand in acquiring those lands for large 'development' projects like mining and indiscriminate industrialization? When the people had none of it, the protest idea was germinated.

Figure 34.2 When the people marched in 2011 over 15 days, they walked toward the city where policies are made, and all along they received the warm welcome of the residents of villages they passed. They wore their identity with pride, in the foreign land of Mumbai city. They danced to the tunes that the forest taught them; they sang with instruments made from the bounty of the forest. They buy only oil and kerosene from the markets, because the forest provides everything else. But this was not the case when they finally arrived in Mumbai. The urban folk looked down upon them, and the Times of India – the largest selling newspaper in India, and in the world – complained about the traffic jam they had caused, which led to many people missing out on their appointments.

– or lack thereof – was always there, but only to be discovered, with every passing season, every thunder of the bulldozer, every new slogan, every new welcome into the home of a revolutionary who believed that my camera and words would change the world. That faith, in the face of my sometimes-wavering voice, has been the home I have been seeking. That is the home where stories are on a pedestal, as they potently flow through my words and lens.

35 Global wildfire and urban development

Blowback from disaster capitalism

Albert S. Fu

Introduction

This chapter examines the social and economic relationship between urban development and wildfire. Wildfires (also called bushfires) are increasing around the world due to global climate change (Westerling and Bryant, 2007; Krawchuk et al., 2009; Liu et al., 2010). In September, 2008, fires ravaged South Africa, Zimbabwe, and Mozambique, destroying hundreds of homes. In July 2009, France, Greece, Italy, Spain, and Turkey saw massive wildfires that resulted in tens of thousands of acres burned. In fact, the Mediterranean basin has seen an increasing number of fires (Pyne, 1997; FAO, 2003, 2011; Pausas and Fernández-Muñoz, 2012; Balaban and Fu, 2014). In the Western USA, California has seen a five-year average of 169,000 acres burned yearly between 2008 and 2013 (CalFire, 2013). In New South Wales, Australia, 290,000 acres burned in October 2013. Climate change also affects fire risk in Latin America and increasingly threatens the Amazon forest (Cochrane and Barber, 2009). This destruction has social consequences as well. For instance, in April, 2014, a wildfire in Valparaíso, Chile led to thousands of homes being destroyed and over 11,000 people left homeless.

This chapter will provide a critical perspective on wildfire's relationship with global urban development in three ways. The first is the problem of urban growth and vulnerability in peri-urban regions. Urban growth facilitates increased consumption, which in turn drives climate change. This creates a dangerous feedback loop, in which climate change harms the livelihoods of rural populations and in turn fuels more urban growth (Rees, 1992). There have always been wildfires and they have always affected cities and other human settlements (Bankoff et al., 2012). However, global climate change, alongside aggressive urban development policies, has dramatically altered the scale and scope of such conflagrations (Mell et al., 2010; Gill et al., 2013). Global warming has increased wildfire risk by worsening drought conditions, creating readily combustible fuel for fire, and longer fire seasons. At the same time, humans are increasingly encroaching upon bush, chaparral, forests, and other ecologies that are particularly prone to wildfire. This encroachment threatens life, property, and the environment.

Fire-prone regions typically have seasonal patterns with regard to bush and wildfire. This fire regime is particularly important in ecosystems in which fire encourages plant germination. However, planners and policy makers believe they can and should design away wildfire. Strategies combating wildfire reflect the belief that safety is an achievable reality through innovative technology and management. Mitigation plans include controlling brush and grass, as well as planning and development strategies (Pence et al., 2003; Mukheibir and Ziervogel, 2007; Rahlao et al., 2009). There are engineering strategies such as fire-resistant architecture, fire breaks, and warning systems (Linares and Linares, 1990; Sorensen et al., 2004; Stephens et al., 2009; Şahin and İnce, 2009). While tactically sound on paper, such strategies do little to address the underlying social and economic drivers of wildfire risk.

The second issue this chapter addresses is disaster capitalism, or the way in which disaster aids the capital accumulation process (see Klein, 2008). There is no doubt that capitalists benefit from real estate development in fire-prone regions, as well as providing the technology to manage disaster. Researchers, insurers, and developers are well aware that wildland–urban interfaces are at greatest risk of fire damage (Radeloff et al., 2005; Gill and Stephens, 2009; Mell et al., 2010; Mann et al., 2014). This, however, does not stop construction and development in the peri-urban fringe. For capitalists, disaster risk and management can be profitable. As lives, wild spaces, and property are destroyed by fire, capitalists profit from construction in deforested areas, new insurance policies, and the sales of new fire-resistant (but not fireproof) technologies.

Wildfire mitigation and defense are more than a technical or engineering problem. They are equally a social and political problem. Mitigation strategies are not neutral (Stetler et al., 2010; Paveglio et al., 2010). Thus, we need to ask questions such as: Who profits from building in fire-prone geographies? How much does it cost to manage brush? Who will pay for land use management? Which communities are prioritized when deploying firefighters?

Correspondingly, I will discuss the problem of social vulnerability through the notion of blowback (see Johnson, 2001). Typically, blowback is a term that describes the unintended consequences of military intervention or a covert operation. As in war, policy makers excited by technical solutions and the economic benefits of military victory employ strategies that create unexpected and harmful outcomes (Hewitt, 1983, 1997; Wisner et al., 1994; Freudenburg et al., 2009). A product of hubris, such policies ignore the social, political, and economic problems of *growth* – a Western capitalist ideology – resulting in blowback that exacerbates existing problems. In particular, we see that consequences fall upon the intersectional matrices of race, class, and gender. As an intersectional social problem, wildfire risk cannot be understood without examining the economic and political dimensions of this so-called natural disaster.

Urban growth and global warming

In Greek mythology, Prometheus stole fire from Olympus for humanity, which in turn inspired art, culture, and civilization. In response, to Prometheus' theft, Zeus ordered Hephaestus to craft Pandora, who later released the evils of humanity upon the opening of her box. Thus, fire became not merely a source of enlightenment and illumination, but also a weapon and source of destruction thanks to humanity's vices. The story of Pandora connects the story of fire to human failings. Outside of warfare, 'natural' disaster such as wildfire represents the greatest challenge to the physical nature of cities (Wisner et al., 1994; Coward, 2006, 2008). Fires damage homes and infrastructure, in addition to threatening life. Despite the role of human negligence, people treat wildfire as a force of nature – a natural element alongside earth, wind, and air. Arson, camp fires, cigarettes, electrical equipment failure, fireworks, waste incineration, and of course global climate change cause or exacerbate wildfire risk (FAO, 2003, 2011; Balaban and Fu, 2014).

According to Pyne, "fire ecology is human ecology" (2001, p. 115). Humans, fire, and nature are interconnected. City and countryside are environments that change over time thanks to human intervention. In turn, so do their fire regimes – the pattern in which fires occur in a region. Cities are thought of as the antithesis to nature and the rural countryside despite its reliance on natural resources (Heynen et al., 2006). However, developments in the rural countryside (including fire) shape cities. Global climate change exacerbates the risk of drought and food insecurity (Bohle et al., 1994). In turn, food insecurity encourages rural–urban migration, leaving a landscape that is at greater risk of wildfire due to its abandonment (Crush, 2013). Piñol and colleagues (1998), for instance, have noted land use changes, such as the abandonment of fields and grazing lands, as affecting the fire regime of Eastern Spain. Due to climate change, vulnerable biomes experience changes in vegetation that fuel risk (Adams, 2013). This creates a vicious feedback loop in which growing cities exacerbate the climate change that makes rural areas vulnerable – which in turn drives rural-to-urban migration.

As a product of deruralization, cities and urban growth are a part of a region's fire regime (also see Bankoff et al., 2012, pp. 10–11). Increased fire risk exacerbated by urban growth is not merely a social and environmental problem, but is also blowback for political and economic policies. Although, as Steinberg (2006) notes, policy makers typically assume that natural disasters are "Acts of God," this is not the case. Humans make voluntary or forced decisions to build homes and live in fire-prone wildland–urban interfaces. Humans build roads. Humans run power lines and burn fuel. Humans push other humans into vulnerable areas. Much of this is driven by the capitalistic drive for growth.

Urban growth, as an economic engine, is an anthropocentric ideology bound to contemporary capitalism (Molotch, 1976). However, the material and population growth of cities contributes to the global warming that increases wildfire risk (Westerling and Bryant, 2007; Coumou and Rahmstorf, 2012). The foundation of modern cities is literally cement and concrete. As one of the world's most con-sumed substances, concrete generates approximately 7 percent of all global carbon emissions (Ali et al., 2011). Ironically, concrete was touted in the early twentieth century as a fire-resistant material to counteract fire risk and encourage more con-struction (Fu, 2013). In addition to concrete consumption, cities are major pol-luters that contribute to climate change (Pataki et al., 2006; Grimm et al., 2008). Indeed, urban dwellers consume less than their suburban counterparts (Owen, 2009). Yet, *more efficient* consumption does not negate the problems associated with growth (Princen et al., 2002). This is why there has been a push for anti-sprawl and smart/no-growth initiatives (Scott, 2007). Half the world's population lives in cities – where consumption and growth thrives. As such, they are on the front line with regard to both the production of anthropogenic climate change and natural disaster.

Disaster capitalism

Capitalist-driven growth has a tendency to produce solutions to problems it created – and these solutions, in turn, create new problems, fueling a profitable cycle. However, part of our hubris is the belief that whatever blowback may occur will be minor, solvable, and profitable, or will not be in one's backyard. The aforementioned relationship between city and countryside is an example of this. Vulnerability to disasters and catastrophes is a part of what Gotham and Greenberg (2014, p. xi) call "crisis driven urbanization." Such urbanization creates a cycle of development strategies which encourage blowback that creates new disasters.

The literature on natural disaster promotes the notion of resilience in pro-tecting lives and property (Pelling, 2003b; Goldstein, 2008; Dooling and Simon, 2012). While resilience refers to a long-term and sustainable approach to dis-aster, it appears that the de facto response has been to mitigate short-term harm from wildfire and other calamities. In other words, disaster industries such as real estate and insurance have figured out ways to shift financial burdens and encour-age management (rather than reduction) of risk.

For instance, wildfire mitigation is often reframed as "resource management" (Gill and Stephens, 2009; McCaffrey and Rhodes, 2009). Given the profits to be made by developing such areas, disaster mitigation strategies are less about slowing growth than reducing damage to encourage growth (Balaban and Fu, 2014). Like Icarus, human fallibility places cities on the edge of disaster. Icarus, excited by his father's technological solution of imprisonment – wax wings – flew too close to the sun. Thus, many technocratic solutions not only reveal an anthropocentric hubris, but the consequence of overconfidence.

Calamity tied to capitalist urban growth is different from the idea that there are contradictions within capitalism that undermine sustainability, which threatens society (Schnaiberg, 1980; O'Connor, 1998; Foster, 1999; Moore, 2000; York and Mancus, 2009; Givens and Jorgenson, 2011). Rather, engrained within the economic system of capitalism is disaster and crisis. For instance, economists Skidmore and Toya have controversially noted a positive correlation between natural disasters and capital accumulation (Skidmore and Toya, 2002; Toya and Skidmore, 2007). While their explanation for the relationship is market oriented – emphasizing innovation and adaptation – critical scholars have compared natural disaster to the subprime mortgage bust. Natural disaster allows for developers to liquefy assets and move investments to more valuable regions of the city or elsewhere (Gotham, 2009). This is also apparent in the way the insurance industry has commodified risk and worked alongside disaster management firms to maximize the returns on urban growth (Steinberg, 2001; Sturm and Oh, 2010).

Around the world, nations, regions, and cities increasingly mitigate disasters through a number of solutions. Responses to wildfire range from brush management to evacuation, secure design, and warning systems (Pence et al., 2003; Mukheibir and Ziervogel, 2007; Rahlao et al., 2009; Paveglio et al., 2010; Fu, 2013). While the politics of resource management shape wealthy nations' responses to wildfire, the lack of organization and resources intensify wildfire risk in the developing world. As Sorensen and colleagues (2004) point out, protective action is resource dependent. Decisions as to where early warning systems should be installed, who is evacuated and to where, and funding of fire suppression are all policy decisions which reveal the inequality that exists in society.

Due to the correlation between physical and social vulnerability, those most affected by wildfire are those who have the least power in society. Who lives where, and therefore has access to what kinds of resources, are affected and shaped by race, class, and gender (Klinenberg, 2003; Hartman and Squires, 2006). Those who are most vulnerable may have limited access to insurance, resources to build, or even protection from the fire-storm. A European Investment Bank (EIB) working paper discussing strategic adaptation to climate change in Europe asks a fundamental question, "who should adapt?" (Fankhauser and Soare, 2012). In an ideal situation, public and private actors *should* work together to protect those who are vulnerable to disasters such as wildfire. However, in general it is the public who suffers from blowback. For example, political aversion to public expenditure on wildland management and fire suppression has intensified wildfire risk in the USA. This is a product of policy that shifted the burden of fire protection from state agencies to local communities. The consequence is privatized protection for those who can afford it. These contractors are often employed not just by the state agencies, but on behalf of insurance companies working to protect their clients located in wealthy areas (Funk, 2009). Fire suppression has become a profitable enterprise as private contractors are increasingly called upon to fight fires for the super-wealthy (Tao, 2011).

The aforementioned responses are seen as profitable opportunities included in the various economic calculations of insurance companies, builders, and financiers (Beck, 1992; Wisner et al., 1994; Mileti, 1999; Tierney, 1999). This is why Balaban and I (2014) have observed that wildfire does little to slow urban development in Turkey and California. In my own work, I have seen that wildfire risk or damage has a negligible influence on property values in wealthy neighborhoods. On the other hand, California wildfires are more likely to affect property values of poorer minority neighborhoods than wealthier, less diverse neighborhoods (Fu, 2013). In Turkey, wildfire 'conveniently' destroys wildlands, allowing for the re-designation of land for development. The result of such policies is the endless feedback loop of disaster and growth. In both cases the wealthy either benefit or suffer far less than the poor.

Blowback and social consequences of wildfire: race, class, and gender

To use the words of Chalmers Johnson: "Because we live in an increasingly interconnected international system, we are all, in a sense, living in a blowback world" (2001, p. 17). Disaster capitalism creates blowback. The myopia of politicians results in urban planning, mitigation strategies, and other policies that address economic complications, but worsen existing social problems. Urban growth as a product of disaster capitalism produces and reproduces a cycle of blowback. It is those who are most vulnerable who bear the social brunt of natural disaster – thus making such catastrophes social disaster.

The aforementioned defunding of active wildfire defense due to anti-tax initiatives and privatization has resulted in blowback that accentuates race, class, and gender inequalities (Davis, 1998; Fu, 2013). This is tied to the very uneven and unequal ways in which the social interacts with geography and ecology. For instance, Wyman and colleagues (2012) have noted differences in land use patterns among white and black landowners in Florida that affect wildfire mitigation strategies (see also Jarrett et al., 2009). In Arizona, Collins (2008) examined the political ecology of fire risk and mitigation for the region's indigenous and Hispanic populations. Wildfire causes casualties, but it also impacts human health in other ways. In California and other areas, agricultural workers living in the wildland–urban interface are often right on the fire line. In past fire-storms, Hispanic farmers were either not warned or were not given evacuation orders in Spanish (Benavides and Arlikatti, 2010). Moreover, the police presence deterred people of color from seeking help at evacuation sites (Martinez et al., 2009). Consequently, many suffered smoke inhalation and did not receive adequate medical treatment due to lack of access to medical facilities; nor were they counseled for trauma and stress. Moreover, clinics and organizations that support vulnerable groups such as immigrants are particularly under strain amidst fire-storms due to their already limited staffing and resources in the chaos created by the disaster (Rahn, 2010).

In addition to race and ethnicity, Eriksen (2013) has noted the gendered nature of emergency management and the lack of initiatives targeting women. As such, given the gendered responses to wildfire, there have been calls to consider gender in warning communications (Tyler et al., 2012; Tyler and Fairbrother, 2013). As Cox writes of bushfire in Australia, "women know that bushfire will occur ... and most of the town's men folk will go out to fight it" (1998, p. 135). Firefighting, like many occupations, is gendered (Desmond,

2008, pp. 38–41) This affects civilians as well. For civilians, men are more likely to be killed defending their homes, while women and children are more likely to die sheltering in a place or during evacuations (Haynes et al., 2010). The gendered nature of firefighting shifts the attention away from women in fire-fighting and evacuation. This, in turn, creates gendered consequences. For instance, female evacuees were more likely to report mental and physical health issues following the 2007 wildfires in San Diego (Jenkins et al., 2009).

Perhaps the most disconcerting intersection of race, class, and gender is California's use of inmates to fight wildfire. In the 2014 Bully Fire in Shasta County, the state deployed hundreds of low-level offenders to fight the conflagration. The vast majority are men of color, and are paid $1.45 to $3.90 a day, while saving the state $80 million in "labor costs" (Connor, 2014; Helmick, 2014). The California Department of Corrections and Rehabilitation reports that approximately 4000 offenders and 200 crews are part of 42 adult and two juvenile conservation camps that are managed alongside the Division of Juvenile Justice Conservation Camps, the California Department of Forestry and Fire Protection, and the Los Angeles County Fire Department (Department of Corrections and Rehabilitation, 2014). According to Goodman (2012), while female offenders are deployed to fight fires, only 297 are women, and the racial composition of male conservation camps reflects that of the general prison population – 29 percent African American, 41 percent Latino, 23 percent white, and 6 percent other – which is largely drawn from urban centers.

In the developing world, social change such as globalization, changes in agriculture, development patterns, and climate change have exacerbated fire risk (Eriksen and Watson, 2009; Nath and Behera, 2011; Osman-Elasha et al., 2011). This is particularly the case in nations that are dependent on agriculture and those which lack resources to combat large-scale conflagrations (Opukri and Ibaba, 2008). The general problem of underdevelopment creates unsafe conditions for the poorest of the poor (Pelling, 2003a). In September 2008, more than 100 bushfires in South Africa killed over 40 people, and in Mozambique and Zimbabwe another 30 lives were lost (BBC, 2008). The amplitude of this fire-storm strained regional fire defense resources – including personnel and water. The consequence in such scenarios is inevitably displacement and forced migration (Tschakert and Tutu, 2010; Arthur and Arthur, 2011).

At the same time, there is a push by state officials in such countries to prevent farmers from burning material. Among indigenous and traditional rural communities in Africa and Latin America, fire has long been a part of land use (Miller and Davidson-Hunt, 2010). In fact, traditional land use and burning helped control flammable brush. As such, the intervention by state officials has resulted in conflicts. Kull (2004) has looked at disputes between those seeking to pre-empt wildfire risk and communities engaged in traditional pasture burning in Madagascar. Eriksen (2007) has seen similar conflicts over burning in Zambia. This disruption of traditional life ultimately encourages migration that increases fire risk rather than reducing the threat.

Conclusion

This chapter is by no means a conclusive survey of global wildfire's relationship with urbanism. There is no doubt that wildfire is a complex problem that is influenced by human and environmental variables. Ecological, geological, climatic cycles, and human rhythms are interconnected. As Foster's reading of Marx suggests, nature is not only defined by the social, but nature in turn affects humans (Foster, 1999). As Gill and colleagues (2013) argue, minimizing harm from wildfire involves both understanding the regional fire regime and dealing with social-economic dimensions of fire risk. Yet, despite the call of researchers for an integrated approach to wildfire, under-examined are the connections between fire and global climate change, urbanization, and economic growth. Moreover, intersectional approaches to wildfire are even less studied.

The blowback from capitalist growth-oriented economic and urban policy is felt greatest by the most vulnerable. Hamza and Zetter (1998) remind us that migrants in poor nations often have no choice but to live in vulnerable areas. As such, there are those who note that there's no such thing as a "natural" disaster (Smith, 2006). Rather, they are social disasters (Klinenberg, 2003; Alagona, 2006; Brunsma and Picou, 2008). This is why events such as earthquakes, extreme temperatures, floods, volcanic eruptions, and wildfire are only considered disasters if they cause loss of life or adversely affect property, commerce, or other economic activities. Given the centrality of cities in the contemporary world, it is important to examine the role of cities and the consequences for urban populations amidst such catastrophes.

References

Adams, M.A., 2013. Mega-fires, tipping points and ecosystem services: Managing forests and woodlands in an uncertain future. *Forest Ecology and Management*, 294, pp. 250–261.

Alagona, P.S., 2006. What makes a disaster 'natural'? *Space and Culture*, 9(1), pp. 77–79.

Ali, M.B., Saidur, R., and Hossain, M.S., 2011. A review on emission analysis in cement industries. *Renewable and Sustainable Energy Reviews*, 15(5), pp. 2252–2261.

Arthur, J.L. and Arthur, I.A., 2011. *Movement Under Environmental Disasters: The case of flooding and bushfires for selected periods in Ghana*. Universitat Bielefeld: Center on Migration, Citizenship and Development, No. 97.

Balaban, U. and Fu, A.S., 2014. Politics of urban development and wildfires in California and Turkey. *Environment and Planning A*, 46(4), pp. 820–836.

Bankoff, G., Lübken, U., and Sand, J., 2012. *Flammable Cities: Urban conflagration and the making of the modern world*. Wisconsin: University of Wisconsin Pres.

BBC, 2008. New deaths in S Africa bush fires. BBC, September 3.

Beck, U., 1992. From industrial society to the risk society: Questions of survival, social structure and ecological enlightenment. *Theory, Culture and Society*, 9(1), pp. 97–123.

Benavides, A. and Arlikatti, S., 2010. The role of the Spanish-language media in disaster warning dissemination: An examination of the emergency alert system. *Journal of Spanish Language Media*, 3, p. 41.

Bohle, H.G., Downing, T.E., and Watts, M.J., 1994. Climate change and social vulnerability: Toward a sociology and geography of food insecurity. *Global Environmental Change*, 4(1), pp. 37–48.

Brunsma, D.L. and Picou, J.S., 2008. Disasters in the twenty-first century: Modern destruction and future instruction. *Social Forces*, 87(2), pp. 983–991.

CalFire, 2013. Fire statistics: Incident information. *CA.gov*. Available at: http://cdfdata. fire.ca.gov/incidents/incidents_stats?year=2013 (accessed August 13, 2014).

Cochrane, M.A. and Barber, C.P., 2009. Climate change, human land use and future fires in the Amazon. *Global Change Biology*, 15(3), pp. 601–612.

Collins, T., 2008. The political ecology of hazard vulnerability: Marginalization, facilitation and the production of differential risk to urban wildfires in Arizona's White Mountains. *Journal of Political Ecology*, 15(1), pp. 21–43.

Connor, T., 2014. Firefighting felons: Hundreds of inmates battling the Yosemite blaze. *NBCNews.com*, August 30.

Coumou, D. and Rahmstorf, S., 2012. A decade of weather extremes. *Nature Climate Change*, 2(7), pp. 491–496.

Coward, M., 2006. Against anthropocentrism: The destruction of the built environment as a distinct form of political violence. *Review of International Studies*, 32(3), pp. 419–437.

Coward, M., 2008. *Urbicide: The politics of urban destruction*. London: Routledge.

Cox, H., 1998. Women in bushfire territory. In E. Enarson and B. Morrow (eds), *The Gendered Terrain of Disaster: Through women's eyes*. New York: Praeger, pp. 133–142.

Crush, J., 2013. Linking food security, migration and development. *International Migration*, 51(5), pp. 61–75.

Davis, M., 1998. *Ecology of Fear: Los Angeles and the imagination of disaster*. New York: Metropolitan Books.

Department of Corrections and Rehabilitation, 2014. CDCR conservation (fire) camps. *CA.gov*. Available from: www.cdcr.ca.gov/Conservation_Camps/ (accessed November 13, 2014).

Desmond, M., 2008. *On the Fireline: Living and dying with wildland firefighters*. Chicago, IL: University of Chicago Press.

Dooling, S. and Simon, G., eds, 2012. *Cities, Nature and Development: The politics and production of urban vulnerabilities*. Aldershot: Ashgate Publishing.

Eriksen, C., 2007. Why do they burn the 'bush'? Fire, rural livelihoods, and conservation in Zambia. *The Geographical Journal*, 173(3), pp. 242–256.

Eriksen, C., 2013. *Gender and Wildfire: Landscapes of uncertainty*. Abingdon: Routledge.

Eriksen, S.E.H. and Watson, H.K., 2009. The dynamic context of southern African savannas: Investigating emerging threats and opportunities to sustainability. *Environmental Science and Policy*, 12(1), pp. 5–22.

Fankhauser, S. and Soare, R., 2012. *Strategic Adaptation to Climate Change in Europe*. EIB Working Papers.

FAO, 2003. *Community-based Fire Management: Case Studies from China, The Gambia, Honduras, India, the Lao People's Democratic Republic and Turkey*. Bangkok, Thailand: Food and Agriculture Organization of the United Nations, RAP PUBLICATION 2003/08.

FAO, 2011. *Wildfire Prevention in the Mediterranean*.

Foster, J.B., 1999. Marx's theory of metabolic rift: Classical foundations for environmental sociology. *American Journal of Sociology*, 105(2), pp. 366–405.

Freudenburg, W.R., Gramling, R., Laska, S., and Erikson, K.T., 2009. Organizing hazards, engineering disasters? Improving the recognition of political-economic factors in the creation of disasters. *Social Forces*, 87(2), pp. 1015–1038.

Fu, A.S., 2013. The façade of safety in California's shelter-in-place homes: History, wildfire and social consequence. *Critical Sociology*, 39(6), pp. 833–849.

Funk, M., 2009. Too big to burn: AIG plays God in a man-made firestorm. *Harper's Magazine*.

Gill, A.M. and Stephens, S.L., 2009. Scientific and social challenges for the management of fire- prone wildland–urban interfaces. *Environmental Research Letters*, 4, 034014.

Gill, A.M., Stephens, S.L., and Cary, G.J., 2013. The worldwide 'wildfire' problem. *Ecological Applications*, 23(2), pp. 438–454.

Givens, J.E. and Jorgenson, A.K., 2011. The effects of affluence, economic development, and environmental degradation on environmental concern: A multilevel analysis. *Organization and Environment*, 24(1), pp. 74–91.

Goldstein, B.E., 2008. Skunkworks in the embers of the cedar fire: Enhancing resilience in the aftermath of disaster. *Human Ecology*, 36(1), pp. 15–28.

Goodman, P., 2012. 'Another second chance': Rethinking rehabilitation through the lens of California's prison fire camps. *Social Problems*, 59(4), pp. 437–458.

Gotham, K.F., 2009. Creating liquidity out of spatial fixity: The secondary circuit of capital and the subprime mortgage crisis. *International Journal of Urban and Regional Research*, 33(2), pp. 355–371.

Gotham, K.F. and Greenberg, M., 2014. *Crisis Cities: Disaster and redevelopment in New York and New Orleans*. Oxford: Oxford University Press.

Grimm, N.B., Faeth, S.H., Golubiewski, N.E., Redman, C.L., Wu, J., Bai, X., and Briggs, J.M., 2008. Global change and the ecology of cities. *Science*, 319(5864), pp. 756–760.

Hamza, M. and Zetter, R., 1998. Structural adjustment, urban systems, and disaster vulnerability in developing countries. *Cities*, 15(4), pp. 291–299.

Hartman, C.W. and Squires, G.D., 2006. *There is No Such Thing as a Natural Disaster: Race, class, and Hurricane Katrina*. London: Taylor & Francis.

Haynes, K., Handmer, J., McAneney, J., Tibbits, A., and Coates, L., 2010. Australian bushfire fatalities 1900–2008: Exploring trends in relation to the 'Prepare, stay and defend or leave early' policy. *Environmental Science and Policy*, 13(3), pp. 185–194.

Helmick, A., 2014. California leans heavily on thousands of inmate firefighters. *KQED.org*, July 25.

Hewitt, K., 1983. The idea of calamity in a technocratic age. In K. Hewitt (ed.), *Interpretations of Calamity*. Boston, MA: Allen & Unwin, pp. 3–32.

Hewitt, K., 1997. *Regions of Risk: A geographical introduction to disasters*. New York: Longman.

Heynen, N.C., Kaika, M., and Swyngedouw, E., (eds), 2006. *In the Nature of Cities: Urban political ecology and the politics of urban metabolism*. New York: Routledge.

Jarrett, A., Gan, J., Johnson, C., and Munn, I.A., 2009. Landowner awareness and adoption of wildfire programs in the southern United States. *Journal of Forestry*, 107(3), pp. 113–118.

Jenkins, J.L., Hsu, E.B., Sauer, L.M., Hsieh, Y.-H., and Kirsch, T.D., 2009. Prevalence of unmet health care needs and description of health care–seeking behavior among displaced people after the 2007 California wildfires. *Disaster Medicine and Public Health Preparedness*, 3(S1), pp. S24–S28.

Johnson, C., 2001. *Blowback: The costs and consequences of American empire*. London: Macmillan.

Klein, N., 2008. *The Shock Doctrine: The rise of disaster capitalism*. London: Macmillan.

Klinenberg, E., 2003. *Heat Wave: A social autopsy of cisaster in Chicago*. Chicago, IL: University of Chicago Press.

Krawchuk, M.A., Moritz, M.A., Parisien, M.-A., Van Dorn, J., and Hayhoe, K., 2009. Global pyrogeography: The current and future distribution of wildfire. *PLoS One*, 4(4), p. e5102.

Kull, C.A., 2004. *Isle of Fire: The political ecology of landscape burning in Madagascar*. Chicago, IL: University of Chicago Press.

Linares, A.Z. and Linares, H.A., 1990. Burn prevention: The need for a comprehensive approach. *Burns*, 16(4), pp. 281–285.

Liu, Y., Stanturf, J., and Goodrick, S., 2010. Trends in global wildfire potential in a changing climate. *Forest Ecology and Management*, 259(4), pp. 685–697.

Mann, M.L., Berck, P., Moritz, M.A., Batllori, E., Baldwin, J.G., Gately, C.K., and Cameron, D.R., 2014. Modeling residential development in California from 2000 to 2050: Integrating wildfire risk, wildland and agricultural encroachment. *Land Use Policy*, 41, pp. 438–452.

Martinez, K.M., Hoff, A., and Nunez-Alvarez, A., 2009. *Coming Out of the Dark: Emergency preparedness plan for farmworker communities in San Diego County*. National Latino Research Center, California State University, San Marcos.

McCaffrey, S.M. and Rhodes, A., 2009. Public response to wildfire: Is the Australian 'stay and defend or leave early' approach an option for wildfire management in the United States? *Journal of Forrestry*, 107(1), pp. 9–15.

Mell, W.E., Manzello, S.L., Maranghides, A., Butry, D., and Rehm, R.G., 2010. The wildland–urban interface fire problem – current approaches and research needs. *International Journal of Wildland Fire*, 19(2), pp. 238–251.

Mileti, D.S., 1999. *Disasters by Design: A reassessment of natural hazards in the United States*. New York: National Academies Press.

Miller, A.M. and Davidson-Hunt, I., 2010. Fire, agency and scale in the creation of Aboriginal cultural landscapes. *Human Ecology*, 38(3), pp. 401–414.

Molotch, H., 1976. The city as a growth machine: Toward a political economy of place. *The American Journal of Sociology*, 82(2), pp. 309–332.

Moore, J.W., 2000. Environmental crises and the metabolic rift in world-historical perspective. *Organization and Environment*, 13(2), pp. 123–157.

Mukheibir, P. and Ziervogel, G., 2007. Developing a Municipal Adaptation Plan (MAP) for climate change: The city of Cape Town. *Environment and Urbanization*, 19(1), pp. 143–158.

Nath, P.K. and Behera, B., 2011. A critical review of impact of and adaptation to climate change in developed and developing economies. *Environment, Development and Sustainability*, 13(1), pp. 141–162.

O'Connor, J.R., 1998. *Natural Causes: Essays in ecological Marxism*. New York: Guilford Press.

Opukri, C.O. and Ibaba, I.S., 2008. Oil induced environmental degradation and internal population displacement in Nigeria's Niger Delta. *Journal of Sustainable Development in Africa*, 10(1), pp. 173–193.

Osman-Elasha, B., Chidumayo, E., and Donfack, P., 2011. Socio-economic and gender related aspects of climate change in Africa. In E. Chidumayo, D. Okali, G. Kowero, and M. Larwanou (eds), *Climate Change and African Forest and Wildfire Resources*. Nairobi, Kenya: African Forest Forum, pp. 176–191.

Owen, D., 2009. *Green Metropolis: Why living smaller, living closer, and driving less are the keys to sustainability*. London: Penguin.

Pataki, D.E., Alig, R.J., Fung, A.S., Golubiewski, N.E., Kennedy, C.A., Mcpherson, E.G., Nowak, D.J., Pouyat, R.V., and Romero Lankao, P., 2006. Urban ecosystems and the North American carbon cycle. *Global Change Biology*, 12(11), pp. 2092–2102.

Pausas, J.G. and Fernández-Muñoz, S., 2012. Fire regime changes in the Western Mediterranean Basin: From fuel-limited to drought-driven fire regime. *Climatic Change*, 110(1–2), pp. 215–226.

Paveglio, T.B., Carroll, M.S., and Jakes, P.J., 2010. Alternatives to evacuation during wildland fire: Exploring adaptive capacity in one Idaho community. *Environmental Hazards*, 9(4), pp. 379–394.

Paveglio, T., Prato, T., Dalenberg, D., and Venn, T., 2014. Understanding evacuation preferences and wildfire mitigations among Northwest Montana residents. *International Journal of Wildland Fire*, 23(3), pp. 435–444.

Pelling, M., 2003a. *Natural Disaster and Development in a Globalizing World*. Hove: Psychology Press.

Pelling, M., 2003b. *The Vulnerability of Cities: Natural disasters and social resilience*. London: Earthscan.

Pence, G.Q., Botha, M.A., and Turpie, J.K., 2003. Evaluating combinations of on- and off-reserve conservation strategies for the Agulhas Plain, South Africa: A financial perspective. *Biological Conservation*, 112(1–2), pp. 253–273.

Piñol, J., Terradas, J., and Lloret, F., 1998. Climate warming, wildfire hazard, and wildfire occurrence in coastal eastern Spain. *Climatic Change*, 38(3), pp. 345–357.

Princen, T., Maniates, M., and Conca, K., 2002. *Confronting Consumption*. Cambridge, MA: MIT Press.

Pyne, S.J., 1997. *World Fire: The culture of fire on Earth*. Seattle: University of Washington Press.

Pyne, S.J., 2001. *Fire: A brief history*. Seattle: University of Washington Press.

Radeloff, V.C., Hammer, R.B., Stewart, S.I., Fried, J.S., Holcomb, S.S., and McKeefry, J.F., 2005. The wildland–urban interface in the United States. *Ecological Applications*, 15(3), pp. 799–805.

Rahlao, S.J., Milton, S.J., Esler, K.J., Van Wilgen, B.W., and Barnard, P., 2009. Effects of invasion of fire-free arid shrublands by a fire-promoting invasive alien grass (Pennisetum setaceum) in South Africa. *Austral Ecology*, 34(8), pp. 920–928.

Rahn, M., 2010. *Initial Attack Effectiveness: Wildfire staffing study*. San Diego, CA: San Diego State University, No. 2.

Rees, W.E., 1992. Ecological footprints and appropriated carrying capacity: What urban economics leaves out. *Environment and Urbanization*, 4(2), pp. 121–130.

Şahin, Y.G. and İnce, T., 2009. Early forest fire detection using radio-acoustic sounding system. *Sensors*, 9(3), pp. 1485–1498.

Schnaiberg, A., 1980. *The Environment, from Surplus to Scarcity*. New York: Oxford University Press.

Scott, J.W., 2007. Smart growth as urban reform: A pragmatic 'recoding'of the new regionalism. *Urban Studies*, 44(1), pp. 15–35.

Skidmore, M. and Toya, H., 2002. Do natural disasters promote long-run growth? *Economic Inquiry*, 40(4), pp. 664–687.

Smith, N., 2006. There's no such thing as a natural disaster. *Understanding Katrina: Perspectives from the social sciences*. Available at: http://understandingkatrina.ssrc.org/ Smith (accessed December 15, 2008).

Sorensen, J.H., Shumpert, B.L., and Vogt, B.M., 2004. Planning for protective action decision making: Evacuate or shelter-in-place. *Journal of Hazardous Materials*, 109(1–3), pp. 1–11.

Steinberg, T., 2001. The secret history of natural disaster. *Global Environmental Change Part B: Environmental Hazards*, 3(1), pp. 31–35.

Steinberg, T., 2006. *Acts of God: The unnatural history of natural disaster in America*. Oxford: Oxford University Press.

Stephens, S.L., Adams, M.A., Handmer, J., Kearns, F.R., Leicester, B., Leonard, J., and Moritz, M.A., 2009. Urban–wildland fires: How California and other regions of the US can learn from Australia. *Environmental Research Letters*, 4, pp. 1–5.

Stetler, K.M., Venn, T.J., and Calkin, D.E., 2010. The effects of wildfire and environmental amenities on property values in northwest Montana, USA. *Ecological Economics*, 69(11), pp. 2233–2243.

Sturm, T. and Oh, E., 2010. Natural disasters as the end of the insurance industry? Scalar competitive strategies, Alternative Risk Transfers, and the economic crisis. *Geoforum*, 41(1), pp. 154–163.

Tao, D., 2011. The big business of battling wildfires. *Fortune*, June 22.

Tierney, K.J., 1999. Toward a critical sociology of risk. *Sociological Forum*, 14(2), pp. 215–242.

Toya, H. and Skidmore, M., 2007. Economic development and the impacts of natural disasters. *Economics Letters*, 94(1), pp. 20–25.

Tschakert, P. and Tutu, R., 2010. Solastalgia: Environmentally induced distress and migration among Africa's poor due to climate change. In T. Afifi and J. Jäger (eds), *Environment, Forced Migration and Social Vulnerability*. New York: Springer, pp. 57–69.

Tyler, M. and Fairbrother, P., 2013. Bushfires are 'men's business': The importance of gender and rural hegemonic masculinity. *Journal of Rural Studies*, 30, pp. 110–119.

Tyler, M., Fairbrother, P., and Phillips, R., 2012. Gender and bushfire. *Fire Note*, 101.

Westerling, A.L. and Bryant, B.P., 2007. Climate change and wildfire in California. *Climatic Change*, 87(S1), pp. 231–249.

Wisner, B., Blaikie, P., Cannon, T., and Davis, I., 1994. *At Risk: Natural hazards, people's vulnerability and disasters*. Oxford and New York: Routledge.

Wyman, M., Malone, S., Stein, T., and Johnson, C., 2012. Race and wildfire risk perceptions among rural forestland owners in North-Central Florida. *Society and Natural Resources*, 25(12), pp. 1293–1307.

York, R. and Mancus, P., 2009. Critical human ecology: Historical materialism and natural laws. *Sociological Theory*, 27(2), pp. 122–149.

36 As the World Melts

Phoebe Godfrey

Figure 36.1 As the World Melts by Phoebe Godfrey, oil on canvas, 6'×8'. As the World
Melts is my version of the anonymous French painting titled Gabrielle
d'Estrées and One of Her Sisters (1594). It explores the self-indulgent denial
of GCC on the part of the American white elite by showing icebergs melting
over into the bathtub where Ladies of Leisure are bathing. So self-consumed
they are oblivious to the world around them and yet a full moon indicates
the end of the current cycle and a new cycle is about to begin with renewed
potential for positive social and ecological change.

37 A personal tale from the environmental wetback

Rethinking power, privilege, and poverty in a time of climate change politics

Soraya Cardenas

Special note: I refer to myself as the environmental wetback because I want to empower my presence through a socially derogatory term, a tactic used during the Chicano Movement (Meier and Ribera, 1994). I want to take back what was taken away from me when I was first labeled a 'wetback' when I was nine years old.

I am not going to try to convince anyone that climate change is real; I am assuming that this reality has already been assessed. My story is about power, privilege, and poverty, and how climate change impacts marginalized peoples. Ultimately, do power and privilege affect the methods and processes used to help marginalized communities? What are alternative considerations?

In this chapter, I cite my experiences as a child and then as a social scientist. I lived and conducted field research in Mexico for about two years in Queretaro while in my late twenties and then another three years in Oaxaca while in my mid-thirties. During my residence as a child, and then as a social scientist, I observed the effects of climate change on the environment and marginalized communities. I lived in communities experiencing drought and temperatures that tested human tolerance, which were exacerbated by the lack of shade trees and man-made concrete structures – what I call 'manufactured environments' – building on the term 'manufactured landscapes' popularized by renowned photographer Edward Burtynsky.

On a personal reflection, I can still remember transporting my two-week-old baby in the car through the narrow and sometimes cobbled streets of the historic downtown Oaxaca because my older son's school was on the other side of the city. It was an almost unavoidable condition of the city's rapid urban growth, most likely fueled by an influx of residents who moved from the rural communities to the city in search of work, which was a consequence of the globalization of labor and agriculture. Poor farmers could not compete in a global subsidized market, and the demand for cheap labor in the United States often left the elderly and children behind in their communities (Davis, 2007).

As I fiercely drove my son home, I felt my nerves crawl as I heard his screams of agony. His body turned into a fireball as his skin was covered with a thick, inflamed rash. When I finally arrived home, I quickly took him inside and opened the faucet, filling my bucket with water and dumping it into my son's bath tub. I took off his onesie and soaked his little, frail body. I later learned that my son had a skin condition, which was severely aggravated by the sun, but I was lucky because I had access to healthcare, a car, and water. My house had a cistern below the ground which pumped water to a tank that rested on my roof; the tank served as a storage unit because water was delivered only twice a week. This system allowed me to have running water 24 hours a day, but if the delivery service failed, I had enough storage for about a week (if used conservatively). Of course, I could always purchase more water for about $10 to $15, which could sustain me for another week. This type of water-delivery system is practiced throughout Oaxaca and much of Mexico. Mexico is a water-scarce country. When I lived in Queretaro, I learned through the government agencies that three of Mexico's major cities are in water-scarce regions and the two largest cities, Mexico City and Guadalajara, share the same basin, namely Cuenca Lerma Chapala. Ironically, the largest consumer of water in Mexico is not the municipalities, but agriculture, which accounts for about 80 percent of all water use. This is partially a consequence of water policies in Mexico. Water is considered a national resource and farmers do not pay for their water use (Cardenas, 2002).

Water scarcity and the politics behind it are only one factor affecting marginalized populations. The reality is that climate change is happening, and we have exceeded our ability to go back in time, though we should make efforts to at least stop any further degradation put forth by 'progress' and CO_2 emissions (Heinberg, 2010); but two important questions remain: (1) What are the disparities between marginalized peoples and those in privileged positions with regard to the fallout of climate change? Are they experiencing climate change disasters equally? In some instances I would say yes, fires, floods, hurricanes, but in some instances poor peoples may be building shanty towns along mud-slide corridors or forced to live in other areas that are more sensitive to the ravages of climate change. For example, in the Chicago heatwave of 1995 about 465 people died in one week and an additional 521 in the month of July. Most of those who died were poor, and elderly African Americans in the urban ghetto, who didn't have air conditioning in their high-rise homes (Klinenberg, 2002). (2) What are the differences in the ability to recover from the fallout of climate change or even have other agencies respond to their needs after devastating losses? For example, when I lived in Maine there was a 100-year flood which flooded many homes and wiped out poor rural housing. Many of those affected lost all their belongings, but the wealthier residents with insurance were able to recover their material losses more easily.

So this is where my story begins. I want to take you through my personal journey as a child and then as a researcher to help you understand how marginalized peoples are affected by climate change with regard to the lack of power, privilege, and resources.

My awakening to privilege

Despite living through a very uncomfortable situation with my son, I had resources. If I did not have access to healthcare, water, and a car, my son – who weighed only about nine pounds – could have potentially been hospitalized or even perished. Yet, many people in Mexico do not have access to basic resources like water. Seven years earlier, while living in Queretaro, I witnessed communities that did not have access to water for two weeks at a time. These already financially strapped families were forced to purchase 19-liter jugs of water, which cost about a dollar, to be used for drinking, cooking, bathing, cleaning, flushing the toilet, and washing clothes and dishes. This price for water may sound like a steal to you, but keep in mind that the minimum wage in Queretaro was about $4.20 per day at the time of my research (Cardenas, 2002). Those employed through the informal sector could be making as little as one or two dollars a day; thus, paying $1 a day for water was a significant sacrifice. Now, this was not my first exposure to poverty and water scarcity in Mexico.

I am the daughter of immigrant parents. Although I was born in the USA and was predominantly raised in Nebraska, I spent many summers as a child in Mexico living with my grandmother, my father's mother. She was a poor woman, living near downtown Colima, Mexico. Her living room area and bedroom were the only two major spaces, enclosed in an old cement structure that shared the wall with a neighbor. The floors were of cracked cement. Her kitchen was semi-covered with a tin-like roof, which was held up by round wooden beams that usually had ants crawling up and down them. I often found dead floating ants in my soup; though I complained, my grandmother would retort in her cranky voice, "Eat it! It's not going to kill you."

Because the roof in the kitchen was shoddy, water dripped onto the dirt floor. The restroom was outside, within four or five steps from the kitchen. It was not attached to the house, but was its own small cement structure. It consisted of a *pila*, a rectangular, cement holding tank about three feet deep with little fish swimming inside, which I now understand were used to keep the tank clean because they ate the slime that grew on the sides. Next to the *pila* was a sink. This was where my grandmother washed dishes and clothes. There were grooves on the flat surface that could be used to scrub clothes back and forth, like a washboard. Inside the structure were two rooms: a toilet and bathing station. When using the toilet, we had to carry a small bowl of water with us as our flushing mechanism. In the bathing room, we could access the *pila*. When cleaning ourselves, we had our bowl, soap, and shampoo to hand. We would take the cool water and pour it over our bodies, followed by lathering suds from soap and shampoo. Although you may think the ants in my soup and the cold showers would be the most vivid of my memories, instead it was the poverty and the limited access to resources. My grandmother received water daily for a few hours, and it was imperative that someone be home during that period so that the *pila* would always have water. I was mortified by the possibility of not having water and being inconvenienced by the city's rudimentary services. Despite my grandmother having limited resources, her neighbors across the street were in even worse shape; they didn't have a *pila* or a sink, but they had something that my grandmother did not: a running stream where they bathed and washed their clothes.

I can still remember being invited by the children to swim in the stream. Because I was so young and small I felt a sense of danger; but still I reveled in the excitement of splashing along the edges as I played with the other children, until I learned my place. One day, the torrential summer rains were coming down, as they did every year. It rained heavily for about 20 to 30 minutes and then the sun would come out for the rest of the day. I was in my aunt's house, who lived two doors down from my grandmother; my aunt's family was equally as poor as my grandmother. My eight cousins, along with their mother, shared three beds. They often slept with their feet hanging off the side of the bed, so that they all could fit. I had just taken a bath and was in my white, ruffled, Tinker Bell undies (now bear in mind that I was only four years old). The boys from across the street came knocking at my aunt's door and invited me to play with them in the puddles, so off I went in my underwear with a pack of boys, running wild, splashing, and laughing with glee. Minutes later my mother came storming out, grabbed me by the arm, spanked my bottom, and told me that I was not allowed to play with these poor children anymore and never should I be seen in public wearing just my underwear. Although I might agree with her second point today, I do not agree with her first; regardless, the class barrier was raised. This idea of class would haunt me for the rest of my life. I knew from that juncture that I was prohibited from engaging with impoverished people, whether at an intimate or informal level. I also knew that, by virtue of being American, I had access to basic amenities like water 24 hours a day – hot and cold. I would never have to wash my clothes by hand in a stream. I was afforded a life of convenience. Ultimately, I knew that I was privileged. This theme would resonate throughout my life and research.

End of nature not privilege

On Saturday, December 1, 2012, I attended a 350.org event, an organization started by Bill McKibben and students from the prestigious Middlebury College. This organization is dedicated to fighting for climate change policy that will reduce our global carbon dioxide output. This organized event was attended by people with middle-of-the-road politics, avant-garde activists, older citizens, students, and educators. The highlight of the event featured Bill McKibben, author of many environmental books, including *The End of Nature*. They even invited a rancher, Randy Thompson, who has been the voice against the TransCanada pipeline that would drag oil from the tar pits of Canada through the USA.

The following Monday, two of my students – who had also been at the event – asked me what I thought about it. At first, I said it was good, then okay, and finally I asked if they really wanted to know what I thought. I proceeded to discuss why I hated it. Although I believe that the event served its purpose and had its place, I see a much bigger dilemma that continues to be neglected and mitigated. I asked my students, "How many people of color did you see on stage?" I said, "I think that I might remember one participant, but the leaders of this movement are white." This was no surprise. Yet, these advocates did make an effort to play the race card.

At the beginning of McKibben's presentation, he showed some impressive artwork of 350.org around the world. I will admit that was cool – but that's just it, it's just cool artwork with symbolic meaning, but no true forcible action. It is trying to impassion people into joining the climate change movement, but the fact remains that we are a global society dependent on fossil fuels for food production, heating our homes, and making our clothes, medical supplies, plastics, etc.

When McKibben began his discourse at the beginning of his presentation, he said something very telling: that most of the world is not white. It is predominantly occupied by Africans, Asians, Indigenous peoples – essentially, nonwhite peoples. He talked about how some of the biggest demonstrations are attended by these same peoples. Many of them are from poor, less developed countries. Yes, I was moved by these comments, and thought to myself, "beautiful," but something inside felt unsettled.

I started thinking about Peter Dauvergne's book, *The Shadows of Consumption* (2008). In his book, Dauvergne points out three important ideas in addressing environmental degradation processes: (1) Global health and environments face risks from corporate and government actions. (2) Strong states and corporations circumvent environmental progress unless it is profitable to go green. (3) Less developed countries face the greatest negative consequences from environmental externalities and slow environmental progress.

He makes a strong argument that powerful corporations and nations influence less powerful countries' consumption patterns. They use poor nations' resources, forcing them to accept outdated products, denying them access to better technologies, and keeping them financially deprived and less competitive. Undeniably, wealthier nations make up the bulk of the consumption and negatively influence others, but less industrialized nations also produce externalities. They contribute to excess by participating in the global economy, adopting Western culture, and demanding more resources to meet the needs of their growing populations (Cardenas, 2010). An example of this is China, the largest consumer of energy in the world, while the USA is one of the largest consumers of Chinese products. In addition, the USA is still the largest consumer of energy per capita (Koch, 2010).

Okay, back to my conversation with my students. I told them that I was annoyed that, once again, white privileged peoples in the USA are the leaders of these movements and that there continues to be an absence of leadership among people of color. It is the privileged who have exploited the world to build their empires and create a society of super-consumers, and now the same peoples who created the problem are our saviors and are going to fix it. My students' response was, "Well, if they created the problem then they should have the responsibility to fix it." I retorted with an analogy.

Let us think back to the Native American experience. Whites came to this continent, essentially destroying Native people's way of life by taking their lands, killing their food sources and culture, placing them on reservations, and making them dependent on the USA. Since the whites devastated Native people's existence, should whites now fix it without regard for Native people's needs or input because whites created the mess? Should not the Native people instead be given support (power and resources, not just lip-service) to make decisions that will rebuild their own tribes? If whites continue to step up and be the leaders, fixing all the problems they created, then they still remain in power. They continue to parallel their imperialist behaviors but with a different agenda, with a noble cause. Again, people of color continue to be asked to follow and become tokens of the cause. This reminds me of black feminist theorist Patricia Hill Collins, and how she discussed white knowledge validation. She points out that whites are able to maintain their authority and demonstrate inclusiveness by allowing a few black women to participate, but that the women they choose continue to validate the white man's processes and their authority (2008).

The 350.org event showed their 'people of color faces' to demonstrate their inclusive nature, but on that stage and behind the scenes, whites continued to be the leaders. And yes, McKibben cited South Africa's apartheid as an example of how to fight the system, stating that Apartheid was able to be dismantled because of divestment, meaning people around the world – predominantly Europe and the USA – had started to pull their monies out of South Africa's economy. Now think about this; investors from Europe and the USA were able to make a difference economically. What does this tell you? A significant amount of capital comes from white-dominated nations.

McKibben later told us that Nelson Mandela visited the USA, specifically the students in the colleges that pushed for this agenda, and thanked them for their support, which helped change the tide of Apartheid. Here's the irony: white political control over South Africa was dismantled through the efforts of white-dominated countries – Europe and the USA. These same countries helped re-establish political balance in South Africa, but institutions in the USA continue to subjugate peoples of color, and South Africa still continues to be economically governed by the white elite; thus, it is only free by outward appearances (Goodenough, 2012).

McKibben then showed his token video of Nobel Prize recipient Desmond Tutu spouting about the importance of climate change advocacy. Ahh, relief; McKibben has managed to cover his bases and reinforce his keen awareness of inclusiveness in his discourse, so no one can say he is ignorant to peoples of color around the world because he has presented his token video. This may sound harsh and you may be placing your hands over your ears, but you cannot escape the truth. People of color around the world are slaves to the predominantly white, global capitalist system, and we as a global community cannot change the environmental locomotive disaster we are riding. McKibben's energies are feel-good efforts with no bite. Now it's my turn to share my ideas on the issue.

Revisiting Mexico

I do not take the stance that I am the authority on climate change or sustainable solutions, but I want to impart my research experiences in Mexico, along with my many years as an academic. These experiences have shaped my views on rethinking and redesigning sustainable climate change models. Globally, we cannot turn back the hand of time and change future climate trends, but how do we provide resources and voice to marginalized peoples?

I would like to share my research experience in Queretaro, Mexico. I first came to Queretaro by way of the organization Groundwater Guardian, a part of the larger Groundwater Foundation organization, which was founded by Susan Seacrest, as a consequence of her concern about elevated nitrates in the Plattsmouth River Valley of Nebraska (Groundwater Foundation, n.d.). Bob Kuzelka, who was a professor at the University of Nebraska-Lincoln, was affiliated with this organization and introduced me to the Director of the State of Water Commission of Queretaro, Mexico at a conference in Illinois. The Director would later give me access to important data about their aquifer, El Valle de Queretaro, and their operations.

Thanks to a Garcia-Robles Fulbright and a University of Nebraska-Lincoln Presidential Fellowship, I was able to conduct water scarcity research in Queretaro for two years. I conducted over 50 interviews with mostly government officials and some community activists and scholars. Not only did I gain a perspective from the people with whom I spoke, but I also lived in the community, which gave me an observer-participant experience.

I witnessed the passing of everyday life in an industrial metropolis, zooming cars, grocery stores, fast-food chains, industrial parks, and much more. The landscape was covered with houses, buildings, historical sites, paved roads, sprawling communities, and smog. Despite the appearance of a water-rich community with lush gardens, elaborate fountains, and a flowing river, the outskirts of the city told the tale of a semi-arid community, thirsting for a drink of water from the sky. The rolling mountains were covered in cacti and the flowing river was actually a dried river bed which was being artificially pumped sewage water within the city limits to give the illusion of a flowing river, the former Rio de Queretaro.

When I first arrived I resided in southern Queretaro, which had access to water 24 hours a day. El Centro and northern Queretaro suffered severe water scarcity. Some parts of the city at times didn't receive water for over a week, as drought conditions weakened an already poor delivery system. These residents would normally receive water either a few times a week or once a day for a few select hours. I remember taking a shower and having to finish bathing with bottled water because I ran out. Keep in mind that I was earning an American salary, so this was more an inconvenience than a financial hardship.

I came to personally experience and witness that not everyone received water equally. While in the field with a representative from the State Water Commission to the wealthier neighborhoods of Jurica and Juriquilla, which are situated in the water-strapped northern region of the city of Queretaro, I saw some of the first egregious signs of inequity. On our way to these wealthier communities, we had to cross some of the poorest neighborhoods of Queretaro. While on our way to Jurica, we ran across a massive leak in the middle of the road. Wasteful municipal practices were rampant throughout the city, which I had been told about by state and federal officials, and I had also read about in the local newspaper. As we stopped to observe the massive leak, among many others on our tour, a local woman confronted us. She was very irate, telling us that she had contacted the State Water Commission, but that no one came. According to her, the leak was several days old. She angrily grunted that in her own home she cannot get even a drop of water, but outside her home the government allows the water to go to waste.

Meanwhile, in the upscale neighborhoods of Jurica and Juriquilla, there were large homes with lush green gardens which were watered regularly. Residents' house servants washed their fancy cars with hoses. According to the state representative escorting me, these wealthier communities had access to water 24 hours a day made possible by their water towers, which were not available to the less privileged neighborhoods. This feeling of being forgotten or being less entitled because of their low economic status echoed among many poor residents. I remember a particular conversation with a low-income homemaker and local activist. She made a reference to the uneven distribution of water, while explaining what she perceived to be the "culture of water."

> It is our culture. If we only receive water for one or two hours then we learn to live with that situation. But if we received water like the wealthier people, I think we too would use more.... It is our culture of water.

What is interesting about this perspective is that the more access members of society have to goods, the more likely they are to consume. This makes me think back to the research I conducted with my students regarding consumption and biomass. One student, Sandy Mosquera, found an article that examined whether environmental attitudes and education influenced the outcomes of behavior. The author, Carmen Tanner (1999), uncovered the finding that, though education appears to affect people's attitudes and beliefs – meaning they may be sympathetic to the environmental cause – their behaviors showed otherwise. She pointed out that a majority of attitudinal studies have shown that environmental concern or attitudinal variables fail to correspond with behavior. In fact, the opposite appears to be true. People with higher educational levels usually have higher incomes; thus, they have greater access to acquiring goods (United States Department of Labor, 2011). It is no surprise that the USA makes up 5 percent of the population, but consumes 20 percent of the world's fossil fuel resources (The Worldwatch Institute, 2016). Meanwhile, America is consciously becoming more green; according to Dunlap and McCright (2008, p. 1057), "environmentalism has diffused throughout most sectors of U.S. society." In essence, mainstream America supports environmentalism, and women are now more active than men in the movement (Dunlap and McCright, 2008). So how has America reacted? By green-washing. According to Peter Dauvergne (2008), companies are using green rhetoric to sell their goods, fueling more consumption.

I returned to Queretaro in 2010. There were several interesting findings. First, after speaking to my State Water Commission informant, he told me that everything was like ENRON. It looked good on the outside but was falling apart from the inside. The State Water Commission had realized by going green that they could obtain global support, which meant investments could be used to line their pockets. According to my informant, acknowledging publicly that there was a water scarcity issue meant that their state office could gain access to global funds.

Another interesting finding arose. The local public was no longer in the dark about water scarcity. The community had learned from a researcher that the dominant aquifer, El Valle de Queretaro, was not a rechargeable aquifer – it was a fossil aquifer. This meant that their water sources were limited. Meanwhile, the water table continued dropping three to four meters annually, destabilizing the ground as parts of the city's sidewalks and exterior walls cracked.

I also spoke with a local resident, an activist with whom I had stayed in contact. She commented that water scarcity was no longer an issue because it was raining all the time and their dams were capturing the water. She neglected to recognize that the aquifer was still losing water, and from 2000 to 2010 the population had increased, due mostly to migration, by over 150,000 – reaching 801,940 residents (INEGI, 2000, 2010). The Aquifero de Queretaro was not recharging. Water usage was not being curtailed and disparities continued.

Reinventing the wheel? Not really

So what does this all mean? And why is this relevant to power, privilege, and resources? I have witnessed that people with money and power are unlikely to give up their privilege and usually only benefit themselves. Although people often turn to education as a method of choice to stop unwanted behavior, the fact of the matter is that education has less of an impact than socio-economic factors, so education alone cannot be the only tactic. In a study by Verhallen and Van Raai (1981), they reveal that socio-demographic factors have a greater impact than attitudes when it comes to environmental advocacy; in other words, money talks.

I propose using education and investments that help impoverished communities. Although educating everyone about climate change is not going to change behaviors, I still believe that it has value, but it cannot be the only factor. Using education to improve the livelihoods of the marginalized people's children and also providing them with business opportunities, such as supporting small businesses, may help alleviate the gaps. Ultimately, we need to stop per-petuating poverty. We must also invest in assisting people who want to recon-nect with their rural roots and find opportunities that help them move back to their land. We need to promote other economic sectors, beyond service. We must consider primary and secondary economic sectors, such as small-scale farming. We should revisit sustainable living communities. Finally, there needs to be voice and leadership. Leadership must be advocated within communities. Marginalized people should be allowed to participate and lead in decisions that impact their own futures, and outsiders may serve as consultants and advocates, but not as decision-makers.

I leave you with one last story to mull over my ideas. While living in Oaxaca I would run my errands prior to picking up my son from school, which consisted of going to the butcher, the tortilla woman, and the local corner store vendor which provided fresh fruit and vegetables and other cooking essentials (I do not use their names to protect their identity). Unlike the USA, where I could buy my food for the week, in Mexico it is often customary to buy one's food for just that day. As I visited each post, every community member wanted to have a conversation with me. I remember telling the grocery vendor of fruit and veget-ables that I was in a hurry, and I needed to go pick up my son. He totally ignored my polite plea and continued to ask me about my day and my life. I soon realized that my community vendors were not going to change their behav-iors for me, so I acquiesced. I started to leave 20 minutes early for my grocery needs. What I learned about this experience was that I was selfish and that these members of my community were worth getting to know. I learned that the grocery vendor had left his home as an adolescent in the rural mountains of Oaxaca to find work. He would later, as an adult male, return to his community for a year to become the mayor. The tortilla lady – who was very poor and living in an open, one-room cement structure – had received a grant to help local women start their own businesses, and these funds made it possible for her to start her tortilla stand. The butcher had once lived in the USA but returned home because he loved living in Mexico; he was able to open his butcher shop with the extra money he made while working in the USA.

I left in the summer of 2006 during a time of severe turmoil in Oaxaca. The National Guard had arrived to help control a conflict between the teachers' union and the government. Seven years later I returned for a visit. During that brief time I learned an important lesson about community. As I walked through my old neighborhood I visited all of my community members. Everyone remem-bered me and gave me hugs and free food. It felt good inside, because despite leaving, they still considered me to be part of their memories, life, and com-munity. A very strong lesson indeed!

References

Cardenas, S., 2002. *The Government of Queretaro's Response to an Impending Water Crisis.* PhD thesis, University of Nebraska-Lincoln.

Collins, P.H., 2012. Black feminist epistemology. In C. Calhoun, J. Gertais, J. Moody, S. Pfaff, and I. Virk (eds), *Contemporary Sociological Theory*, 3rd edn. West Sussex: Wiley-Blackwell, pp. 323–331.

Dauvergne, P., 2008. *The Shadows of Consumption: Consequences for the global environment*. Cambridge, MA: MIT Press.

Davis, M., 2007. *Planet of Slums*. New York: Verso.

Dunlap, R.E. and McCright, A.M., 2008. Social movement identity: Validating a measure of identification with the environmental movement. *Social Science Quarterly*, 89(5), pp. 1045–1065.

Goodenough, T., 2012. Jacob Zuma says South Africa is controlled by white power-brokers from 'Apartheid-Era.' *Daily Mail*. Available at: www.dailymail.co.uk/news/article-2164961/Jacob-Zuma-says-South-Africa-controlled-white-power-brokers-Apartheid-era. html (accessed June 26, 2012).

Groundwater Foundation, n.d. Our history. Available at: www.groundwater.org/who/history.html (accessed January 3, 2014).

Heinberg, R., 2010. *The Post Carbon Reader: Managing the 21st century's sustainability crisis*. Healdsburg, CA: Watershed Media.

Instituto Nacional de Informacion Estadistica y Geografica (INEGI), 2011. Available at: www.inegi.gob.mx/inegi/default.aspx (accessed November 14, 2011).

Klinenberg, E., 2002. *Heat Wave: A social autopsy of disaster in Chicago*. Chicago, IL: University of Chicago Press.

Koch, W., 2010. China surpasses U.S. in energy consumption. *USA Today*. Available at: http://content.usatoday.com/communities/greenhouse/post/2010/07/china-surpasses-us-in-energy-consumption/1#.Vo1T3fkrJD8 (accessed January 5, 2016).

Meier, M.S. and Ribera, F., 1994. *Mexican Americans/American Mexicans: From Conquistadors to Chicanos*. New York: Hill and Wang.

Tanner, C., 1999. Constraints on environmental behavior. *Journal of Environmental Psychology*, 19(2), pp. 145–157.

United States Department of Labor, 2013. *United States: Bureau of Statistics: Consumer Expenditure: 2011*. Available at: www.bls.gov/cex/ (accessed June 12, 2013).

Verhallen, T.M.M. and Van Raaij, F.W., 1981. Household behavior and the use of natural gas for home heating. *Journal of Consumer Research*, 8(3), pp. 253–257.

Worldwatch Institute, The, 2016. The state of consumption today. Available at: www.worldwatch.org/node/810 (accessed January 5, 2016).

38 Climate Action Planning (CAP)

An intersectional approach to the urban equity dilemma

Chandra Russo and Andrew Pattison

The leading international climate research institution, the Intergovernmental Panel on Climate Change (IPCC), has demonstrated that climate change will have devastating impacts on communities the world over if there is not substantive policy intervention. Those who face the intersections of racism, classism, and sexism have long borne the brunt of environmental hazards, both globally and in the USA (Agyeman et al., 2003; Bullard, 2005; Chavis and Lee, 1987). This is no different when it comes to climate change which, under existing social conditions, puts the global poor, people of color, and women at the greatest risk (Nagel, 2012; Shearer, 2011). While such a dynamic is often discussed as it impacts the Global South, classism, racism, and patriarchy have intersected to immiserate certain urban communities in the USA, rendering them vulnerable not only to climate change itself but also to the collateral damage of the very policies intended to prevent environmental destruction.

While the issue of climate change has drawn immense attention from policymakers the world over, the USA, one of the greatest national consumers of natural resources and carbon emitters, has refused to advance meaningful climate legislation. In the vacuum of federal action, cities and states have taken up the mantel for meaningful progress, most commonly through the development of climate action planning (Ramaswami et al., 2012; Boswell et al., 2012; Finn and McCormick, 2011). A Climate Action Plan (CAP) lays out a program to reduce greenhouse gas emissions through programs to generate renewable energy, reduce buildings' energy use, cut transportation and land use-related emissions, and reduce emissions from waste management (Boswell et al., 2010).

The proliferation of CAPs in US cities is an exciting development for the environmental movement. However, there is growing evidence that local climate planning has neglected issues of social equity (Finn and McCormick, 2011; Pearsall and Pierce, 2010). Many scholars have pointed out the impossibility of the capitalist ethos that promises perpetual economic growth. On a planet with finite space, resources and resiliency, continual development, even if it is green, is not merely stratifying, it is also built upon a flawed premise (Noble, 2012). Moreover, for those living at the intersections of raced, classed, and gendered disadvantages, CAPs may not signal an improvement but, ironically, a deterioration in their potentially sustainable livelihood. Modeled in a long tradition of urban planning that prioritizes profitable growth at the cost of equity and fairness (Mollenkopf, 1983; Harvey, 1973), CAPs may actually deepen gendered, raced, and classed inequities of wealth and opportunity in US cities.

In what follows, we briefly explain how CAPs have emerged in cities and municipalities throughout the USA as a foremost policy tool for fighting climate change. We then examine the intersections of race, gender, and class to discover who does and does not have access to wealth in the urban spaces where the vast majority of CAPs are enacted. We conclude with an intersectional assessment of the most common features in municipal CAPs. We suggest that CAPs render the needs of poor communities of color and a disproportionate number of women invisible while doing little to ameliorate cavernous wealth divides in US cities. CAPs thereby do not successfully address what many scholars agree to be one of the root causes of environmental degradation and climate change: systemic inequality (Wilkinson et al., 2010; Agyeman et al., 2003).

Global climate change and the promise of municipal CAPs

Beginning in the industrial age, the burning of fossil fuels from coal and natural gas plants as well as tailpipe emissions has released enough greenhouse gas into the Earth's atmosphere to impact a change in the planet's climate (IPCC, 2014). While some continue to argue that human-caused climate change is not scientifically verified, the Intergovernmental Panel on Climate Change confirms anthropogenic "warming of the climate system" (IPCC, 2014), which has spurred global action. In February 2005, the Kyoto Protocol, an international agreement to address climate change, went into effect, setting emission reduction targets for all signatory countries. While nearly 200 countries have ratified the protocol to date, a complex of corporate and political interests have prevented the USA from participating in the treaty or the promulgation of its own climate legislation (Layzer, 2011). Those interested in curbing environmental destruction have thus turned to target policy choices at the city, county, and state levels.

As one example, on the same day that the Kyoto Protocol went into effect in 2005, Seattle Mayor Greg Nickels helped launch the US Mayors Climate Protection Act to advance the goals of the Kyoto Protocol through local government action (USMCPA, 2013). As of April, 2013 there were more than 1000 cities that had signed the Act, committing to employ a variety of climate-related policies, typically taking the form of a Climate Action Plan (CAP). In the years since, CAPs have become *the* climate change prevention strategy throughout the USA, becoming ever more integrated into traditional city planning processes, complete with formal adoption by city councils (Boswell et al., 2010). However, as greater amounts of monetary capital and political will go toward planning and implementing CAPs, some have warned about the pitfalls of fashioning climate change prevention strategies in the mainstream urban planning tradition (Finn and McCormick, 2011). Critics suggest that the priorities of economic growth have long trumped those of equity in municipal politics (Pearsall and Pierce, 2010). CAPs are thus ill suited for pursuing any version of environmental protection that might constrain the interests of capital (Lutzenhiser and Hackett, 1993; Logan and Molotch, 1987). With this in mind, we now turn to the dynamics governing access to capital in urban centers.

An intersectional assessment of urban wealth inequality

An intersectional analysis suggests that racism, classism, and sexism cannot be disaggregated when considering the lived experiences of individuals and communities who face overlapping oppressions and, contrastingly, overlapping privileges (Collins, 1991; Crenshaw, 1991; hooks, 1984). Applied to a study of US cities, an intersectional lens reveals how race, class, and gender work together to largely determine which communities have access to capital, desirable real estate, and political clout (Peake, 1997; Sze, 2007). For the purposes of assessing the impact of CAPs, we are predominantly interested in who constitutes the urban, low-income population, as we argue that these communities disproportionately bear the costs of current climate change prevention strategies. An intersectional approach demonstrates that urban low-income communities are forged by the overlapping systems of not only class stratification, but also racism and patriarchy. Drawing upon this insight allows us to examine differential access to capital as well as the spatial segregation of US cities, both important factors in analyzing the impacts of CAPs.

Harvey's (1973) analysis of how class functions in the urban environment is a useful starting point. For Harvey, as for Marx, social class is not merely a measure of income or wealth, but a delineation of how individuals and social groups relate to capital and the means of production. In other words, how people make their money is class stratified and relevant in the urban environment. When it comes to sustainable urban planning, for instance, the capitalist class is able to accrue wealth through investment in new, green technologies and energy-efficient property. The working classes and urban underemployed, on the other hand, depend on their labor, state assistance, and informal economies to earn an income.

Furthering his application of Marx to the study of urban development, Harvey argues that those who own property and rent or sell it to others, such as landlords and realtors, are able to use housing and property "as a means of exchange – housing services are exchanged for money" (Harvey, 1973, p. 164). For renters, on the other hand, housing serves predominately as a use value, a place in which to reside. This means that those who rent are investing their incomes in the use of their houses as homes without gaining monetary profit or access to capital from this property. In juxtaposition, those who own property are able to accrue wealth on that property. Whether different groups are able to relate to property as having use or exchange value determines the possibility of earning profit and developing assets.

Class as a relationship to capital is racially segregated, with communities of color facing generations of asset stripping, barriers in access to property, and lower wages (Massey and Denton, 1993; Lipsitz, 2006). Scholars have argued that the wealth divide between white communities and people of color is largely due to accumulated assets, not earned income, and Americans hold most of their wealth in their homes (Oliver and Shapiro, 1997). For generations, people of color have been excluded from owning property and today face redlining, reverse redlining, predatory subprime mortgage lending, and rejection by insurance companies to protect property (Squires, 2003).

Moreover, legacies of subsidized white flight to the suburbs along with the dominant strategy of urban growth means that "decades of runaway sprawl have resulted in a geographically segregated society" (Chen, 2007, p. 299). This has disproportionately trapped people of color into asset-poor inner cities (Noble, 2012; Wilson, 1987; Mollenkopf, 1983). Industrial restructuring under neoliberalism has also had its most devastating impact on the wages and work opportunities of people of color who have been forced to turn to government assistance and informal economies to make ends meet (Wacquant, 2009). These communities are today over-policed and hyper-incarcerated, further damaging the earning potentials of families of color (Lipsitz, 2006).

The neoliberal policies that have restructured the US labor force, shrunk the state's safety net, and further segregated US cities have also converged to create the "feminization of poverty" in America, meaning that more women than men are poor (Peake, 1997). As low-paying service work has replaced middle-income jobs, women's per capita income has decreased. Women work longer hours for lower wages. The federal government has systematically dismantled the assistance programs for families and children upon which single mothers depend (Amott, 1993). It is no wonder that poverty rates are highest for single-headed families, a fact that is exaggerated for women (DeNavas-Walt et al., 2009). Nevertheless, while women, and specifically mothers, are disproportionately represented in low-income communities, this must be contextualized in terms of both race and class. In fact, being a white woman still confers relative income benefits, as white women continue to earn more than men of color (McCall, 2000).

For decades, activists and scholars in the environmental justice tradition have demonstrated that those who face the intersections of racism, classism, and sexism bear the brunt of environmental threats. Their neighborhoods are most often sited for toxic waste facilities, and they face the most dire environmental health dangers (Taylor, 2014; Bullard, 2005; Chavis and Lee, 1987). Because women of color are multiply disadvantaged in urban spaces and most often responsible for the care of children and the elderly (UNDP, 2007), they have also been at the forefront of confronting the environmental decay that threatens their families and communities (Sze, 2007; Di Chiro, 2008). However, their voices and actions are all but invisible in much of the public discourse informing municipal climate change policy (Shrader-Frechette, 2002). While urban policy-making may have the potential for dismantling environmental injustice if envisioned from an intersectional perspective, today's proposed solutions do not achieve this. Rather, CAPs, as a foremost version of municipal efforts to curb climate change, may in fact reproduce environmental injustice. We now apply an intersectional assessment of urban wealth inequality to an analysis of three key features in municipal CAPs.

CAPs and the (re)production of intersectional inequalities

CAPs generally contain provisions to: (1) increase renewable energy generation; (2) decrease energy consumption by built structures, and (3) reduce transportation emissions (Betsill and Rabe, 2009; Boswell et al., 2010, 2012; Ramaswami et al., 2012). Although water and waste reduction are often addressed and are important for comprehensive sustainability planning, they are not usually the major features of CAPs. For this reason we do not include them in our analysis. In what follows we look at examples of each of these to show that, as currently structured, the majority of CAPS provide wealth accumulation opportunities for those with access to capital while continuing to disinvest in urban women, communities of color, and the working and under-employed poor.

Increasing renewable energy generation

CAPs typically seek to increase the amount of energy generated from renewable sources in three major ways: the renewable portfolio standard, fees and taxes on energy use, and incentives for clean energy generation in building construction. All of these are structured in a regressive manner as large purchasers and those with investment capital are granted opportunities to increase their savings and profits while low-income communities bear disproportionate hardships.

The "renewable portfolio standard" mandates that utility companies use more renewable fuels, such as wind or solar energy, in order to reduce the GHG emissions that come from burning "dirtier" fuel sources like coal and gas. Renewable portfolio standards have been shown to dramatically increase renewable energy production in multiple parts of the country and, in this sense, effectively curtail climate change emissions at the local level (Ramaswami et al., 2012). However, such polices are also associated with at least some increase in energy costs for customers (Wiser et al., 2007). Large corporations that consume greater resources are given a kind of wholesale discount, as the more energy a single buyer purchases, the less it has to pay per unit. In this way, those with greater capital displace increased energy costs onto smaller scale consumers, such as home owners and renters. Further, as lower income US residents spend a higher proportion of their income on energy, they are the most negatively impacted by the energy rate increases in renewable portfolio standards. While state-funded energy assistance programs have the potential to alleviate these impacts, such programs have been de-funded since the energy crises of the late 1970s (see Higgins and Lutzenhiser (1995) for an excellent history of the decline of federal funding to states supporting low-income energy assistance programs and the attempts to replace these programs with traditional welfare assistance and non-profit or civic organizations).

A second and similar policy tool to increase renewable energy production includes fees and taxes on energy consumption. "Carbon taxes" are applied either to energy consumption or to the volume of emissions generated. Akin to the price increases associated with renewable portfolio standards, these taxation schemes are structured so that major consumers pay less tax per energy unit. Such programs have been used more extensively outside of the USA, where researchers have found that they disproportionately impact those with less wealth. In Denmark, for instance, carbon taxes have been found to be more regressive than income and even sales tax, which is usually considered to be the most regressive tax (Wier et al., 2005).

A third common means for increasing renewable energy generation provides enticements for producing energy at the building site itself, such as through monetary incentives for property owners who install solar panel roofing. An example of this is the California Solar Initiative, which promises financial bene-fits for on-site solar energy generation installations on existing residential homes and other kinds of properties. This program has been enormously successful by traditional environmental measures, allowing California to become the first state to install more than one gigawatt of customer-generated solar energy. However, of the more than 100,000 customers who were able to take advantage of the program in 2011, only 1 percent of those were families making less than $50,000 in annual income (CPUC, 2012). This meager number of low-income family participation exists despite government gestures at making the program broadly accessible. The Single-family Affordable Solar Homes (SASH) program was implemented alongside the California Solar Initiative to help include those who could most benefit from cost-saving solar installations in the program. However, that only 1 percent of participants were doing so through SASH suggests that the program's subsidies for low-income families have had little benefit for low-income communities and people of color who wish to take advantage of the California Solar Initiative. While laudable in intention, SASH appears impotent in practice.

Decreasing energy consumption by built structures

The second major category in CAPs seeks to reduce the amount of energy consumed by built structures through 'green building policies.' Green building policies range from improving the quality of building insulation for better temperature regulation, to installing roofs that capture rainwater, to establishing stricter limits for energy use in buildings. Green building policies are promising for curbing carbon emissions, but tend to offer profitable investment opportunities for those who can afford new property or upgrades to what they already own. Meanwhile, by way of increased rental pricing, the cost of housing is transferred to those who do not own their homes. Thus, Harvey's (1973 [2009]) critical assessment of the city as a nexus for reproducing income inequality through property ownership under neoliberal capitalism resonates. Those who do not own are forced to spend a growing percentage of their income on housing that is not accruing equity while contributing less to personal savings in the hope of future ownership.

Green building measures can be implemented either in the form of new construction or as retrofits to existing buildings. This translates into up-front costs for developers and owners but significant long-term savings owing to the decrease in energy expenditures. As one indicator of the financial benefits of green construction, the 2013 Building Energy Efficiency Standards in California are predicted to increase the cost of constructing a new home by $2290 (CEC, 2012). However, home owners are expected to save triple this amount over a 30-year mortgage. On the other hand, rents increase as property owners are able to displace increased costs on renters. What monetary benefits renters see in decreased utilities expenditures are dwarfed by the cost of increased rents. This inequity is not merely an unfortunate arrangement of the financial system but actually a wealth generator for those who own property. Indeed, the fact that energy-efficiency standards increase rental prices for commercial real estate is a point used to tout the 'business-friendly' nature of these policies (Fuerst and McAllister, 2011). There are further barriers for low-income communities in seeing the financial benefits of making their homes more energy efficient. Even in cities and towns where energy-efficiency measures for homes are not mandatory, but rather voluntary or subsidized, those without wealth are often unable to access capital to make energy cost-saving retrofits due to economic status or lack of "credit worthiness" (Golove and Eto, 1996). Again, while government assistance and grant programs could alleviate this inequity if aggressively implemented, these have been downsized and receive little political support in today's budget constrained economy.

Reducing transportation emissions

As transportation generates a large percentage of carbon emissions, an important category of CAPs includes efforts to decrease vehicle miles traveled (VMT). There are two major ways in which this is done: taxes or pricing schemes that raise the cost of driving, and the redesign of urban space for maximum transportation efficiency. We treat these in order.

1 Pricing schemes

There are three ways in which CAPs propose to raise the cost of automobile travel: taxes on fuel; taxes on driving during peak traffic flow; and incentives for hybrid or electric vehicle owners. The first two techniques, similar to other taxation proposals in CAPs, are regressive in nature. Low-income communities spend a greater proportion of their income on energy costs, not only for the home but also in the form of gasoline for vehicles (Lutzenhiser and Hackett,

1993). Thus, fuel and congestion taxes impact low-income communities more than they affect those with greater wealth (Sevigny, 1998). The third technique – enticements for the owners of non-traditional automobiles – again benefits those with greater disposable incomes while neglecting the transportation needs of the working classes, people of color, and women. Better parking spots or highway lanes are really the tip of the iceberg as hybrid and electric vehicles are but another green technology which benefits those with access to capital. More expensive than traditional vehicles, hybrid and electric cars require a high up-front cost. For instance, based on a cost comparison of these three automobiles conducted online at www.toyota.com, the 2012 Toyota Prius hybrid costs 50 percent more, and the 2012 Chevrolet Volt, an electric car, costs twice as much as the comparably sized, traditional fuel 2012 Toyota Matrix. Still, hybrid and electric cars come with the promise of significant savings in fuel prices over time, a promise that offers little for those who cannot afford these cars in the first place.

2 Land use policy

The organization of urban space through different land use policy is another manner of reducing emissions from the transportation sector. Distinct from policies intended to raise the cost of car travel, we understand the emerging category of 'transit-oriented development' to have the potential to remedy legacies of raced, classed, and gendered spatial segregation in US cities. Transit-oriented development locates high-density, mixed-use urban space close to low-cost, public transportation, such as light rails or major bus routes. These mixed-use developments incorporate a diversity of residential and commercial zones, allowing housing, employment, and retail opportunities to be positioned close together (Boarnet and Compin, 1999).

Community leaders have argued that transit-oriented development could actually benefit low-income communities the most because of its dramatic potential to reduce transportation costs. For instance, while not attending to distinctions based on race, class, or gender, the Center for Transit-Oriented Development nevertheless predicts that families residing in 'transit-rich neighborhoods' will spend less than one-tenth of their income on transportation. This is in comparison to families in automobile-dependent neighborhoods who spend an entire quarter of their income on transportation (Belzer and Poticha, 2009).

The problem is that these projects are rarely implemented in a manner that integrates low-income communities into policy considerations or the planning process itself (Wood and Brooks, 2009). There are a few reasons for this. First, transit-oriented development is a relatively new way of arranging urban space. Historically, US policy has encouraged home ownership, especially for white families, as a means for wealth accrual. However, providing a house and land for each nuclear family has long meant pushing sectors of the population into the suburbs, far away from most public transport (Markusen, 1980). Transportation-oriented development intends to achieve the opposite of the sprawling urban spaces that these earlier policies encouraged. Working against the historical gradient by which US cities have been organized, urban densification is still seen as experimental by many in the private and public sector. Those with surplus capital tend to be the ones investing in, and hence envisioning the design of, transit-oriented development. At the very least, this raises concerns regarding procedural equity, as middle- and certainly lower-income communities are rarely part of planning these new communities (Belzer and Poticha, 2009). The state's role in promoting transit-oriented development by incentivizing the private sector through tax credits further presupposes that those with wealth will be the ones to guide transit-oriented development.

Evidence also suggests that transit-oriented development prioritizes new construction over the preservation of already existing neighborhoods, endangering the affordable housing that does exist near transportation hubs. A 2008 report found that contracts on nearly two-thirds of the privately owned, subsidized housing within walking distance of public transportation in US cities were to expire by 2013, at which point private owners might choose to replace affordable housing with more profitable development (National Housing Trust and Reconnecting America, 2008). In the San Francisco Bay area, affordable housing is increasingly being located away from low-cost transportation toward the periphery of the city (Chapple et al., 2007), demonstrating that those with investment capital guide transit-oriented development to the exclusion of low-income communities.

Further, a number of studies demonstrate that without a substantial expansion of public transport in nearly every major US city, low-income communities cannot benefit much from these transportation opportunities even if they can afford nearby housing. Most low-wage and entry-level positions that employ these communities are not accessible by public transport (Coulton et al., 1996). Moreover, those receiving state assistance – disproportionately low-income women – are often forced to juggle geographically dispersed, entry-level shift work with childcare, education, and state-mandated job training. In lacking reliable and accessible public transportation, society's most vulnerable risk missing their various state-mandated obligations, resulting in serious penalties (Chen, 2007).

Thus, TOD creates a bit of a paradox for low-income communities. Housing closer to affordable transportation has become prime real estate, forcing the working poor further away from public transport to where housing is affordable. Still, in the absence of adequate public transport, households face longer commutes, wiping out any intended savings in housing costs. A recent study (Haas et al., 2006) demonstrates that lower income households are constantly budgeting transportation against housing expenditures. Averaging expenses across 28 metropolitan areas in the USA, the study finds that for the wealthy urban dweller earning between $100 and $250,000 annually, only 22 percent of annual income is spent on housing and transportation combined. For those making $35,000 to $50,000 a year, this total expenditure jumps to 39 percent. Making less than $20,000 a year means that affording housing and transportation is, in fact, impossible. Those in this bracket spend 115 percent of their income – more than they have – on transportation and housing alone.

The challenges that those without access to capital face in making ends meet are not alleviated by the current model of transit-oriented development with its foundations in privatized development with limited state intervention. Of course, land use patterns and social inequity pre-date CAPs. Nevertheless, as long as CAPs do not ensure the creation and maintenance of affordable housing in conjunction with an expanded public transportation system, current divisions in urban wealth and property ownership are only exacerbated.

Conclusion

This analysis shows some of the current inequities inherent in climate action planning. At best, subnational CAPs can help reduce carbon emissions as a meaningful bulwark against climate change. On the other hand, CAPs appear to be the latest incarnation of urban policies that deepen inequality along gendered, raced, and classed lines. CAPs often deploy regressive pricing structures. They benefit those who own property and have access to investment capital while shifting the costs of technological advancements and energy savings onto the poor. As women, communities of color, and the urban underemployed and working classes have less access to capital and property ownership, they are often merely left out of the financial benefits that other communities receive. More often, they are actually further disadvantaged by CAPs. The few programs that have sought to offer access to these environmentally beneficial practices have proven largely ineffective.

We argue that CAPs are front and center in the urban equity dilemma. It is beyond the scope of this chapter to assess whether the CAP, forged in the classical urban planning tradition and leveraged on a bedrock of neoliberalism, is capable of overcoming the raced, gendered, and classed inequities that plague our cities. It does seem, however, that small assistance programs for households and incentives for private developers have been insufficient to counter legacies of social injustice. If we are to begin to make the emission reductions necessary to address climate change, a fundamental transition to a more redistributive economy, especially as it relates to household energy expenditures, is required. Indeed, creating a climate action agenda that simultaneously addresses environmental destruction while dismantling the raced, gendered, and classed axes of exclusion and disadvantage in American cities is the only environmentally and socially just option.

References

Agyeman, J., Bullard, R.D., and Evans, B., eds, 2003. *Just Sustainabilities: Development in an unequal world*. New York: Earthscan.

Amott, T., 1993. *Caught in the Crisis: Women and the US economy today*. New York: Monthly Review Press.

Belzer, D. and Poticha, S., 2009. Understanding transit-oriented development: Lessons learned 1999–2009. In D. Wood and A. Brooks (eds), *Fostering Equitable and Sustainable*

Transit-oriented Development. New York: Center for Transit-oriented Development, pp. 4–11.

Betsill, M.M. and Rabe, B.G., 2009. Climate change and multilevel governance: The evolving state and local roles. In D. Mazmanian and M. Kraft (eds), *Towards Sustainable Communities: Transition and transformations in environmental policy,* 2nd edn. Cambridge, MA: MIT Press, pp. 201–226.

Boarnet, M.G. and Compin, N.S., 1999. Transit-oriented-development in San Diego County: The incremental implementation of a planning idea. *Journal of American Planning Association,* 65(1), pp. 80–95.

Boswell, M.R., Greve, A.I., and Seale, T.L., 2010. An assessment of the link between greenhouse gas emissions inventories and climate action plans. *Journal of American Planning Association,* 76(4), pp. 451–462.

Boswell, M.R., Greve, A.I., and Seale, T.L., 2012. *Local Climate Action Planning.* Washington, DC: Island Press.

Bulkeley, H. and Betsill, M., 2003. *Cities for Climate Change: Urban sustainability and global environmental governance.* London: Routledge.

Bullard, R., ed., 2005. *The Quest for Environmental Justice.* San Francisco, CA: Sierra Club Books.

California Energy Commission (CEC), 2012. Energy Commission approves more efficient building for California's future. Available at: www.energy.ca.gov/releases/ 2012_releases/2012–0531_energy_commission_approves_more_efficient_buildings_nr.html (accessed August 1, 2012).

California Public Utilities Commission (CPUC), 2012. About the California Solar Initiative. Available at: www.cpuc.ca.gov/puc/energy/solar/aboutsolar.htm (accessed August 1, 2012).

Chapple, K., Spade, E., and Lester, B., 2007. Shaping a mixed-income future: Lessons from the San Francisco bay area. Center for Community Innovation Working Paper, Institute for Urban and Regional Development.

Chavis, Jr., B.F. and Lee, C., 1987. *Toxic Wastes and Race in the United States.* United Church of Christ Commission for Racial Justice.

Chen, D., 2007. Linking transportation equity and environmental justice with smart growth. In R.D. Bullard (ed.), *Growing Smarter: Achieving livable communities, environmental justice, and regional equity.* Cambridge, MA: MIT Press, pp. 299–320.

Collins, P.H., 1991. *Black Feminist Thought: Knowledge, consciousness and the politics of empowerment.* New York: Routledge.

Coulton, C.J., Verma, N., and Shengyang, G., 1996. Time-limited welfare and the employment prospects of AFDC recipients in Cuyahoga County: A baseline technical report. Center on Urban Poverty and Community Development.

Crenshaw, K., 1991. Mapping the margins: Intersectionality, identity politics, and the violence against women of color. *Stanford Law Review,* 43(6), pp. 1241–1299.

Davis, M.D. and Weible, C.M., 2011. Linking social actors by linking social theories: Towards improved GHG mitigation strategies. *Carbon Management,* 2(4), pp. 483–491.

DeNavas-Walt, C., Proctor, B., and Smith, J., 2009. Income, poverty and health insurance coverage in the United States, 2006. *Current Population Reports,* U.S. Census Bureau, Washington, DC, pp. 60–233.

Di Chiro, G., 2008. Living environmentalisms: Coalition politics, social reproduction and environmental justice. *Environmental Politics,* 17(2), pp. 276–298.

DuBois, W.E.B., 1903. *The Souls of Black Folk.* Mineola, NY: Dover Publications.

Ellwood, D., 1998. *Poor Support: Poverty in the American family*. New York: Basic Books.

Finn, D. and McCormick, L., 2011. Urban climate change plans: How holistic? *Local Environment*, 16(4), pp. 397–416.

Friedan, B., 1963. *The Feminine Mystique*. New York: W.W. Norton.

Fuerst, F. and McAllister, P., 2011. Green noise or green value? Measuring the effects of environmental certification on office buildings. *Real Estate Economics*, 39(1), pp. 45–69.

Golove, W.H. and Eto, J.H., 1996. Market barriers to energy efficiency: A critical reappraisal of the rationale for public policies to promote energy efficiency. Energy and Environment Division of Lawrence Berkeley National Laboratory, Berkeley, CA.

Haas, P.M., Makarewicz, C., Benedict, A., Sanchez, T.W., and Dawkins, C.J., 2006. Housing and transportation cost trade-offs and burdens of working households in 28 metros. The Center for Neighborhood Technology.

Harvey, D., 1973. *Social Justice and the City*. Athens: University of Georgia Press.

Higgins, L. and Lutzenhiser, L., 1995. Ceremonial equity: Low-income energy assistance and the failure of socio-environmental policy. *Social Problems*, 42(4), pp. 468–492.

hooks, b., 1984. *Feminist Theory: From margin to center*. Boston, MA: South End Press.

Intergovernmental Panel on Climate Change (IPCC), 2014. *Climate Change 2014: Synthesis Report*. Geneva, Switzerland: IPCC.

Kraft, M.E. and Mazmanian, D.A., 2009. Conclusions: Towards sustainable communities. In D. Mazmanian and M. Kraft (eds), *Towards Sustainable Communities: Transition and transformations in environmental policy*, 2nd edn. Cambridge, MA: MIT Press, pp. 317–334.

Layzer, J., 2011. *The Environmental Case: Translating values into policy*, 3rd edn. Washington, DC: CQ Press.

Lipsitz, G., 2006. *The Possessive Investment in Whiteness: How white people benefit from identity politics*. Philadelphia, PA: Temple University Press.

Logan, J. and Molotch, H., 1987. *Urban Fortunes: The political economy of place*. Berkeley: University of California Press.

Lutzenhiser, L. and Hackett, B., 1993. Social stratification and environmental degradation: Understanding household CO_2 production. *Social Problems*, 40(1), pp. 50–73.

Markusen, A., 1980. City spatial structure, women's household work, and national urban policy. *Signs*, 5(3), pp. S22–S24.

Marx, K. and Engels, F., 1978. *The Marx-Engels Reader*, 2nd edn, ed. R.C. Tucker. New York: W.W. Norton.

Massey, D. and Denton, N., 1993. *American Apartheid: Segregation and the making of the underclass*. Cambridge, MA: Harvard University Press.

McCall, L., 2000. Explaining levels of within-group wage inequality in U.S. labor markets. *Demography*, 37(4), pp. 415–430.

Mollenkopf, J.H., 1983. *The Contested City*. Princeton, NJ: Princeton University Press.

Molotch, H., 1976. The city as a growth machine: Toward a political economy of place. *American Journal of Sociology*, 82(2), pp. 309–332.

Nagel, J., 2012. Intersecting identities and global climate change. *Identities: Global Studies in Culture and Power*, 19(4), pp. 467–476.

Noble, D.W., 2012. *Debating the End of History: The marketplace, Utopia, and the fragmentation of intellectual life*. Minneapolis: University of Minnesota Press.

National Housing Trust and Reconnecting America, 2008. *Preserving Opportunities: Saving affordable homes near transit*.

Oliver, M.L. and Shapiro, T.M., 1997. *Black Wealth/White Wealth: A new perspective on racial inequality*. New York: Routledge.

Peake, L., 1997. Toward a social geography of the city: Race and dimensions of urban poverty in women's lives. *Journal of Urban Affairs*, 19(3), pp. 335–361.

Pearsall, H. and Pierce, J., 2010. Urban sustainability and environmental justice: Evaluating linkages in public planning/policy discourse. *Local Environment*, 15(6), pp. 569–580.

Ramaswami, A., Bernard, M., Chavez, A., Hillman, T., Whitaker, M., Thomas, G., and Marshall, M., 2012. Quantifying carbon mitigation wedges in U.S. cities: Near-term strategy analysis and critical review. *Environmental Science and Technology*, 46, pp. 3629–3642.

Sevigny, M., 1998. *Taxing Automobile Emissions for Pollution Control*. Cheltenham: Edward Elgar.

Shearer, C., 2011. *Kivalina: A climate change story*. Chicago, IL: Haymarket Books.

Shrader-Frechette, K., 2002. *Environmental Justice, Creating Equality, Reclaiming Democracy*. Oxford: Oxford University Press.

Squires, G., 2003. Racial profiling, insurance style: Insurance redlining and the uneven development of urban America. *Journal of Urban Affairs*, 2(4), pp. 391–410.

Sze, J., 2007. *Noxious New York: The racial politics of urban health and environmental justice*. Cambridge, MA: MIT Press.

Taylor, D., 2014. *Toxic Communities: Environmental racism, industrial pollution, and residential mobility*. New York: New York University Press.

The Sentencing Project, 2010. *Racial Disparity*. Available at: www.sentencingproject.org/template/page.cfm?id=122 (accessed July 30, 2012).

United Nations Development Programme (UNDP), 2007. *Fighting Climate Change. Human solidarity in a changing world*. New York: UNDP.

US Mayors Conference of Mayors Climate Protection Agreement (USMCPA), 2013. Available at: www.usmayors.org/climateprotection/agreement.htm (accessed March 15, 2012).

Wacquant, L., 2009. *Punishing the Poor: The neoliberal government of social insecurity*. Durham, NC: Duke University Press.

Wier, M., Birr-Pedersen, K., Jacobsen, H.K., and Klok, J., 2005. Are CO_2 taxes regressive? Evidence from the Danish experience. *Ecological Economics*, 52, pp. 239–251.

Wiser, R., Namovicz, C., Gielecki, M., and Smith, R., 2007. The experience with renewable portfolio standards in the United States. *The Electricity Journal*, 20(4), pp. 8–20.

Wilkinson, R.G., Pickett, K.E., and De Vogli, R., 2010. Equality, sustainability, and quality of life. *BMJ*, 341.

Wilson, W.J., 1987. *The Truly Disadvantaged: The inner city, the underclass, and public policy*, 2nd edn. Chicago, IL: University of Chicago Press.

Wood, D. and Brooks, A., 2009. Overview of briefing papers. In D. Wood and A. Brooks (eds), *Fostering Equitable and Sustainable Transit-oriented Development*. New York: Center for Transit-Oriented Development, pp. 1–3.

39 Dear future generations

Sorry

Prince Ea

Dear future generations,
I think I speak for the rest of us when I say,
sorry, sorry we left you our mess of a planet.
Sorry that we were too caught up in our own doings to do something.
Sorry we listened to people who made excuses,
to do nothing.
I hope you forgive us,
we just didn't realize how special the earth was,
like a marriage going wrong,
we didn't know what we had until it was gone.
For example,
I'm guessing you probably know what is the Amazon Desert, right?
Well believe it or not,
it was once called the Amazon Rain Forest,
and there were billions of trees there,
and all of them gorgeous and just um.
Oh, you don't know much about trees, do you?
Well let me tell you that trees are amazing,
and I mean, we literally breath the air
they are creating, and they clean up our pollution,
our carbon, store and purify water,
give us medicine that cures ours diseases, food that feeds us.
Which is why I am so sorry, to tell you that,
we burned them down.
Cut them down with brutal machines, horrific,
at a rate of 40 football fields every minute,
that's 50% of all the trees in the world all gone
in the last 100 years.
Why? For this.

And that wouldn't make me so sad,
if there weren't so many pictures of leaves on it.
You know when I was a child,
I read how the Native Americans had such consideration,
for the planet that they felt responsible,
for how they left the land for the next 7 generations.

Which brings me great sorrow, because most of us today,
don't even care about tomorrow.
So I'm sorry, I'm sorry that we put profit over people,
greed over need, the rule of gold above the golden rule.
I'm sorry we used nature as a credit card with no spending limit.
Over drafting animals to extinction,
stealing your chance to ever see their uniqueness,
or become friends with them.
Sorry we poison the oceans so much that you can't even swim in them.
But most of all, I'm sorry about our mindset,
cause we had the nerve to call this destruction,
"Progress."

Hey Fox News, if you don't think climate change is a threat.
I dare you to interview the thousands of homeless people in Bangladesh,
see while you was in your penthouse nestled,
their homes were literally washed away
beneath their feet due to the rising sea levels,
and Sara Palin, you said that you love the smell of fossil fuels,
well I urge you to talk to the kids of Beijing
who are forced to wear pollution masks just to go to school.
You see you can ignore this, but the thing about truth is,
it can be denied, not avoided.
So I'm sorry future generations,
I'm sorry that our footprints became a sinkhole and not a garden.
I'm sorry that we paid so much attention to ISIS,
and very little to how fast the ice is melting in the Arctic.
I'm sorry we doomed you
and I'm sorry we didn't find another planet in time to move to.
I am s…

You know what, cut the beat, I'm not sorry.
This future I do not accept it,
because an error does not become a mistake,
until you refuse to correct it.
We can redirect this, how?
Let me suggest that if a farmer sees a tree that is unhealthy,
they don't look at the branches to diagnosis it,
they look at the root, so like that farmer,
we must look at the root,
and not to the branches of the government,
not to the politicians run by corporations.
We are the root, we are the foundation, this generation,
it is up to us to take care of this planet.
It is our only home, we must globally warm our hearts
and change the climate of our souls
and realize that we are not apart from nature,
we are a part of nature.
And to betray nature is to betray us,

to save nature, is to save us.
Because whatever you're fighting for:
Racism, Poverty, Feminism, Gay Rights,
or any type of Equality.
It won't matter in the least,
because if we don't all work together to save the environment,
we will be equally extinct.

Sorry.

40 All Yours

Mr. Fish

Figure 40.1 All Yours by Mr. Fish.

Part V
Water

Figure V.1 Water by Tina Shirshac.

Flooding   Drought   Emotions   Spirituality   Turmoil

41 The fluidity of identity and the crisis of material reality

Denise Torres

Water is the embodiment of metamorphosis capable of transforming from solid to liquid to gas in a potentially endless cycle that is essential to life on Earth. As a section and intersection, water simultaneously represents the progressive rapidity of these transformations and the resultant turmoil in our physical, environmental, and social worlds. Materially, some of the manifestations of global climate change (GCC) addressed in Part V include increased flooding, drought, and other water-based disasters. Experientially, we examine the substrates of our fluid identities, those such as our emotions and spirituality that may remain below the surface, submerged, or contained, and yet may nevertheless inform our actions and attitudes.

We begin with Jose Gonzalez's artwork entitled *El Agua es la Vida*, or water is life, which exhorts us to 'cuidala,' or care for it. The centrality of water to life is captured in the adage that humans, like the Earth, comprise 75 percent water. While most adults comprise no more that 60 percent water, there is variability within and among us: individual water composition is dependent on variables such as age, sex, percentage of fat versus lean muscle, and fluid intake (United States Geological Survey, 2015). This context dependency is the focus of the first half of Part V as our contributors interrogate which variables become salient when we experience disasters. Hamad Sindhi, in 'Citizenship: environmental disasters, intersectional vulnerabilities, and changing citizenship models,' grapples with the question of an often silent identity – citizenship – using the exemplar of Hurricane Katrina. His work is especially timely as the USA marks Katrina's tenth anniversary amid calls for ending birthright citizenship. Sindhi's more conceptual piece is solidified by Al Duvernay's 'Race, social class, and disasters: the Katrina version of reality,' a firsthand account of how certain citizens were and continue to be neglected.

We offer a meditative interlude to begin the second half of Part V. Cherese Mathews' poem, 'Poison water blessings,' exposes our 'delusional' disconnection from our actions and is paired with Subhankar Banerjee's photo, 'Sea ice,' allowing us to pause and reflect on how our actions are impacting our ice cover and oceans. Of note, solitary, floating icebergs have been used to symbolize GCC; they have also been used as a cultural model, where the visible portion represents the behaviors of a society and the larger, invisible part represents values and thought patterns that are implicitly learned and difficult to change. We offer that while parts of us remain invisible, we are intra-acting and negotiating above and below surface simultaneously. Indeed, as our environments change, aspects of us may gradually melt away, while others are rapidly shaken off; or we may expand, taking in and hardening the water around us as rain and snow settle upon us. In addition, we may also grow or shrink as we softly – or abruptly – bump up against others. To demonstrate the power of examining what is below, Lisa Corrigan and Holly Rawn chip through the surface layers of US evangelical attempts to engender responsibility for the Earth, uncovering the perpetuation of 'white' male, Christian privilege in 'Evangelical Environmental-ism: an analysis of gender and ideology.' In Victoria Team and Eyob Hassen's 'Climate change and complexity of gender issues in Ethiopia' they examine the theme of women's voicelessness while linking water to global food, health, and migration crises. To reinforce the fluidity and turmoil which GCC provokes, we close out Part V with a poem by Anthony Richardson, 'How climate change makes me feel,' which exposes the emotions colliding within him. Hopefully, by giving voice to the flood of feelings which many experience, we can gently connect, recognizing that despite our differences our very human emotions bind us.

Reference

United States Geological Survey (July, 2015). The water in you. Available at: http:// water.usgs.gov/edu/propertyyou.html (last modified: July 27, 2015) (accessed August 20, 2015).

42 El Agua es la Vida

José G. González

Figure 42.1 El Agua es la Vida (Water is Life) by José González. Water may now be a commodity and consumer product. But it has long been a culturally integrated component in the conservation practices of our communities – it is life, and should be treated as such.

43 Citizenship

Environmental disasters, intersectional vulnerabilities, and changing citizenship models

Hamad Sindhi

Disasters, whether man-made or environmental, have played a major part in our understanding of what citizenship means in modern Western society. This chapter will focus on the development of the idea of citizenship in the shadow of major disasters. I will first make the claim that citizenship is a concept in flux, and more so in the aftermath of major crises. How a state responds to protect its citizens after environmental disasters, I argue, redefines what it means to be a citizen in that state. And because the capitalist state doles out resources differentially across axes of class, race, gender, and disability, I will argue that we need to apply an intersectional lens when we study citizenship. I will then explore whether the state acknowledges lived intersectional experiences of its most marginalized citizens during times of disaster by looking at Hurricane Katrina as an example. Finally, given the perpetual nature of environmental threats as a result of anthropogenic global climate change (GCC), I will advocate for the use of an intersectional lens when analyzing and thinking about citizenship in a disaster-prone world.

But first, what do we mean by 'citizenship'? Citizenship is more than simply an abstract idea or concept about where and how we belong. It is also a social force, and like any other social forces it is largely hidden, and it constrains and determines social actions in particular contexts. Citizenship informs our relationships with other people within our actual and imagined territorial vicinity, it shapes our actions and responsibilities toward the state, and it realizes the protections and rights to which we can lay claim whenever we are harmed or threatened.

Scholars who have studied citizenship usually approach it from one of two perspectives: a top-down view that sees citizenship as a set of rights given out by states, or a bottom-up view that sees citizenship as a set of negotiated claims, where the power of negotiation often depends on the perceived race, class, gender, and disability status of the citizen. For instance, the prevailing scholarship on citizenship in the 1950s contended that the concept gained more prominence as states kept giving their subjects more rights (Marshall and Bottomore, 1992). Contemporary scholars who follow this perspective still privilege the role of the state but may also include citizens as actors who can and do demand more rights (Bulmer and Rees, 1996; Hindess, 2002; Pakulski, 1997). The alternative perspective privileges the actions of citizens in their negotiated claims for more recognition by the state. Scholars who follow this perspective look at the social actions of individuals or groups of citizens, usually marginalized, who start using the language of rights to articulate claims to recognition and protection by the state (Gilbert and Phillips, 2003; Isin and Nielsen, 2008).

I propose that both of the aforementioned perspectives, and the theories that derive from them, can benefit from the explicit application of an intersectional lens – a theoretical tool developed by feminist thinkers Kimberle Crenshaw and Patricia Hill Collins. Intersectionality, according to Collins (2000), allows us to rethink our theories and practice in social science by applying an analysis of the intersecting effects of race, class, gender, and other identities. Hence, the subjects we study and theorize around can no longer be thought of as occupying only one social position, but multiple positions with differential relationships to power. The application of an intersectional lens would mean that one would have to analyze how a state would respond to a citizen who is, for example, a black, transgender, disabled woman living in a shelter in New Orleans, where the intersecting identities of such a citizen would be analyzed simultaneously (race, gender, ability, and class). Although one would expect the use of intersectionality in the work of scholars who follow the bottom-up perspective (where citizens lay claims to rights previously denied them), this is not apparent in the literature. Most scholars have only used singular identities (such as race only) to analyze claims-making behavior, and not an intersection of multiple identities (see, e.g., Lister, 2007).

When Kimberle Crenshaw (1989) first proposed the theory of intersectionality, she was intrigued by a particular legal case (*DeGraffenreid* v. *General Motors*) which involved black women plaintiffs who had recently been fired from their jobs by General Motors. On the face of it, the case was a simple matter of discrimination against the plaintiffs because of their status as black women. However the anti-discrimination laws at the time did not recognize 'black women' as a class distinct from either 'black' or 'women.' In other words, the law only allowed the plaintiffs to seek redress as *either* 'black' *or* 'women,' not both. When viewed in this way, no discrimination was obvious because if the plaintiffs used racial non-discrimination laws, there was evidence that GM had not discriminated against black men, and when they used gender non-discrimination laws, there was no evidence that GM had discriminated against white women. Hence, it was the intersecting identities of these black women plaintiffs that made them vulnerable to discrimination, and the law refused to recognize such discrimination. The case went in the favor of General Motors.

State recognition and using the law for protection vis-à-vis anti-discrimination laws was the impetus for thinking through intersectionality as a theoretical framework. And though legal studies, race and gender theory, and other scholarly fields have greatly benefited from applying an intersectional lens to their research questions and analyses, citizenship studies has room for improvement. This does not mean that no citizenship scholars are using intersectionality in their analyses. Nira Yuval-Davis (2006), for example, asks scholars of citizenship to pay attention to the effects of multiple social positions on each other and on the status of citizenship itself. Following this intervention in citizenship scholarship by Yuval-Davis and scholars of intersectionality before her, it is no longer adequate to create models of citizenship that assume one-dimensional citizens.

Citizenship may, of course, also be viewed as an identity or a status itself. Many people living in a state do not possess the official status of citizenship in terms of legal documents stating that they reside in or belong to the state in which they are living. Scholars should certainly include citizenship status (as in does one have it or not) as an axis of identity when doing intersectional analyses. However, for the purpose of this analysis, I will treat citizenship as a process rather than as an identity. In order to tease out how citizenship is reconfigured, enacted, and differentially applied, I will be forced to limit the analysis to those people who are at the very least recognized as citizens by the state, albeit unequally. Because I want to focus on this inequality between citizens, including those categorized as non-citizens would detract from the purpose of this study.

In the next section I will look at the prevailing model of citizenship and some of the recent corrections to it, and focus on how disasters have usually been a backdrop to and sometimes even an impetus for rethinking citizenship models. This makes sense once we realize that citizenship, in terms of claims to benefits and protection from the state, becomes most visible during times of widespread crises. Consequently, in the following section I will look at how disasters have helped shape the evolving structures of citizenship, and why applying an intersectional lens to the study of both disasters and citizenship can be beneficial. Next, I will look at some evidence from the US federal government's response to one of the deadliest disasters in its history to demonstrate that the state still has some way to go in acknowledging its citizens' lived intersectional identities and experiences. Finally, I will close out this chapter by looking at the potential impact which global climate change may have on our understanding of citizenship, and how including time and space in our intersectional lenses can be useful for this context.

Liberal ideologies and the contributory model

Liberal ideologies stemming from the English Enlightenment (a precursor to the advent of industrial capitalism) have for some time now provided the framework for any and all discourses around citizenship. In particular, Thomas Hobbes' concept of the 'state of nature' has been useful for citizenship scholars. Hobbes assumed that all people in prehistory (i.e., before any social arrangements existed) had equal capacities and capabilities to compete for and obtain the scarce resources they needed to survive (Macpherson, 1962). The concept assumes that social constraints like inequality and oppression did not play any role in the competition for those scarce resources, and so the 'natural' condition of all humans is to compete with each other on equal terms.

Scholars expanded this idea to theorize the rise of nation-states: basically, to regulate the peace and trust between equally capable men, a more powerful entity was needed (Pelczynski, 1984). Accordingly, scholars following this tradition have thought of citizens as independent contractors who come together in mutual interest to form a state which, in turn, protects its citizens against each other, foreign enemies, man-made and natural disasters, etc. (Turner, 2006). Theories and discussions about citizenship that ground themselves in this framework may be thought of as *contractarian*; they center the contract between the citizen and the state as a foundational principle. It is well worth pointing out here that the 'state of nature' assumes that all citizens have equal capacities, and that social forces such as patriarchy, racism, ableism, and other forms of social oppression do not exist. Hence, when scholars, and society at large, assume that the 'natural' state that informs human behavior and motivation is fair and equal competition for scarce resources, they privilege propertied white men (who for a long time were the only ones able to compete in the marketplace). Consequently, citizenship and its protections, when built on the logics of a 'state of nature,' in effect also privileged only white men of a certain class and ability to prosper.

Shortly after World War II, T.H. Marshall was thinking and writing about citizenship both as a social structure and an ongoing process in Great Britain. Marshall observed that citizenship worked in a 'contributory' way. Citizens would contribute their labor, reproductive capacities, and service (especially during times of war) in exchange for equal rights and benefits from the state. He also theorized that citizenship followed a trajectory of progressive inclusion in England. From about the twelfth century to around the early 1940s, citizenship in Great Britain slowly came to encompass different sets of rights: first, civic rights (liberties and freedoms), then political rights (representation in decision making), and finally social rights (social and economic safety nets) (Marshall and Bottomore, 1992). In the final stage (social rights), the theory assumes that since citizens would contribute unequally based on their socio-economic positions, citizenship could diminish the inequalities of social class through a redistribution of resources (Turner, 2009). In a sense, citizenship would function as a safety net for those who could not contribute as much.

There are many critiques of this progressive trajectory (Bulmer and Rees, 1996), but I would like to focus on Marshall's theory about the 'contributory' nature of citizenship. As a general model this may have accounted for much of how citizenship was set up in Great Britain after World War II; however, it missed many people who could not contribute in any of the three ways Marshall indicated (work, reproduction, or service) due to long-term disabilities, sexual orientation, homelessness, etc. More often, though, those who were left out of the citizenship equation were those who *did* contribute, but were not seen as legitimate citizens worthy of equal protection by the state. Women (before getting the right to vote and other equal protections and benefits), immigrants, and the colonized subjects of the British Empire made up this category. Citizenship, in Great Britain, was designed to exclude these people (Tyler, 2010). Instead of becoming an equalizer as Marshall theorized, in practice citizenship was set up to create hier-archies along the lines of race, class, gender, and disability.

One of the most recent critiques of Marshall's theories comes out of an ana-lysis of citizenship by Margaret Somers (2008), who observed the failure of the state after Hurricane Katrina devastated New Orleans in 2005. Writing some 60 years after Marshall, Somers challenges the contributory model and comes close to positing a more realistic model for citizenship, a *conditional* model (my term, not Somers'). Like the contributory model, the conditional model still holds that citizens enter into a 'contract' with the state to exchange their labor, repro-ductive capacities, and service for protection. Unlike the contributory model, it takes into account the reality that the state offers protection and benefits only to some. The more useful the citizen is perceived to be, the more rights and benefits they can get from the state. And in a patriarchal, racist, and ableist society like the USA, the perception of 'usefulness' is inflected through race, gender, class, and (dis)ability. In other words, the protection offered by citizen-ship in the USA is distributed unequally based on race, class, gender, and disa-bility status. This corrects for the traditional contributory relationship, especially in the context of a neoliberal state that is less and less interested in protecting many of its citizens.

According to David Harvey (2006), neoliberalism is a pervasive ideology that re-centers individualism, free market assumptions, private property, and other ideologies from the English Enlightenment. Neoliberalism may be seen as a reversion to the original liberal ideologies (hence the term *neo*liberal) of the English Enlightenment. The role of the state in a neoliberal context is to privi-lege the logic of the free market over and above all other considerations (Harvey, 2006). Neoliberal ideologies advocate for the shift of state responsibil-ities to the market, because they assume that all individuals can compete on equal terms in the marketplace. Hence, during disasters, the state shifts its responsibility to protect its citizens to the market (Klein, 2008). Somers observes this shift in responsibility for particular citizens in the aftermath of Hurricane Katrina – the state offered little to no help, and many poor black people had to contend with private contractors to survive and rebuild. Those who could not access the market did not survive.

Disaster citizenship, intersectional citizenship

Although Marshall himself focused more on the benefits, or rights, that citizens ought to receive from the state, his theory has been extended to include the duties and obligations on the part of the citizens to maintain claims on those rights. Indeed, much of Marshall's work revolves around the claims making done by workers in an industrialized Great Britain, culminating in more social and economic benefits for the proletariat. Working in a post-disaster context (i.e., post-World War II), Marshall may also have been cognizant of disabled war veterans and civilians, and the failure of traditional liberal ideologies in accounting for them. Thus, the emphasis in his model on the state's obligations (rather than the citizens' contributions) may have been supported by his observations of a disaster-disabled citizenry, as much by the working-class citizen, both of whom needed protection by the state. Turner (2009) critiques this by focusing on the ableist implications of these duties, observing the exclusionary consequences of these responsibilities for disabled, elderly, mentally ill, and otherwise incapacitated citizens. Indeed, there is no direct evidence in Marshall's writings linking war-induced disability to state protection, and critiques that this model of citizenship ultimately imagines and privileges an able-bodied, white, male, and self-sufficient citizen are still largely valid.

Somers starts to correct for the assumptions in the contributory model by examining the effects of Hurricane Katrina on the black and poor residents of New Orleans. The ability of a disaster to expose the contours of internal exclusions in American citizenship is not surprising (Tierney, 2007), and, of course, Somers' main point is to excavate the long social and historical process by which poor and working-class black New Orleanians were made stateless. Somers looks to the history of slavery, social exclusions stemming from the institutionalization of segregation, and the deliberate and consistent decline of welfare programs as indicators of the internal exclusion faced by the poor and black population throughout the history of the USA. Such exclusions left these groups exposed to the violence of both nature and the market, as evidenced by the stories of devastation and deprivation precipitated by Hurricane Katrina. For Somers, the questions that immediately surfaced after Katrina struck – why did the state not step in sooner, why was the response of the state (an otherwise resourceful and powerful entity) so miserable, and how did so many citizens become 'refugees' in their own country? – required an analysis of how citizenship is actually structured in the USA. The implications are clear for those who study citizenship: unless we take multiple and intersecting identities into account, the questions Somers poses won't make sense.

Both the contributory and conditional models of citizenship are based on shifting thinking about what rights and protections citizens can claim once they are affected by disasters. The reconceptualization of what it means to be a citizen following disasters is intriguing, at least at face value. Both Marshall and Somers look to citizenship as a form of protection, bestowed either by the state or a combination of the state and civil society, against market forces in a capitalist society. Whereas classical liberal (and now neoliberal) ideologies would imagine citizens as independent, self-sufficient, and able-bodied, the experiences of disasters often expose the fallacy of such liberal assertions. Since notions of citizenship were conceptualized and reconceptualized within the framework of liberal ideologies as well as in the context of disasters, they often embody the tension between imagined self-sufficiency and lived realities. Citizenship scholars must work to move in the direction of the latter, exposing, as much as possible, the inequalities inherent in the experience of citizenship by those who occupy different social locations.

A study of claims making by citizens in post-disaster contexts can demonstrate that when citizens claim protection from the state they do it from an intersectional standpoint (i.e., from a lived experience that is informed by their multiple and coextensive identities). Even if we demonstrate this, though, it is not at all clear that the state recognizes the complex intersectional realities that its citizens live in, and many times falls back on notions of flat, similar, and one-dimensional citizens perpetuated by (neo)liberal ideologies. Whether seen through the lens of class struggles and post-World War II interdependency, or through the history of slavery, racism, and the declining welfare state, the lived realities and experiences of people who occupy multiple and intersecting identities must be attended to whenever ideas about citizenship are reformulated in post-disaster contexts. They must not just be implied, but be explicitly drawn out.

Example: the congressional report on Hurricane Katrina

When citizens seek protection and benefits from the state after disasters, they do so from a place of intersecting identities and their related oppressions. To reiterate the hypothetical example from this chapter's introduction, how would a black, transgender, disabled woman living in a homeless shelter in post-Katrina New Orleans seek protection from the State of Louisiana or the US Federal government without invoking the multiple oppressions of racism, sexism, transphobia, and homelessness? Alternatively, how will the state respond to provide the most comprehensive protection if it cannot read the multiple oppressions that have converged for this woman, and have now been exacerbated by the disaster? Will the state only consider the disabled status of the woman, ignoring her homelessness, her race, and her gender identity? Or will it provide her with shelter, but in a men's facility? At this point, I do not have the data to fully answer these questions. However, I can start to explore a more basic question: Does the state acknowledge the lived, intersecting realities of its marginalized citizens in a post-disaster context? In this section, I look at the Congressional report on Hurricane Katrina to find evidence (or lack thereof) of whether and how the state reads the intersecting identities of the hurricane's victims.

The 2006 Congressional report on Hurricane Katrina, entitled *A Failure of Initiative: The Final Report of the Select Bipartisan Committee to Investigate the Preparation for and Response to Hurricane Katrina* (U.S. Congress, 2006), was completed some six months after the hurricane devastated New Orleans, as well as other parts of Louisiana and neighboring states. This report is one of the most comprehensive archives of the response by the federal government, the State of Louisiana, local governments, and charitable organizations. The report, as its name suggests, concluded that the state, at all levels, not only failed to fulfill its obligation to protect, "an abdication of the most solemn obligation to provide for the common welfare" (U.S. Congress, 2006: x), but also impaired the effectiveness of local organizations, associations, and individuals in providing help. It is important to note that though some of the language used in the report promotes the idea of governmental duties and obligations toward its citizens, most of the spirit of the document is informed by Nozick's (1974) notion of the 'nightwatchman' state, or, a state that only passively provides support through the market or through local requests. Ultimately, the report did not call upon the failure of 'duties' or 'obligations' – terms that presume a priori responsibilities of the state – but rather the failure of 'initiative,' a term that denotes action only called upon in specific circumstances.

When reading the document, one is hopeful that, finally, the committee will consider an intersection of various identities in its analysis of the evidence. Indeed, one of the opening statements in the report states,

> There was little question that Katrina had sparked renewed debate about race, class, and institutional approaches toward vulnerable population groups in the United States. In the aftermath of the storm, a wide array of media reports, public statements, and polls underscored this reality.
>
> (U.S. Congress, 2006, p. 19)

This introduction of the report lists at least five studies and newspaper articles that analyze the confluence of race and class during the storm. Indeed, there is language that comes close to promising a fuller analysis of the intersection of (at least) race and class: "At the very least, the Select Committee determined it should further explore at this hearing how socioeconomic factors contributed to the experiences of those directly affected by the storm" (U.S. Congress, 2006, p. 23).

The introduction of the report does note that the intersection of race and class presented unique challenges for many poor black New Orleanians, especially when it came to communicating messages around evacuation and the severe nature of the hurricane. But it stops there. There is no follow-through, and a similar analysis is nowhere to be found when the report discusses medical need, shelter, and other aids after the hurricane. Gender is not once commented upon, except to acknowledge that about the same proportion of women died as in the general population. The following is an account of the dead in the report:

> In terms of ethnicity, the dead in New Orleans were 62 percent black, compared to 66 percent for the total parish population. The dead in St. Bernard Parish were 92 percent white, compared to 88 percent of the total parish population. The percentage of the dead by sex was approximately the same as the overall population. In terms of wealth, the analysis found that the percentage of dead bodies found in poorer New Orleans and St. Bernard Parish neighborhoods – as measured by poverty rates and median household incomes – was roughly equivalent to their percentage in the overall population.
>
> (U.S. Congress, 2006, p. 115)

No statistics are presented on how many black women, or poor disabled folk, or poor, black, disabled women were affected or died during the storm.

In its conclusion, the report again reiterates the importance of considering the overlapping racial and socio-economic contexts with regard to messaging and communicating the importance of the evacuation. However, no conclusions are made as to the impact of the intersections of race, class, gender, and disability when it came to state response and aid. It appears that the report only imagines differences among citizens before the hurricane, when everything was stable, but suddenly flattens the citizen after the disaster. Some of the last words in the report are: "The failure of initiative was also a failure of empathy, a myopia to the need to reach more people on their own terms" (U.S. Congress, 2006, p. 362). "People on their own terms" are usually located at the intersections of varying identities and social contexts. By failing to adequately acknowledge the multiple and interacting oppressions of black, poor, disabled, and sometimes homeless New Orleanians, the report itself seems to have failed in the task of reaching those people on their own terms.

Whereas the text of the report provides an account of the institutional failures of the state, the images in the report give another account – that of the perceived failures of particular citizens. Lakshmi Fjord (2007) analyzes the images reproduced in the Congressional report and connects them to narratives about race and disability in a post-disaster context. Fjord relies on the central concepts of intersectionality theory, and particularly on theoretical insights about the intersection of race and disability developed by disability studies scholar Douglas Baynton. Baynton observes: "disability has been used to justify discrimination against other groups by attributing disability *to them*" (cited in Davis, 2013, p. 17, emphasis added). Thus, when discriminating against black people, popular discourse usually reiterates prejudices of inability, irrationality, or incapacities prevalent in black bodies, families, or culture. Consequently, 'black' often comes to mean 'dependent' or 'disabled' and therefore excluded from citizenship in a state that privileges independent and able-bodied individuals.

When Fjord applies this specific intersectional lens on the images depicted in the Congressional report, she finds evidence of the conflation of race and disability that neither separates the oppressions of racism and ableism, nor presents them as an intersecting identity. For example, 'black families refusing to leave their homes' reinforced racist and ableist stereotypes (blacks as irrational), but did not elicit any questions or investigations into the reasons for this refusal by the Select Bipartisan Committee. Similarly, images of black people securing food or carrying guns to protect their own neighborhoods were perceived by the media as marauding gangs instead of responsible citizens trying to protect their families. In its concluding remarks, the Congressional report does criticize the media for its irresponsibility in perpetuating such images: "If anyone rioted, it was the media. Many stories of rape, murder, and general lawlessness were at best unsubstantiated, at worst simply false" (U.S. Congress, 2006, p. 360). However, by uncritically inserting the same images as those used by the media into its pages, and not following up with inquiries about the multiple barriers faced by people at vulnerable intersections, the report failed yet again to acknowledge their lived realities.

Both the text and the images reproduced in the Congressional report on Hurricane Katrina give us a glimpse into how the state imagines its citizens. It appears that the state starts to acknowledge some of the multiple oppressions people face, at least in terms of race and class, but it largely balks at analyzing this further. The report indicates that the state does seem to completely ignore the fact that oppressions like racism exists. However, akin to the anti-discrimination laws in *DeGraffenreid* v. *General Motors*, the report is also unable to fully comprehend its subjects as living with multiple identities and experiencing multiple oppressions. It seems that there is still a way to go for the state to recognize the multiple and intersecting components of people's identity and social location, and to really meet people where they are.

Conclusion: global climate change and intersectional vulnerabilities

As environmental and social threats gain prevalence and become a constant worry for both citizens and governments, notions of citizenship have developed in interesting and theoretically challenging directions. We can no longer imagine a citizen in the same way that Marshall did half a century ago: as someone whose identities and social locations do not matter. Somers' critique of the figure of the liberal citizen, so prevalent in citizenship theories (that is, a citizen who has equal capacities and capabilities, and equal opportunities), is important and moves us forward to a more intersectional analysis of citizenship. Evidence from the Congressional report on Hurricane Katrina supports the hypothesis that the state still does not take into account the lived intersectional realities of its citizens. Somers' analysis also provides us with an alarming diagnosis about the incapacity of the state to fulfill its obligations in upholding the social and economic rights of its vulnerable citizens. This is all the more important when we begin thinking about global climate change (GCC) and its impact on the relationship between the state and its citizens.

Sociologists who study the effects of GCC have theorized that, together with globalization, climate change has the potential to adversely impact everybody regardless of citizenship or other social statuses (Giddens, 2009). In other words, the consequences of GCC will be democratized and felt by everyone. Other scholars have questioned this assumption and theorized the opposite: climate change will not affect everyone equally, and marginalized groups will bear the burden of its consequences (Roberts, 2009). Given the negligence of governments (at least in the USA) toward intersectionality and multiple oppressions faced by their marginalized citizens, one may presume that multiply marginalized communities will be the most affected. Numerous studies of environmental degradation and disasters, from toxic waste disposal to heatwaves to hurricanes, have shown evidence of inequalities in adverse impact on non-privileged groups (Beamish, 2002; Erikson, 1995; Grinde and Johansen, 1995; Mies and Shiva, 2014). It makes sense that such inequalities will continue as GCC unfolds in different ways in various parts of the world.

Most environmental scientists and sociologists agree that GCC has become a world system phenomenon. Scholars are beginning to think about what this could mean for a world that is divided into nation-states, and what it could mean for the people who inhabit and seek protection from those states. Yes, there have been dozens of United Nations meetings, protocols, resolutions, and reports, all of which started as an effort to bring countries together to solve a global phenomenon. However, details from many of these meetings and resolutions show that UN ambassadors largely disagree with each other, are mostly concerned about their own nations' economic well-being, and usually reflect the anxieties stemming from the local politics of their countries when making decisions (Roberts, 2009). In sum, scholars have wondered whether the current system of nation-states is adequate to resolve the problems that will result from global climate change. Given this apprehension, it is also important to think about the role of citizens in this situation. How can citizens claim protection from states that shrug off their responsibilities toward their national and the global environments? What kinds of citizens will states protect and distribute resources to when disasters related to climate change hit? And ultimately, what will happen to the concept of citizenship if and when trust erodes in the protective capacities of the state in the face of GCC?

We can start answering such questions by building on work done by citizenship scholars from Marshall to Yuval-Davis. Given the dislocation, sense of powerlessness (among both citizens and state institutions), and anxieties of inescapability brought on by global climate change, I believe that we will have to consider including other factors in the intersection of citizens' identities. In fact, we may have to go beyond race, gender, class, and disability identities to also include space, time, and perhaps also transnational status in our analyses.

For example, Rosemary Garland-Thomson's (2011) concept of 'misfitting,' derived from disability studies, may be generalized as juxtaposing embodied ability and spatial architecture (including technologies on which we depend) and calls for studying an embodied interaction of a variety of differently bodied and abled people (and their intersecting identities) with everyday structures, objects, and technologies. This intersection is useful in thinking about citizenship during GCC because space and location become significant as climate change unfolds differentially across different places. Time is also an important part of intersecting vulnerabilities during GCC. According to Jasbir Puar (2009), including time into the intersectional lens can make us view identities such as disability as 'encounters,' since a disability in one context (place and time) may not be a disability in another. Following this thinking, one's multiple vulnerabilities or multiple resiliencies may interact differently before, during, or after the effects of climate change are realized. Finally, the ability to gain access to different countries, in the form of multiple intersections between transnational status, class, occupation, and resourcefulness, would also place one at a different intersectional vulnerability than those who will find it harder to migrate. As scholars, it is our task to understand and meet people where they are. The state seems to fail in this, but we must not.

References

Beamish, T.D., 2002. *Silent Spill: The organization of an industrial crisis*. Boston, MA: MIT Press.

Bulmer, M. and Rees, A. (eds), 1996. *Citizenship Today: The contemporary relevance of T.H. Marshall*. London: UCL Press.

Collins, P.H., 2000. *Black Feminist Thought: Knowledge, consciousness and the politics of empowerment*. New York: Routledge.

Crenshaw, K., 1989. Demarginalizing the intersection of race and sex: A black feminist critique of antidiscrimination doctrine, feminist theory and antiracist politics. *The University of Chicago Legal Forum*, 139.

Davis, L.J. (ed.), 2013. *The Disability Studies Reader*. New York: Routledge.

Erikson, K., 1995. *A New Species of Trouble: The human experience of modern disasters*. New York: W.W. Norton.

Fjord, L., 2007. Disasters, race, and disability: (Un)seen through the political lens on Katrina. *The Journal of Race and Policy*, 3(1), p. 46.

Garland-Thomson, R., 2011. Misfits: A feminist materialist disability concept. *Hypatia*, 26(3), pp. 591–608.

Giddens, A., 2009. *Politics of Climate Change*. Cambridge: Polity Press.

Gilbert, L. and Phillips, C., 2003. Practices of urban environmental citizenships: Rights to the city and rights to nature in Toronto. *Citizenship Studies*, 7(3), pp. 313–330.

Grinde, D.A. and Johansen, B.E., 1995. *Ecocide of Native America: Environmental destruction of Indian lands and peoples*. Santa Fe: Clear Light Books.

Harvey, D., 2006. Neo-liberalism as creative destruction. *Geography*, 88 B(2), pp. 145–158.

Hindess, B., 2002. Neo-liberal citizenship. *Citizenship Studies*, 6(2), pp. 127–143.

Isin, E. and Nielsen, G. (eds), 2008. *Acts of Citizenship*. New York: Zed Books.

Klein, N., 2008. *The Shock Doctrine: The rise of disaster capitalism*. New York: Picador.

Lister, R., 2007. Inclusive citizenship: Realizing the potential. *Citizenship Studies* 11(1), pp. 49–61.

Macpherson, C.B., 1962. *The Political Theory of Possessive Individualism*. London: Clarendon Press.

Marshall, T. and Bottomore, T., 1992. *Citizenship and Social Class*. London: Pluto Press.

Mies, M. and Shiva, V., 2014. *Ecofeminism (Critique. Influence. Change)*. New York: Zed Books.

Nozick, R., 1974. *Anarchy, State and Utopia*. New York: Basic Books.

Pakulski, J., 1997. Cultural citizenship. *Citizenship Studies*, 1(1), pp. 73–86.

Pelczynski, Z.A. (ed.), 1984. *The State and Civil Society: Studies in Hegel's political philosophy*. Cambridge: Cambridge University Press.

Puar, J., 2009. Prognosis time: Towards a geopolitics of affect, debility and capacity. *Women and Performance: A Journal of Feminist Theory*, 19(2), pp. 161–172.

Roberts, J.T., 2009. Climate change: Why the old approaches aren't working. In K. Gould and T. Lewis (eds), *Twenty Lessons in Environmental Sociology*. New York: Oxford University Press, pp. 191–208.

Somers, M.R., 2008. *Genealogies of Citizenship: Markets, statelessness, and the right to have rights*. Cambridge: Cambridge University Press.

Tierney, K.J., 2007. From the margins to the mainstream? Disaster research at the crossroads. *Annual Review of Sociology*, 33, pp. 503–525.

Turner, B.S., 2006. *Vulnerability and Human Rights*. University Park, PA: The Pennsylvania State University Press.

Turner, B.S., 2009. T.H. Marshall, social rights and English national identity. *Citizenship Studies*, 13(1), pp. 65–73.

Tyler, I., 2010. Designed to fail: A biopolitics of British citizenship. *Citizenship Studies*, 14(1), pp. 61–74.

United States Congress, 2006. *A Failure of Initiative: The Final Report of the Select Bipartisan Committee to Investigate the Preparation for and Response to Hurricane Katrina*. House Report 109–377.

Yuval-Davis, N., 2006. Belonging and the politics of belonging. *Patterns of Prejudice*, 40(3), pp. 197–214.

44 Race, social class, and disasters

The Katrina version of reality

Alvin DuVernay

The following is an account of my Hurricane Katrina experience. It is a snap-shot of my sentiment immediately prior to, during, and after the Federal levee failures caused intense flooding and destroyed most of New Orleans. It was written upon special request from the editor of the New Orleans Geological Society (NOGS) and was later published in their rag, *The NOGS Log*, August, 2006, Vol. 47, No. 2. I offer it here as a glimpse into both the horror and the humanity birthed from natural and man-made disasters.

Anatomy of a storm

Katrina, August 29, 2005
As a species, we must react and adapt to our environment. The alternative is extinction. The list of human inventions for protecting ourselves from the ele-ments is endless and dates back to our earliest ancestors: caves, huts, animal skins, foul weather gear, storm- and quake-proof structures, mechanically condi-tioned air, etc. Then one day Mother Nature approaches gently and methodi-cally, but seemingly with purpose and resolve. This time she must clearly demonstrate her status in the scheme of things. She enters your life and dares you to stand toe-to-toe with your puny man-made adaptations. "Go ahead," she urges. "Put on your slicker suit. No, not that one, the expensive one you bought from Sharper Image for 'protection in extreme conditions.' I dare you!" She whispers. "Use your Swiss Army knife and release that fine garment from its her-metically sealed, vacuum-pack."

"Let's dance!" she demands.

Preparation for Hurricane Katrina proceeded like so many other storms in my memory of a half-century or so. I collected, compiled, and spamodulated in anticipation that the worst might happen. Every storm is the same drill, the notion being that with proper anticipation and equipment, personal harm can be mitigated and aid to those less prepared can be given. Staring down Mother Nature had never felt arrogant or foolish; it is simply what I have always done.

I had been monitoring the storm's projected path for days on the Web and did so one last time before leaving work on Friday. No worries for us; it would seem that poor Florida was to be hit by another monster. Saturday morning however, the neighborhood was abuzz with storm and evacuation talk. Had I missed something? My back door neighbor suggested I have a fresh look at the weather sites. Sure enough, most of the models had her pointing right at New Orleans with Category 5 winds and surge in tow. Damn – time to gear up in earnest.

It's a familiar exercise but a chore nonetheless: charge up the boat batteries; top off gas tanks (boat, truck, bike, chain-saw, extra tanks); fill containers with water (plastic bottles, drums, ice chests, bath tubs, buckets) – some for drinking, some for sanitation; provision the boat (canned food, 12V TV, radio, chain-saw, ropes, axe, first aid, rain gear, clothes); police the grounds for potential missiles (seemingly heavy or even fixed objects become destructive in 100 mph winds); fill several iceboxes with block ice; and finally, stage some perishables in the refrigerator and freezer so that they can be moved quickly into iceboxes when the power outage is imminent.

Saturday came and went; Sunday brought more bad news (i.e., Katrina was still bound for the Delta). Rob Sloan, my Chef Menteur camp partner, wanted to lighten the load in his freezer before he evacuated, so we met at the camp and baited the crab traps. Blue Crabs always run great after a storm. There's a better than average chance you'll lose your freezer contents and this year we had a lot to lose. Within the three weeks prior, two events had filled our freezers: we slammed the trout at Grand Isle, and I had hit the Lake Pontchartrain Pogy run just right – three throws with my 7 ft cast net filled an 80 qt icebox. I had a freezer full of the best eating fish on the planet and enough bait to get me through the spring catfish run and summer crab season for the coming year.

I spent that Sunday monitoring the news; moving my dad from his house in Metry to mine in Lakeview; and staging supplies, provisions, tools, and equipment where they would be most needed. The local news was filled with scenario building and animated desperation as the leadership floundered with suggestions and eventually demands on what we should and should not be doing. Something was missing – indeed, a crisp implementation plan.

That evening and the following morning was the single most impressive thing I had ever experienced. Speaking as a geoscientist, this is a fairly bold declaration. We've seen some things. We've peered down the throats of active volcanoes, and canoed glacier lakes and rapids. We've kicked through miles of cactus in Mexico and have even survived vino and grappa overdose on the Mediterranean. Many of us have grappled for hours wrestling the death grip of the Louisiana flotant marsh. Ours is a life of passion for the extreme.

For hours the house shuddered with each gust of wind and there was a continuous din of snapping trees and limbs. I'd monitored the rainwater street flooding in front of my house all night and morning. It was flowing (as it should have been) from south to north. By daybreak the storm winds began to subside, so I walked the neighborhood to assess the damage before going down for a well overdue nap. Gratefully, I whispered a prayer of thanks that we had dodged yet another deadly bullet. Simultaneously I observed that the flow regime had made a 180-degree change and was now flowing north to south – my prayer changed to one of mercy. Relentlessly, doggedly, the water crept up the lawn. To the sidewalk, over the porch and into the house it came with ambivalence yet malicious results. Something was wrong I remember thinking. It had to be a break in one of the levees. Surely the peak surge had passed so the levees could not have been topped. The levees failed? How could this be? The storm had weakened to Cat 2 when it passed our area and the US Army Corps of Engineers had assured us the levees were designed and built to withstand a Cat 3 storm. No time to meditate on that – It was about 9 a.m. and time to dance – no time for a nap.

There's no way to describe what it feels like to watch everything you own being slowly digested by putrid urban flood waters. Further, no one could have imagined what an ominous and persistent shadow this event would cast on our lovely city. No one could have believed how ineffectual our local and federal leaders would prove to be; and equally, how impotent their emergency planning. No one could have anticipated the inequities, fraud, deception, and incompetence that would soon and continuously be exposed. I didn't at the time. I was operating on pure inspiration and adrenalin – no humanity, no sense of past or future, no blame, guilt, or remorse. It was fight-or-flight time – instinctive and primal.

My dad and I got the mutts and all of our provisions from their staging area up to the camelback rooms above the garage. My neighbor Shannon had evacuated and called for an update; the best I could come up with was, "Your house is fine so far, but I can't talk now because the water is rising!" Within an hour, the water was waist deep and it was time to float the boat out of the garage and prepare for our escape. We tied the boat off to the upstairs window and monitored the radio for information while measuring the water's rate of rise. Fortunately WWL radio was broadcasting and doing their best given the breakdown in typical communication tools. Frustrated at the information (or lack thereof) coming across, I tried calling in to provide a real-time account. Alas cell phone usage was dismal.

We watched and listened for a few hours more and decided to motor off toward dry land in Metry. The water was steadily rising and showed no signs of stopping. From a lucky phone connection I'd made earlier, we knew that just west of the 17th St Canal was relatively dry. That would be our destination if we could navigate through the labyrinth of broken trees, power lines, and neighborhood debris. Immediately, we realized that this would not be a simple exercise in navigation. Our modest fishing vessel, a 16-ft aluminum flat, was quickly transformed into a lifeboat. I knew why I had not evacuated but why hadn't so many others; so many without proper means of escape; so many elderly and infirm? Too many years of dumb luck and blind faith in the federal levees had given us all a false sense of security.

We picked up as many people as we safely could that afternoon and shuttled them to the Vets Hwy, 17th St Canal Bridge. If the boat was full we promised others that we would come back for them – and did. Darkness came and it was too dangerous to continue. I remember wondering while tying my boat to a tree for the night – where is the real cavalry? I left the keys and spare gas tanks in the boat in case it might be useful for others. Mentally, I'd written my boat off – another casualty of the storm. The extended adrenalin surge was over. My brain and muscles burned and I could finally acknowledge my breaking heart.

My good friend Richard Thurman picked us up in his truck and brought us to my dad's house a mile or so away. It was high and dry, suitable for catching up on some news from my 12V TV, and getting a few hours of restless sleep before reigniting the adrenalin jets. We had plenty of food, but eating was the furthest thing from my mind. I opted for a couple of cold Abita beers on the front stoop instead – they wouldn't be cold for long.

That evening was unforgettable. There was no power, running water, people, birds, bugs, nor wind. There was no sound. The night sky was beautiful but wrong – too many stars for suburban America. The silence was deafening – peaceful but frightening.

I was jolted out of bed on Tuesday at 4 a.m. to the rude yet protective barking of our dogs. My (soon-to-be) good friend Dickie Durham was walking the street with his neighbor Maria looking for people and/or resources to help rescue her family who were trapped in the flood. We talked for a bit and planned to meet at the boat at first light. She was desperate to get to them right away, and I did my best to explain how hazardous it would be to attempt such a rescue in the dark. She was not convinced but reluctantly complied.

Unable to go back to sleep, I armed myself with axe, bottled water, and tobacco, and walked back to the boat. It was still tied off where I had left it so I swam out to it and sorted through the tools, ropes, and provisions while waiting for daylight and my helpers.

Dawn broke soon. Dickie and I shoved off for our first mission, and I was thankful that I hadn't much time for reflection. The water was still rising so negotiating the currents under bridges, around trees, and over power lines was a bit tricky. We found Maria's family and easily got them out of a second-storey window into the boat. Four generations were happily alive and well – great grandparents visiting from Italy, Grandma, Mom, and her three children including an angel of only six weeks old. When I took hold of the tiny basket from the window and saw that delicate human child gently breathing, eyes closed, and at peace – well, maybe the storm itself wasn't so impressive after all compared to this.

Maria and her family were reunited and we continued penetrating the neighborhood to pull people, pets, and belongings off of roofs and out of windows. In short order a couple of other civilians launched boats and joined in the ersatz rescue force. Richard joined me in my boat, and Dickie and another good friend Charlie Dominic found other boats to crew with.

After a couple of hours the Coast Guard choppers were working the area as well. We watched them hover over a house and effectively (albeit slowly) haul one person at a time into their vehicle. Meanwhile we would get two or three folks and as many pets from two or three different houses and shuttle them all back to dry land – six to nine per boat load. I remember thinking how efficient this effort could be if those choppers were systematically locating and mapping out where the people in need were, and communicating that information to those of us on the water.

We dropped off our passengers on the Vets Hwy, 17th St Canal Bridge where private citizens from the dry side of the disaster had congregated. One woman had set up a lean-to out of plastic for shade and took care of the orphaned pets. People were helping each other – strangers in life but brothers and sisters in spirit, they shared towels, dry clothes, food, and water. They drove the rescued off in their cars, maybe to their own homes, I don't know. Dickie's wife became the self-appointed site director. She enlisted support from what would have otherwise been a random collection of sightseers. She rallied the congregation with clear but curt direction, solicited gasoline and drinking water for the boats, and with conviction of purpose curtailed inappropriate behaviors. She made me take a break and eat a cheese sandwich (white bread, no mayo) – a mundane yet crystalline act of humanity.

The task was daunting (i.e., too many people with too few resources) and still no cavalry. Richard left me to go get his neighbor's boat and put it into the fray. It was only late morning and I already felt spent and sun-baked – my God it was hot. I kicked back to have a long drink of water and roll up some nicotine before pushing off back into the neighborhood. This time I would be alone.

A guy with black rubber boots and a formidable collection of cameras around his neck approached my boat. He introduced himself as Chris Graythen, a Getty Images photographer, and asked if he could ride with me and get some pictures of the event. My initial reaction was 'no way' because I could afford neither space nor weight in the boat for a photographer. I further allowed that I could use some help, and if in the process he could get some shots, that would be fine. He agreed. Chris proved to be an awesome first mate. He was casually competent in the ways of boating and rescuing, his compassion was inspirational and energizing, and he stayed with me the whole day. A word on this last point: as a professional photographer, Chris could have been traversing the city capturing Pulitzer Prize-winning images. Chris chose humanity and me; for that I shall be forever grateful.

About midday Chris and I learned that something, perhaps FEMA, was set up on the I-610 overpass so we started dropping off our people there. The cavalry had arrived. Say what you want about that organization's leadership but those folks on the ground were kicking butt. They had airboats, chain-saws, emergency medical, food, water, shelter, and buses. And they worked tirelessly.

We continued our civilian efforts until we started losing light. It was time to stand down and leave the mission to the professionals. Richard got his neighbor's boat back home and Charlie rescued my boat with a spare trailer of his. With broken spirits and sore bodies, we proudly limped home. I'd have traded my pension for a shower that evening but had to settle for a towel bath with bottled water and Dr. Tichenor's antiseptic. What evil sorts of microbes and chemicals had we been exposed to? Rash-like blisters were bubbling up on my face, arms, and chest. Paranoia set in.

How many people? How many pets? Unknown. Hundreds, I suspect. It all just ran together and it seemed like it couldn't get worse. After a few minutes watching the TV, I learned that it indeed had. Most of the city and points eastward and south were in similarly dire or worse circumstances – Uptown, Downtown, Mid City, Gentilly, Ninth Ward, Arabi, St. Bernard, North Shore, Mississippi, etc, etc, etc. Horrific reports of panic in the Dome, looting, and carjacking overwhelmed me with a sense of dread as I had no means of transportation out of the area.

The rest of that evening and the next morning I inventoried provisions and laid out rations to last for several weeks. We were armed, prepared, and alive. And like so many others, we were wondering what recovery would look like and when it would begin. Local and national leadership was thin at best, and it was clear that survival would be a personal matter. Salvation arrived mid-morning. Richard came by the house and declared that we could have his truck, he would use his motorcycle, and we should all escape to safety and sanity. White-knuckled and ten hours later, Dad and I arrived at my sister's house in Houston. A year later, I'm still waiting to see some leadership and grateful that I'm not dependent on them for my personal recovery efforts.

Casualties of the storm: they are indeed many and diverse – lives, pets, property, hearts, minds, relationships, futures, etc. What has this event done for us? These experiences have enhanced our personal power and resolve. We share a special bond that others will never understand, never empathize with, and can never relate to nor tolerate. We are more tolerant because we've all gone a little crazy and back (often). We are more patient because we've learned that to behave otherwise will merely enhance the frenzy. We embrace simple and mundane pleasures as if they were priceless treasures because we have been reduced to 'starting over' in the fullest extent of the phrase. Those who were lucky started with a car and a change of underwear, or a boat and some tools, or their pets and a friend's sofa for rest. We cherish the tolerance and charity of others because we understand personal inequities, judgment lapses, and inexplicable reactions brought on by the event. We are comrades of cause and effect.

Compared to nature, we humans are so tiny, yet through simple choices and adrenaline-driven gestures springing from those choices, we have so much power to affect the outcome of events beyond our control. I wonder what the ripple effects of our gestures will be. I wonder how those people we helped will touch the lives of others. I wonder who and what the baby in the basket will grow up to be and how many lives that child will touch through the years. I wonder...

My heart has been shredded in so many ways and at so many levels throughout this event. Oddly, it did not start with the water inundating and destroying my home while my dad and I escaped in the boat. It began days later when I finally got connected to full-time media coverage and witnessed their gross misrepresentation, and the obvious instigation styles, tools, techniques, and agenda (i.e., find the angriest person around, spin the report, and fuel the negative energy with subjective interpretation).

For the few days of the storm my spirit was glowing with the kindness, generosity, and the tender mercies of humankind. The on-demand rescue fleet, sharing of resources, and general concern for each other's well-being was abundant. I saw little of that on the news. Yes, there were extremely bad things happening and indeed very bad people doing those things. There were, however, equally good things happening. Again, I saw little of that on the news. Both sides are news and information, and the public deserves the whole story. Not just the sensational horror that might as well be slapped on a plastic lunch box and sold at the dime store.

I believe in the goodness of people despite some of the media's obsession for the contrary. I believe that my time spent with the civilian rescue fleet validates my belief. I still feel the love, the caring, the perseverance, and the overwhelming camaraderie. I see the bad stories. I see the evil portraits. I choose to believe in what I feel. This I believe.

> In life there are many events,
> Some with cause to pine,
> These aren't they.
> Tomorrow is mine.

Commentary

That was then. It's been eight years since Katrina and a million miles of anti-humanoid behavioral observations. Am I the only one frustrated with knowing that no matter whom we elect – liberals, moderates, conservatives – we get laws and policies that favor wealthy individuals and/or corporations over the public good? I had no idea how prophetic the following one line in the above story would become: "no one could have imagined what an ominous and persistent shadow this event would cast on our lovely city." In the years since the storm, I have observed the dynamics of the privileged class versus the not so fortunate manifest themselves in many ways.

Full disclosure: I am a paleontologist by profession and have had a long, comfortable career with a major oil company. More to the point, I have spent my entire life playing in, observing, and studying ecosystems, species diversity, and the environment. The balance that most organisms achieve with the Earth while exploiting resources is a thing of beauty – it is the stuff of great poetry, song, and art and is infinitely sustainable. The history of human exploitation of resources, however, is profoundly frustrating – the stuff of nightmares and bedwetting. It is this frustration along with my passion for the planet that fuels my incessant editorials to journalists and policy makers. It is what compels me to be seen or heard whenever possible and to whomever will listen on the issue of a healthy biosphere.

On the strength of that passion, I was even involved in a global energy crisis film, *The Age of Stupid*, a great docudrama set in the future, after a fictional (yet plausible) collapse of society as we currently know it. The film features an archivist in the year 2050 or so who looks back through the centuries of modern civilization and meditates on how and why we have come to such a dark existence. He pulls up related news and human-interest stories from the past and opines as to why we were so stupid to ignore historical lessons and conspicuous warnings. The film-makers very cleverly chose several real-life stories around the world to illuminate the real impact of the energy crisis and climate change. There is an octogenarian mountaineer in the Mont Blanc area of France who has witnessed escalating waste-fulness and its direct effects. A young Nigerian woman demonstrates what life is like while trying to make a living and get an education under the cloud of a pol-luted landscape and self-serving politicians. There are two Iraqi children displaced to Jordan after the invasion and occupation of their country. These siblings mourn the killing of their father, and they long for their home and a reunion with other family members. There is also an Indian airline executive, a London wind farmer, and me. I'm the American oil field worker/disaster survivor who happens to be sympathetic to the film's theme, objectives, and style.

I have experienced the proverbial weeping and gnashing of teeth since Katrina and the federal flood. It was, indeed, unbearable. I would not wish that on anyone, but having said that I was comparatively fortunate. By my standards, I am rich and have been since I started working for a buck and a quarter per hour in a boat-yard at the age of 13. Before the storm I owned my house, a modest fishing camp, a small boat for exploring the swamps, and numerous creature comforts; since then I've collected insurance and was able to qualify for grants and loans. Within five years or so I was, for the most part, physically whole. Psychological and spiritual harmony is a story for another time.

Those with less financial prowess in my city, on the other hand, fared quite differently. It is loathsome that in the twenty-first century, racial injustice and intolerance are still issues, and that minority status often correlates to poverty; but there it is, and here we are. In Oklahoma or Alaska, we might be talking about Native Americans. In Arizona or Texas, it would be Latinos/Hispanics. In other places, it might be Irish or Italian or Russian or Vietnamese, etc. New Orleans poor are predominantly black/African Americans. Within that com-munity, females are even harder hit because of the high occurrence of the single mom phenomenon and, unfortunately, the inequities produced by still-persistent misogynistic attitudes.

New Orleans is a unique community. She is laced with layers of ceremony, tra-dition, class, race, heritage, and history. Most of the time these attributes weave and intertwine with each other in a harmoniously geometric fashion like a fine silk web. They manifest themselves in our food, music, and omnipresent celebrations of life and death. They are the marrow of our community and the reasons we love her so and want to live nowhere else. Then there are the times when the webbing is compromised and those on the outer threads, the more vulnerable, suffer dispro-portionately. Humanity should be measured by how it treats those without signi-ficant education, savings, or social and business networks. As I lived and watched our recovery, I have judged that we did not fare well on that grading metric.

For years, scientists have predicted (and continue to warn) that climate change would alter the environmental balance as we know it. They have modeled and demonstrated that naturally occurring weather-related events and the resulting disasters would become both more extreme and more frequent. Forest fires, blizzards, droughts, floods, storms, and the like would become more difficult to forecast and even harder to deal with and recover from.

As a paleontologist, I spent my career studying climate changes and the resulting effects on our planet's organisms through geologic time. Extinctions and the evolutionary emergence of species due to environmental pressure is the natural machine of life. Is the planet currently warming? As a matter of statistical fact, it is. We are in a geologic warming trend and have been for quite some time. The debate, of course, is not whether we are in a geologic glacial or interglacial system. The debate is about anthropogenic global warming. Have we, through our interaction with the environment, exacerbated the rate of climate change? First, mankind is typically not celebrated for its harmonious existence with the Earth and her resources. The human modus operandi is to operate in spite of Mother Nature, not in collaboration with her. Second, who cares! What is the worst thing that could happen if we spent some energy and political will cleaning up our act and reducing emissions and pollution? Clean air? Clean water? Creation of jobs dedicated to friendly exploitation and utilization of our resources? If you like "war for oil," you're going to love "war for water."

I won't say that Hurricane Katrina was a direct result of global warming, anthropogenic or not. I'm not a climatologist and have not personally crunched the numbers. Further, caution must always be used when trying to pin a global climate trend to any single event. I will declare, however, that at the very least it should be used as a model for things to come, both in our physical and socioeconomic world. By definition, a model is a version of reality. It would be prudent to give the Katrina version some attention.

The following are a few short examples of tears in the very fabric of our social structure caused either directly or indirectly by Hurricane Katrina and the federal flood that followed.

In the immediate aftermath of the storm, there seemed to be an endless list of horrors perpetrated on people of limited resources. Monstrous things, as though being poor and/or needy made them somehow less than human, somehow deserving of abuse or worse. Remember the images of the thousands in the Superdome begging for food, water, and security? Those were mostly lowwage working folk who lived in a city with excellent public transportation. They had no need of a personal vehicle, ergo they had no means to evacuate. Or worse, they trusted the guarantees from the US Army Corps of Engineers that our levees would protect us from a Katrina-like storm.

Elsewhere in the city concurrent with the Superdome dramas, your life could have been in danger if you were merely searching for sustenance or shelter. There have been several police officers convicted of shooting and/or killing innocent (unarmed) civilians for such behaviors. Others have been convicted of conspiring to cover up the events.

There is nothing worse than murder of course, but I am hoping for a special kind of afterlife for politicians who create underprivileged assistance programs and then use their power and influence to enrich themselves through those very activities and agencies. Some have already been caught and convicted. One was busted for diverting federal hurricane recovery funds from non-profits, others for pilfering from a teenage pregnancy advocate agency. Many business people have been caught for fraudulent use of recovery resources as well. I suspect we haven't seen the end of that. Stay tuned.

Lack of housing was the most obvious and ubiquitous inequity that I observed. If you owned your home, there were multiple avenues to pursue for recovery. If you were a tenant, there was virtually no assistance available. I went to the French Quarter one evening to treat myself to some fine dining after a particularly grueling day of stink and debris clean-up. It had been a few months since the event and a handful of restaurants were starting to open for business. The paper plates, plastic flatware, limited menu, and skeleton crew of wait staff told the whole story. A city that thrives on food and entertainment is dependent on a healthy working class. A city that had just been 80 percent destroyed and had no room for the labor force to stay is in a world of shit. Rental houses lagged in repairs as the landlords' priorities were getting their own homes up and running. Public housing – solid and structurally sound, with minor water damage – was fenced off. There are those who hold to this day that the powers-that-be were waiting for just such an event in order to get rid of the housing projects. Think prime real estate.

Even the low-wage working-class people who did own their homes couldn't easily recover because of a bizarre accounting method employed by adjusters and disaster grant providers. The money you were awarded was based on the amount of damage, but also the pre-storm appraised value of your home. It seems fair enough until you try to repair a house in a poor neighborhood, which has a lower cost per square foot value than the same house in a different area. The bricks-and-mortar folks don't care where you live; their unit cost is the same for everyone.

Why must the weakest among us suffer the most at the hands of the strongest? Why must the powerful perceive them as targets for their greed, as a predator perceives prey? The closest analogy that I can use to wrap my mind around it is the bully mentality. If a schoolyard bully were to answer honestly as to why he picks on the weak kids, his response would have to be 'because I can.' We've evolved so far since Lucy descended from the trees and stood upright, holding her infant to her left side and a spear to her right. Must we wait much longer for societal evolution to catch up?

Indeed, we must not. We must get involved, volunteer, and donate; we must demand accountability, declare our outrage, pen our thoughts, and give voice to our solutions. Purge the inequities however, whenever, and wherever we can. There are truly righteous people wanting to do the right thing for the right reasons. "I still feel the love, the caring, the perseverance, and the overwhelming camaraderie. I see the bad stories. I see the evil portraits. I choose to believe in what I feel. This I believe."

Related reading

Not Just the Levees Broke, My Story During and After Hurricane Katrina, by Phyllis Montana LeBlanc. LeBlanc is a very powerful and opinionated African American lady with much to say about how she and others were treated after the storm. Her book is raw and deliciously honest. I wish to know her – I look for her in crowds at jazz clubs and such so that I can introduce myself and thank her for sharing herself with us.

Zeitoun, by Dave Eggers. Zeitoun is a Muslim Syrian immigrant – much of the book is a bit over the top with all his spirituality, but his post-Katrina adventure will tear your heart out and simultaneously set you on fire over the enforcement machine's tactics, intolerances, biases, and prejudices.

1 Dead in Attic, by Chris Rose. Rose is a long-time, much-loved New Orleans editorialist who can pen a story better than anyone I know. He can invoke tears and laughter in the same paragraph, and then do it again in the next paragraph. His book is a compilation of his post-Katrina columns.

45 poison water blessings

Cherese Mathews

standing
silent witness
to the slaughter

regal sentries massacred
crash down
to rocks absorbing shock of death
folded
into earths embrace

wombs stripped of precious treasure
left empty
running through
with poison water blessings

toxic nourishment
drawn up by thirsty roots

Commentary

There is deep sadness and anger, mingled with a small but precious bit of hope, when I witness the injustices in our world. That small bit of hope is essential; it is the generous place I *must* give myself when I see the destruction of our planet and the delicate balances that are threatened. Hope is the thing that holds me up when women, children, indigenous people, and our natural world are not simply disregarded but purposely exploited and perpetrated against. And hope is the sliver of light that shines through my despair for all that has been lost and all that is at risk in each moment.

We live in a world of contrasts. I am acutely aware of the power and privilege of my whiteness. I am also the daughter of the downtrodden and persecuted blood that runs through my veins. My own disenfranchised genetic history comes from the poorest of the Scottish hills, the poverty of England, probably the poor immigrants of Germany and, yes, the Trail of Tears. And though I am a child of the 1960s, full of slogans and fight, I am also a part of the larger culture. I sit in simple sacred ceremony of the sweat lodge, and yet drive often hundreds of miles to be in that ceremony. I treasure deep connections to community and family, and sometimes fly half-way across the country to cultivate those relationships. Some of the food I eat has been trucked thousands of miles to my table. I have dedicated myself to nurturing a small piece of land as best I can ... and I use flea spray on my dog. Yet every day I strive to live closer to this Earth. I wish the choices were simpler.

This particular poem is my way of feeling into the painful contrasts and seeming impossible conflicts in our ongoing struggle. We are all a part of this violence. I can only hope to keep the conversation and the hope alive enough for the right action to follow.

Standing, silent witness to the slaughter, regal sentries massacred.... We may stand in solidarity with the regal tree nation, yet we will watch as the world around us is destroyed. If we are not killing by our own hand, then we are killing by our participation in a culture that pays other people to do our dirty work. Where is our power for change? How do we step out of silence? It is not enough to have good bumper stickers on our cars. It is not enough to vote green. It is not enough to just feed the homeless.

Crash down, to rocks absorbing shock of death. Folded into earths embrace.... Gaia opens her arms wide to receive those that are being lost: plant, animal, human, glacial. The losses are enormous and her arms just open wider as they fall. This is what the system is designed to do. It takes the death and regenerates them into new life. But how much can fall into death before we have a movement toward death that is stronger that the movement toward life? When the crashing of beings is a teeming river, a tidal wave of death ... then, at some point we will be pulling the sweet flow of life back in on itself to death before it sees the light of the sun! And we are doing this. We are the rock, unmoving, but somehow absorbing some of the horror of this massacre. Culturally, we close in, we shut down. We watch more television, we distract ourselves, and we buy One More Thing. We wait. We are afraid. We can't go to Afghanistan and take the guns away ourselves. We can't just say stop Monsanto, or Bank of America. We can't make people stop using bleach. It is all so simple. Just stop using the damn bleach! Stop depending on oil. Stop exploiting and destroying the peoples of the earth. So simple.

Wombs stripped of precious treasure, left empty. Running through with poison water blessing.... We are carrying the wombs not only stripped of the sacred treasure but raped, tormented, and disempowered. Women feel this deeply as they are often the victims of this attack on the safety and reverence of life. The womb of culture is also the target of direct dismemberment and destruction. There is little safety in a culture that has had its language, traditions, and place destroyed or taken from it. And in the hole that is left from that swathe cut by "progress" is the poison. What is running through that wound may even look like home, or culture or safety ... but it is not. The roots are too shallow now. And we are drinking from a poison river.

Toxic nourishment, drawn up by thirsty roots. . . . I remember being horrified when infant formula companies sold women around the world the idea that only a factory-made product could feed their babies well. Women gave up the ancient and fundamental right to feed their babies not only the best food for them, but with ease. They replaced it with, if not outright poison, well it is close. This violence against babies, families, and culture was premeditated. The innocent suffer. There is something familiar about the feeling of drinking in the toxic nourishment. What else can we do? We need to take something in, we need to feed and be fed.

You and I may not all be directly responsible for the overt destruction of the rights of individuals, cultures, animals, or ecosystems. But we must all take responsibility for staying awake and participating in solutions. We must be active in educating ourselves and our communities. We must continue to seek a better way to live in this world, a better way to serve the common good. Hope is the feeling that what is desired can be realized. It is essential for us to live with hope holding close our knowledge of what is right and just.

46 Sea ice

Subhankar Banerjee

Figure 46.1 Sea ice, Beaufort Sea, Alaska, July, 2002 (source: photograph by Subhankar Banerjee). One of the most significant indicators of global warming is the melting of Arctic sea ice. The sea ice (along with ice sheets and glaciers) keeps the Earth cool by reflecting back into space a significant part of the solar radiation. But as the sea ice melts due to warming, the darker water absorbs much of the incoming radiation, which then results in further warming. Since the 1970s when recording began, the Arctic sea ice has been on a "death spiral" losing both its extent and thickness: by 2012, the minimum sea ice extent declined by about 50 percent from the 1979–2000 average, and the thickness by about 65 percent from 1975, when it was first measured. The rapid melting of Arctic sea ice is having devastating impacts on ice-dependent species like polar bears, narwhals, seals, and walruses, as well as on the coastal indigenous Inuit people.

47 Evangelical environmentalism
An analysis of gender and ideology

Lisa M. Corrigan and Molly Rawn

The public debate in Washington on climate change has stalled as a major climate change initiative has failed to make it to the floor for debate in the 114th Congress, and climate skeptics continue to imprint their agenda on the House Energy Committee. Challenges to the Obama Administration's green initiatives are due to the influence of evangelical Christians on climate change (Stromberg, 2011; Wang, 2010; Wilkinson, 2012). Evangelical Christians have historically been Protestants that relate born-again conversations, embrace biblical literalism, and commit to evangelizing the 'unsaved' (Bineham, 1993). In 2008, the Pew Forum on Religion and Public Life found that of the 78 percent of Americans who self-identify as Christian, 26.3 percent consider themselves evangelical, making them the largest category of the study (see also Heltzel, 2009). Ninety percent of evangelical Christians self-identify as white (Emerson and Smith, 2000). Former president of the National Association of Evangelicals (NAE) Ted Haggard estimated that the 51 denominations in the NAE represent 30 million evangelicals ready to mobilize and lobby for issues like climate change (Goodstein, 2005).

But these evangelicals hold differing views on climate change. Liberal groups such as the Evangelical Environmental Network (EEN), whose open letter "An Evangelical Declaration on the Creation of Care" (1994) was the first public statement on evangelical stewardship, and the Evangelical Climate Initiative (ECI), propose a prominent role for evangelicals in climate change advocacy. Led by Richard Cizik (former NAE vice president of government affairs), Jim Ball (Executive Director of EEN), Leith Anderson (President of the NAE), and popular pastor Rick Warren, this group of 86 senior evangelicals and a total of 260 signatories issued a full-page public letter in the *New York Times* entitled "Climate Change: An Evangelical Call To Action" (2006), which catalyzed the liberal evangelical movement to accept climate change as fact, see humans as the cause, and to promote environmental stewardship through 'creation care' (Krakoff, 2011; Wilkinson, 2012).

Conservative evangelicals articulate contempt for the science of global warming, interpret the Bible to absolve humans of planetary destruction, and see economic development as a strategy of adaptation to environmental changes. Some even "come close to celebrating the demise of the Earth, enthusiastically citing the decay as proof that the return of Christ is near" (Sider, 2000, p. 44). Led by James Dobson (Focus on the Family) and E. Calvin Beisner (Cornwall Alliance for the Stewardship of Creation), the major conservative evangelical coalition produced and supported the Cornwall Alliance's 2006 manifesto, "A Call to Truth, Prudence, and Protection of the Poor: An Evangelical Response to Global Warming," which elucidated the tenets of conservative evangelical environmentalism and refuted the "Creation of Care" manifesto.

These ideologies demonstrate the growing schism among evangelicals over climate change (Nisbet, 2009). Sadly, scholars have made a paltry attempt "to formulate a synopsis of how evangelical environmentalism has historically developed and what central tenets it affirms" (Simmons, 2009, p. 43). This chapter provides both an historical and an ideological examination of emergent ideologies within evangelical environmental groups (EEGs) through an ideographic analysis of the manifestos of their coalitions to identify how they have clashed over environmentalism (Prelli and Winters, 2009). We examine the open letter and web content of Focus on the Family (FOTF), the Cornwall Alliance, the Evangelical Climate Initiative (ECI), and the Evangelical Environmental Network (EEN), since these organizations have controlled evangelical responses to climate change due to their size and wealth as the largest evangelical organizations in the United States (Medhurst, 2009).

Here, we use an intersectional feminist approach to trace the development of EEG ideographs. Ideographs are the rhetorical devices that build a movement's broader ideology (Charland, 1987). An ideograph is an "abstract word or phrase drawn from ordinary language, which serves a constitutional value for a historically situated collectivity" (Condit and Lucaites, 1993, p. xiii). Ideographs place competing terms in "a *hierarchy*, or *sequence*, or *evaluative series*," that instruct us how to view ideology across time (Burke, 1969, p. 187). Ideographs become meaningful through clusters that build meaning and shape both publics and public discourse as they change over time (McGee, 1975). Ideographs may be studied both synchronically, in terms of how they are used in one rhetorical situation, and diachronically, in terms of how their usage functions over time (Clarke, 2002). We use both methods to understand evangelical environmental ideographs because the ideas we hold about nature dictate how we speak about it, ultimately determining whether we protect or destroy it (DeLuca, 1999). Within the evangelical context, <life>, <Creation>, and <stewardship> have dominated the rhetorical landscape of EEGs.

We begin with the ideograph of <sanctity of life> to understand the gender hierarchies historically employed by the evangelical right that link the ideograph <Creation> with the ideograph <Life>. The intersectional approach highlights how hegemonic power is unequally distributed to create what Patricia Hill Collins (2008) has termed a "matrix of oppression" for women of color across the globe. In charting the intersectional dimensions of the evangelical debate over climate change, we examine contemporary discourse to understand how <Creation> and <Life> were rearticulated as <Creation> and <stewardship> to demonstrate how EEGs utilize <sanctity-of-life> language in new, anti-feminist ways. By tracing the shift from <sanctity of life> to <stewardship>, we uncover how these ideographs carry anti-feminist ideological baggage throughout successive iterations, which is problematic for the poor and for women. Although evangelical environmentalism appears progressive, both conservative and liberal environmentalists harness hierarchies of power that privilege mostly wealthy, white men over women, people of color, and poor people across the globe. Only through an ideographical analysis can we understand the intersectional dimensions of evangelical environmentalism.

<Life> in evangelical discourse

Beginning in the 1970s, the 'sanctity of human life' became a central concern for evangelicals who asserted the a priori nature of human life from conception and asserted the fetus' personhood over the political and biological rights of the mother (Gushee, 2009). The <sanctity-of-life> ideograph grew out of Cold War politics that saw masculinity and aggression as "positive features of political rhetoric and activity" galvanizing conservatives against *Roe* v. *Wade* (Rogers and Heltzel, 2008, p. 413). In the culture wars of the 1980s, <sanctity of life> dominated as conservatives argued that the fetus was a protected human, thus undermining the <choice> ideology advanced by the women's movement in the 1960s (Railsback, 1984). This conservative articulation of <sanctity of life> ultimately linked the Republican Party, evangelicals, and Catholics, despite the GOP's historical support for abortion as a middle-class family planning practice (Williams, 2011). Consequently, <sanctity of life> and later, <life>, became the ideograph defining the conservative agenda and its backlash against feminist sexual politics.

Evangelical President George W. Bush's first executive order, the Global Gag Rule, embraced the <sanctity-of-life> ideograph articulated during the Reagan years, defunding international family planning organizations that informed about or provided abortions. Ostensibly, the Global Gag Rule "tells Third World family planning organizations that getting help must conform to the morality of the Christian Right," thus demonstrating the domination inherent in this conception of <sanctity of life> (Karon, 2012, n.p.). Bush also declared January 22 National Human Sanctity of Life Day to coincide with the anniversary of *Roe*. In his 2009 proclamation Bush explained, "All human life is a gift from our Creator that is sacred, unique, and worthy of protection," adding, "every person waiting to be born, has a special place and purpose in this world. We also underscore our dedication to heeding this message of conscience by speaking up for the weak and voiceless among us" (Bush, 2009, n.p.). Here, <sanctity of life> connects fetal life to <Creation>, providing a divine warrant for protecting the fetus and authorizing evangelicals to speak *for* the unborn, thus authorizing exclusively white men to articulate the importance of various life forms in policy.

Focus on the Family (FOTF), the largest evangelical organization in the USA, articulates the <sanctity-of-life> ideograph by noting that they are "dedicated to defending the sanctity of human life, and by human life we mean God's creation from fertilization to natural death." They continue,

> In the beginning, God created the earth and everything in it, including humans.... We believe that every human, in every condition from the single cell stage of development to natural death, is made in God's image and possesses inestimable worth. Abortion runs contrary to these beliefs.
>
> (Our position, 2010, n.p.)

Here, <sanctity of life> is central to understanding the *scope* of what evangelicals call <Creation>, which extends from conception to death and warrants intervention into the wombs and lives of women.

Tracing competing interpretations of <Creation> and <stewardship>

<Sanctity of life> was elevated in the 1980s with the rise of anti-abortion evangelism in the USA and the popularity of Vatican II in the Catholic Church. <Sanctity of life> frames evangelical legislation focused on scaling back reproductive rights, but it also provides a context to understand how <Creation> marks current evangelical environmental discourse. While often used as a synonym for the "environment," <Creation> is a relational term which articulates beliefs of a "Creator" of the universe and the conception of an omnipotent "God." Gnanakan explains,

> [T]here is an ongoing relationship between God and his creation. In saying God is Creator, we are affirming that it is God who is Lord, and that it is God who is the initiator, the sustainer and therefore continues to graciously relate to a creation of which we are only a part.
>
> (Gnanakan, 2006, p. 111)

Evangelical environmentalism is rooted in a belief in God as the Creator of all life, thus connecting this ideograph to <sanctity of life>. While both liberal and conservative EEGs consider God the sustainer of all creation, conservative evangelicals argue, "God alone, who is Lord and the source of everything, is responsible for all that is created and must not be confused with his creation" (Gnanakan, 2006, p. 112).

Conservatives articulate a hierarchy of <Creation> on Earth that subordinates humans to God and yet indelibly marks them as superior to the rest of creation. In this hierarchy, God is separate from the environment; as Simmons (2009, p. 44) explains, "God is above nature and transcendent from all that God has created." In this perspective, God ordains creation and destruction, including environmental devastation. This debate over dominion as either a vertical or horizontal relationship has been at the center of debates surrounding the ethical imperative to care for nature.

Liberal evangelicals distinguish between 'Creator' and 'creation' but attempt to mitigate hierarchies of power by connecting all creation. For example, "An Evangelical Declaration on the Creation of Care" affirms that "The cosmos, in all its beauty, wildness, and life-giving bounty, is the work of our personal and loving Creator" (Berry, 2000, p. 19). However, the Declaration also states that God's Creation "is a symphony of individual creatures in harmonious relationship," and "we are *creatures*, shaped by the same processes and embedded in the same systems of physical, chemical, and biological interconnections which sustain other creatures" (p. 20, emphasis added). Understanding creation as a "symphony" of interconnected systems linking all creatures provides some pushback to the rigid hierarchy articulated in conservative evangelical writings, demonstrating the flexibility of <Creation> in competing environmental discourses.

Despite minor differences in understanding the relationship between Creator and Creation, the dominion command (found in Genesis 1:26) has been the central theological space for the debate about the hierarchy of power within <Creation> and its relationship to <stewardship> (Nagle, 2008). Certainly, the conservative evangelical perspective dominates theological dominion discourses. For example, Cornwall Alliance spokesman Dr. E. Calvin Beisner (2008, p. 5) argues that Christians must embrace the dominion command because "it gives human beings legitimate authority to subdue and rule the Earth … to serve their needs and the glory of God." He explains, "The dominion mandate to Adam and Eve at the creation makes human responsibility for creation stewardship inescapable" (p. 1). This interpretation of subduing and ruling the Earth justifies environmental destruction through the invocation of "stewardship and absolute property rights in Western civilization" that have "resulted in its mutation into a metaphor of human control and mastery over nature" (Northcott, 1996, p. 180).

Thus, conservative stewardship recognizes no intrinsic value in non-human creation. FOTF justify this ethic of domination contained in their interpretation of <Creation> by enacting the ideograph <stewardship>. Their "A Statement on the Environment" explains, "the very first job God gave humans was to tend and keep their surrounding environment" (A Statement, 2010, n.p.). For example, FOTF asserts that "environmentalists act as if concern for plant or animal trumps all others. This exclusive emphasis is likewise alien to Biblical thought, which cherishes *God* and *humans* above all else" (A Statement, 2010, n.p.). Thus, conservatives establish <Creation> as the property of humans. This view objectifies nature and forecloses the possibility of relationships between humans and nature outside a paradigm of domination (DeLuca, 1999).

Where conservative evangelicals imbue <stewardship> with notions of property ownership and hierarchy, liberals see <stewardship> as a calling to *protect* the Earth from human destruction, arguing, "[t]he earthly result of human sin has been a perverted stewardship. A patchwork of garden and wasteland in which the waste is increasing" (Berry, 2000, p. 20). Liberal evangelicals understand that humans are "embedded in creation" (Bouma-Prediger, 2001, p. 73).

Lowe (2009, p. 44) explains that liberal evangelicals "are called to be stewards of creation as we live out our lives on this earth. We are called to love God and love our neighbor, and to continue moving from being part of the problem to being more of the solution." The liberal articulation of what they call 'Creation care' demonstrates that the successful reinterpretation of ideographs is dependent upon nimble articulations of ideographs for both liberal and conservative evangelicals (Wunderlich, 2004, p. 82). Their website (Evangelical Environmental Network, n.d., n.p.) explains 'Creation care' as the idea that "human beings have a special role and a special responsibility in God's creation since they are created in God's image and have free will. Human beings are called to care for the rest of God's creation, not abuse or destroy it."

Despite the liberal interpretation of 'Creation care,' the link between <stewardship> and <Creation> is nonetheless troubling because <Creation> is unbounded in both political poles of the debate, including not only the environment but *all* of humanity. FOTF claims that "Biblical thinking integrates consideration of the *physical* environment with the *spiritual and moral* climate as well" (2010, n.p.). <Stewardship> functions to mobilize men to be *moral* stewards over other humans, with dangerous implications for women, minorities, and non-Christians in particular. For example, FOTF's "Statement" suggests, "Let us also cherish Creation as our Lord would, guarding it, as faithful stewards, from physical, moral, and spiritual harm alike" (2010, n.p.). This call to steward prompts evangelicals to see others as wayward moral property much like in the Global Gag Rule or in the abortion debate (Wunderlich, 2004, p. 86).

This concept of stewardship over humanity is developed in the Environmental Climate Initiative's (ECI) "A Call to Action," arguably the most significant example of the liberal evangelical position on climate change to date (Wilkinson, 2010). Here, the ECI urges Christians to mobilize around climate change because of the devastating environmental outcomes for 'the poor.' First, they argue, "Millions of people could die in this century because of climate change, most of them our poorest global neighbors" (Climate Change, n.d., n.p.). 'The poor' are used to externally motivate the ECI's audience to care about the changing climate outside of the USA. For example, the ECI states that small changes in global temperature will disproportionately hurt agricultural production "especially in *poor* countries" and that "[t]he consequences of global warming will therefore hit the *poor* the hardest, in part because those areas likely [to] be significantly affected first are in the *poorest* regions of the world" (Climate Change, n.d., n.p., emphasis added). 'The poor' are presented as the only victims of climate change, and since the ECI locates them in particular 'regions' of the world, we can understand the racial politics of talking about places such as Sub-Saharan Africa or East Asia, where scientists predict catastrophe if climate change continues unabated. Here, an intersectional perspective highlights how evangelism is coded as First World and white where 'the poor' are objectified as non-white and described as in need of stewardship. Likewise, the dangers of climate change are externalized from the First World and projected as a catastrophic futurism for the Third World. For example, the ECI claims, "We pray for those who are likely to be most hurt by climate change, both now and in the future" (A Prayer, n.d., n.p.). The ECI believes that wealthy Christians will not be affected by climate because it is a problem for those with less money, or with "fewer resources available to cope" (Climate Change, n.d., n.p.). The consequences of this are precisely the hierarchies of caste and race set up by the dominion command interpreted as 'stewardship.'

Like the ECI, the use of <stewardship> in conjunction with <Creation> in the discourse of the Evangelical Environmental Network (EEN) equates stewardship with mission work and claims that climate change is a result of failed stewardship. The EEN declares,

> Christians, noting the fact that most of the climate change problem is human induced, are reminded that when God made humanity he commissioned us to exercise stewardship over the earth and its creatures. Climate change is the latest evidence of our failure to exercise proper stewardship, and constitutes a critical opportunity for us to do better.
>
> (Climate Change, n.d., n.p.)

They continue, "we must also help the poor adapt to the significant harm that global warming will cause" (Climate Change, n.d., n.p.). Here, climate change represents an 'opportunity' for (white) Christians to exercise benevolent stewardship over 'the (non-white) poor,' who become both scapegoat and victim, displacing collective responsibility for the First World consumption that has prompted climate change, embodying what has been called the 'colonialist' mission of stewardship (Hazelrigg, 1995). Indeed, the EEN explicitly asserts <creation care> as an essential part of the mission of what amounts to white evangelism (Boorse, 2001).

The claims made by the conservative Cornwall Alliance oppose the liberal EEGs, though both use 'the poor' as a rationale for their <stewardship> agendas. The Cornwall Alliance uses <stewardship> to demonstrate how increasing environmental *regulations* will devastate 'the poor.' They argue that carbon dioxide regulations "would cause greater harm than good to humanity – especially the poor – while offering virtually no benefit to the rest of the world's inhabitants" (Climate Change, n.d., n.p.). The Alliance also minimizes the effects of climate change for the poor, describing the consequences as "not catastrophic" (The Cornwall Alliance, 2009, p. 2). As with other EEGs, <stewardship> is linked to <Creation>, establishing the hierarchy that allows stewardship to effectively frame the poor as property. The Cornwall Alliance (n.d., n.p.) explains: "We aspire to a world in which human beings care wisely and humbly for all creatures, first and foremost for their fellow human beings, recognizing their proper place in the created order." This extension of <stewardship> opens the possibility for the steward to decide who needs stewarding, implying that some people (particularly non-Christians, the poor, and women) are incapable of caring for themselves and situating evangelical stewards above other people in a rhetorical move that echoes the abortion wars of the 1980s.

Connecting the ideographic functions of \<stewardship\> and \<life\>

Our assessment of \<Creation\> and \<stewardship\> highlights their flexibility and demonstrates how evangelicals envision hierarchies of power as they enact the dominion command. Only by connecting \<stewardship\> to \<life\> can we understand the material consequences of the debate over dominion as conservative evangelicals seek to define \<stewardship\> as dominion over other *humans*, 'the poor,' and also women. Here, \<stewardship\> is gendered as a masculine benevolence over groups of people such as the poor because \<creation\> suggests that human beings were created in the image of (Father) God. Farley (1976, p. 166) argues that the concept of *imago Dei* is extended only to men, while acknowledging the "devastating refusal by Christian theology to attribute the fullness of the *imago Dei* to women" where "women have been declared lesser than men." Thus, the omission of women from 'creation' is at the root of woman's subordination to man, and the omission of women from EEG discourse demonstrates how women are elided under seemingly gender-neutral language in the hegemonic relationship between \<stewardship\> and \<life\>. Mumby (1997, p. 344) has defined hegemony "as non-coercive relations of domination in which subordinated groups actively consent to and support belief systems and structures of power relations that do not necessarily serve – indeed, may work against – those groups' interests." These ideographs demonstrate the subtle domination in EEG discourse as evangelical women see themselves as part of the stewardship project without seeing themselves as objects of it. Fortunately, the intersectional method "critiques dominance and hegemony using symbols generated by the dominant/hegemonic system" while working "to challenge rather than reproduce power structures" (Howard and Prividera, 2004, p. 91).

This intersectional feminist analysis reveals hegemony within EEG discourse to showcase how \<life\>, with its ideological baggage from the abortion wars, has been harnessed in the climate change debate to elevate fetal 'life.' Condit (1990, p. 61) explains that, beginning in the 1960s, "the major rhetorical effort of the pro-life movement" centered on "constructing and amplifying the verbal linkages between the terms *fetus* and *life*." Thus, references in EEG discourse to stewarding \<life\> are attached to abortion discourse. For example, the liberal EEN uses \<life\> to support its call for environmental \<stewardship\>, essentially positioning the fetus instead of 'the poor' as the warrant for their claims about mercury poisoning as a 'life issue.' In one interview (Little, 2005, n.p.), Richard Cizik outlines how the EEN accomplished this:

> Jim Ball of the Evangelical Environmental Network and I carried a placard at a pro-life rally that said, 'Stop Mercury Poisoning of the Unborn.' I distributed flyers showing that one in six babies is born with dangerous mercury levels, and urged protestors to demand improvements in the Clear Skies Act.
>
> (Little, 2005, n.p.)

Utilizing fetal protectionism in the service of environmentalism places the 'life' of a fetus as superior to other forms of 'life,' demonstrating how ideologies bound to <life> have been redeployed from the abortion wars to the climate change debate. The fact that mercury poisoning affects men, women, and children who have already been born is not enough to warrant improvements to the Clear Skies Act. In another example, EEN President Reverend Mitchell Hescox (2012, n.p.) began his testimony on mercury poisoning to the House Energy and Power Subcommittee in February 2012 by quoting Speaker John Boehner's speech from the March for Life rally and explaining: "Protecting life and providing the opportunity for abundant life must be a matter of principle and morality," particularly in "protecting our unborn children and infants." The unborn become the warrant for environmental action, connecting <life> to liberal environmental discourse.

Consequently, the ideograph <stewardship> relies on an interpretation of <life> that renders all of those stewarded as helpless and in need of protecting in a frame that George Lakoff (1995, pp. 191–193) has called "The Strict Father Model," where the patriarchal, white father figure enacts a benevolent protectionist family rhetoric to maintain control over women, children, and the economy. This paradigm is found in both liberal and conservative discourse. For example, on the website of its "What Would Jesus Drive campaign?" the EEN claims that Christians are called "to love and protect those with less power, such as the poor, children, the unborn, those yet to be born, and our fellow creatures" ("A Statement," 2009, n.p.). The call to "protect" is a call for stewards to exercise power over fetal <life> along with "children," and "the poor." Women are included among those who need to be stewarded because they are the bearers of the 'unborn life.' In EEG discourse, women, children, and 'the poor' each represent <life> akin to a helpless fetus in need of male stewardship. Because she is seen as incapable of making decisions for herself, a woman is equally incapable of making decisions on behalf of the potential 'life' of her fetus. Thus, the narrowness of <life> is indicative of an ideology where women continue to be measured in terms of their reproductive potential. For all EEGs, female sexuality exists for the pleasure of men and for procreation. FOTF (2009, n.p.) claims, "sex is the greatest gift that [a woman] can give [her] husband." And in "On the Care of Creation" (2009, n.p.) the EEN claims, "Our responsibility is not only to bear and nurture children, but to nurture their home on earth." Therefore, <stewardship> means stewardship of the Earth, stewardship of the poor, and stewardship over women to ensure they occupy their proper role in the hierarchy of power that produces children. Abortion represents a woman's violation of her role. Its occurrence is evidence of a woman's sexuality, but procreation, the sole purpose for her sexuality, is eliminated.

In addition to violating her purpose as a procreator, a woman who has had an abortion is also viewed by FOTF as a polluter. FOTF explains that Christian stewards "tremble to consider the consequences to a nation that spends billions for pure air and water, yet whose land ... is polluted by the blood of more than 40 million innocent preborn children" (2010, n.p.). For FOTF the blood of innocent "preborn children" represents the worst type of environmental pollution: moral pollution. Thus, women who have abortions are the greatest polluters the conservative interpretation of <life> and all other calls for environmental <stewardship> pale in comparison to abortion. Although FOTF is the most explicit in connecting <life> to environmentalism to reify patriarchy, the inclusion of <sanctity-of-life> language even in the liberal evangelical organizations cements this anti-feminist logic.

Conclusions and implications

For both liberal and conservative evangelicals framing climate change policy, <life>, <Creation>, and <stewardship> are the dominant ideographs in the cluster, nimbly accommodating ideological shifts across time, though their connections to extremely conservative ideologies in the evangelical community concerning <life> and its attenuating discourses concerning the subordination of women are unshakeable. In this cluster, <Creation> constitutes the reasons for environmental action through the language of property. <Stewardship> provides the warrant for practices encompassing the physical and moral shepherding of the Earth and its creatures, thus creating hierarchies of power that justify ecological destruction in the name of human desires. But in both conservative and liberal rhetorical incarnations, the women and poor are dominated by the logic of <stewardship> and its connection to <life>, justifying the subordination of the poor to the wealthy, the women to the men. Because the liberal and conservative evangelicals deploy *the same* ideographs, these hierarchies of power are familiar to their audience, harnessing the history of conservative ideological power in framing historical iterations of these ideographs.

The power in articulating these same ideographs over time is that a very narrow conception of <life> is articulated by the white men representing the major evangelical organizations in this study. The glaring absence of women's voices is a form of 'sexual fundamentalism' that subordinates women in the articulation of sexual politics within EEG discourse. Thus, as the culture wars re-centered domestic politics on abortion and female sexuality, these concerns infiltrated many aspects of evangelical life, including emergent environmental politics, where man's stewardship is most salient in matters of <life>, for the poor and the 'unborn.' Representing this <life> is a man's greatest call to <stewardship> as men are called to steward both the woman and the fetus through controlling the woman's environment as well as the fetus' environment. Thus the seemingly progressive environmentalism provides rhetorical continuity for the kinds of misogynistic, hierarchical gender relationships that are articulated in the abortion discourse of EEGs across both camps, despite the presence of environmentalist concerns or the absence of explicit abortion rhetoric. The circulation of these ideographs demonstrates how <life> still overshadows often well-meaning evangelical environmental statements regarding climate change with perilous repercussions for ecology and women.

As policies such as the Global Gag Rule make clear, this imposition of the sexual politics of <life> is the primary mechanism of controlling poor populations, particularly of color, across the globe. This has devastating consequences for poor women, particularly of color, across the globe, who are excluded from environmental conversations by <stewardship> and <life>, particularly if they advocate for reproductive justice. For example, at the Rio + 20 environmental summit in June 2012, activists produced almost 24,000 pages of documents in which women were mentioned in less than 0.01 percent of the text, despite the fact that there is no way to understand sustainable development without women (Ehlers and Brune, 2012). For scholars invested in intersectional analyses, an assessment of the ways in which whiteness and masculinity determine evangelical environmental politics, particularly in understanding how environmental concerns interconnect with contraception, family planning, and sustainability, is critical. By blaming the victims, whether they are women who seek to regain legal protection for abortion, or the poor, conservative framing of evangelical environmental discourse mirrors the way in which anti-woman, anti-poor, anti-abortion <life> discourse emerged in the USA.

References

A Prayer Guide for Global Warming, n.d. Evangelical Climate Initiative. Available at: http://pub.christiansandclimate.org/pub/PrayerGuide.pdf (accessed June 1, 2012).

A Statement on Climate Change Legislation: 'What Would Jesus Drive?,' n.d. Evangelical Environmental Network and Creation Care Magazine. Evangelical Environmental Network. Available at: http://whatwouldjesusdrive.org/statement.php (accessed November 10, 2009).

A Statement on the Environment, 2010. Focus on the Family. Available at: http://media.focusonthefamily.com/topicinfo/environmental_statement.pdf (accessed June 1, 2015).

An Open Letter to the Signers of 'Climate Change: An Evangelical Call to Action' and Others Concerned about Global Warming, 2009. The Cornwall Alliance. Available at: www.cornwallalliance.org/docs/an-open-letter-to-the-signers-of-climate-change-an-evangelical-call-to-action-and-others-concerned-about-global-warming.pdf (accessed June 1, 2012).

Beisner, E.C., 2008. What is the Most Important Environmental Task Facing American Christians Today? Mount Nebo Papers, 1, pp. 1–36.

Berry, R.J., 2000. An Evangelical Declaration on the Care of Creation. Reprinted in R.J. Berry (ed.), The Care of Creation: Focusing Concern and Action. Downers Grove, IL: InterVarsity Press, pp. 17–22.

Bineham, J.L., 1993. Theological Hegemony and Oppositional Interpretive Codes: The Case of Evangelical Christian Feminism. Western Journal of Communication, 57, pp. 515–529.

Boorse, D., 2011. Loving the Least of These: Addressing a Changing Environment. National Association of Evangelicals, pp. 1–56. Available at: http://creationcare.org/blank.php?id=97 (accessed July 23, 2011).

Bouma-Prediger, S., 2001. For the Beauty of the Earth: A Christian vision for creation care. Grand Rapids, MI: Baker Academic.

Burke, K., 1969. *A Rhetoric of Motives*. Berkeley: University of California Press.

Bush, G.W., 2009. National Sanctity of Life Day, January 15. Available at: http:// georgewbush-whitehouse.archives.gov/news/releases/2009/01/20090115-1.html (accessed January 4, 2015).

Bush, G.W., 2011. Proclamation of National Sanctity of Human Life Day, January 16. Available at: www.lifesitenews.com/news/archive/ldn/2009/jan/09011601 (accessed January 5, 2011).

Charland, M., 1987. Constitutive Rhetoric: The Case of the peuple quebecois. *The Quarterly Journal of Speech*, 73(2), pp. 133–150.

Clarke, T., 2002. An Ideographic Analysis of Native American Sovereignty in the State of Utah. *Wicazo Sa Review*, 17(2), pp. 43–63.

Climate Change: An Evangelical Call to Action, n.d. Evangelical Climate Initiative. Available at: www.christiansandclimate.org/statement (accessed June 1, 2012).

Collins, P.H., 2008. *Black Feminist Thought*. New York: Routledge.

Condit, C., 1990. *Decoding Abortion Rhetoric: Communicating social change*. Urbana: University of Illinois Press.

Condit, C. and Lucaites, J., 1993. *Crafting Equality: America's Anglo African word*. Urbana: University of Illinois Press.

Cooper, M., 2006. The Unborn Born Again: Neo-imperialism, the Evangelical Right, and the Culture of Life. *Postmodern Culture*, 17(1), n.p.

Cornwall Declaration on Environmental Stewardship, n.d. The Cornwall Alliance. Available at: www.cornwallalliance.org/docs/the-cornwall-declaration-on-environmental-stewardship.pdf (accessed June 1, 2012).

Deluca, K.M., 1999. *Image Politics: The new rhetoric of environmental activism*. New York: Guilford Press.

Ehlers, S. and Brune, M., 2012. Why Women's Needs Must Be Part of the Conversation at Rio. *Grist*. Available at: http://populationaction.org/2012/21/why-womens-needs-must-be-part-of-the-conversation-at-rio/ (accessed June 21, 2012).

Emerson, M.O. and Smith, C., 2000. *Divided by Faith: Evangelical religion and the problem of race in America*. Oxford: Oxford University Press.

Evangelical Environmental Network, n.d. Why Creation Care Matters. Available at: http://creationcare.org/blank.php?id=41 (accessed June 1, 2012).

Farley, M.A., 1976. Sources of Sexual Inequality in the History of Christian Thought. *The Journal of Religion*, 56, pp. 162–176.

Gnanakan, K., 2006. Creation, Christians, and Environmental Stewardship. *Evangelical Review of Theology*, 30(2), pp. 110–120.

Goodstein, L., 2005. Evangelical Leaders Swing Influence Behind Effort to Combat Global Warming. *New York Times*. Available at: http://74.125.155.132/scholar?q=cache:uHLzs299y4YJ:scholar.google.com/+evangelicals+climate+change&hl=en&as_sdt=0,4 (accessed November 3, 2010).

Gushee, D.P., 2009. God and Godlessness in the Environment: Can a Sanctity-of-human-life Ethic Ground Christian Ecological Responsibility? *Notre Dame Journal of Law, Ethics, and Public Policy*, 23, pp. 471–496.

Hazelrigg, L.E., 1995. *Cultures of Nation: An essay on the production of Nature*. Gainesville: University Press of Florida.

Heltzel, P., 2009. *Jesus and Justice: Evangelicals, race and American politics*. New Haven, CT: Yale University Press.

Hescox, M., 2012. A Christian Perspective on the Costs of Mercury to Human Health and Well-being. 112 Congress. Available at: http://republicans.energycommerce.house.gov/

Media/file/Hearings/Energy/20120208/HHRG-112-IF03-WState-MHescox-20120208. pdf (accessed May 27, 2015).

Howard, J.W. and Prividera, L.C., 2004. Rescuing Patriarchy or Saving 'Jessica Lynch'; The Rhetorical Construction of the American Woman Soldier. *Women and Language*, 27(2), pp. 89–97.

Karon, T., 2001. For Bush, Humility and the 'Global Gag Order' Don't Mix. *Time*. Available at: www.time.com/time/world/article/0,8599,96407,00.html (accessed June 1, 2012).

Krakoff, S., 2011. Parenting the Planet. In D. Gordon Arnold (ed.), *The Ethics of Global Climate Change*. Cambridge: Cambridge University Press, pp. 145–169.

Lakoff, G., 1995. Metaphor, Morality, and Politics, or, Why Conservatives Have Left Liberals in the Dust. *Social Research*, 62(3), pp. 177–214.

Little, A.G., 2005. Cizik Matters. Grist.org. Available at: www.grist.org/ cgibin/printthis. pl?uri=/news/maindish/2005/10/05/cizik/index.html (accessed September 20, 2008).

Lowe, L., 2009. *Green Revolution: Coming together for the care of creation*. Downers Grove, IL: InterVarsity Press.

McGee, M.C., 1975. In Search of 'The People': A Rhetorical Alternative. *The Quarterly Journal of Speech*, 61(3), pp. 235–249.

Medhurst, M.J., 2009. Evangelical Christian Faith and Political Action: Mike Huckabee and the 2008 Republican Presidential Nomination. *Journal of Communication and Religion*, 32(2), pp. 199–239.

Mumby, D., 1997. The Problem of Hegemony: Rereading Gramsci for Organizational Communication. *Western Journal of Communication*, 61(4), pp. 343–375.

Nagle, J.C., 2008. The Evangelical Debate over Climate Change. *University of St. Thomas Law Journal*, 5(1), pp. 57–86.

Nisbet, M., 2009. Communicating Climate Change: Why Frames Matter for Public Engagement. *Environment*, 51(2), pp. 12–23.

Northcott, M.S., 1996. *The Environment and Christian Ethics*. Cambridge: Cambridge University Press.

On the Care of Creation, n.d. *Evangelical Environmental Network and Creation Care Magazine*. Available at: www.creationcare.org/resources/declaration.php (accessed November 10, 2009).

Our Position (Abortion), 2010. Focus on the Family. Available at: www.focusonthe-family.com/socialissues/sanctity-of-life/abortion/our-position.aspx (accessed November 3, 2010).

Prelli, L.J. and Winters, T.S., 2009. Rhetorical Features of Green Evangelicalism. *Environmental Communication*, 3(2), pp. 224–243.

Railsback, C., 1984. Contemporary Abortion Debate. *Quarterly Journal of Speech*, 70(4), pp. 410–424.

Religious Landscape Survey. 2008. The Pew Forum on Religion and Public Life. Available at: http://pewresearch.org/pubs/743/united-states-religion (accessed November 10, 2009).

Rogers, R. and Heltzel, P.G., 2008. The New Evangelical Politics. *Society*, 45(5), pp. 412–414.

Sider, R.J., 2000. Biblical Foundations for Creation Care. In R.J. Berry (ed.), *The Care of Creation: Focusing concern and action*. Downer's Grove, IL: InterVarsity, pp. 43–49.

Simmons, J.A., 2009. Evangelical Environmentalism: Oxymoron or Opportunity? *Worldviews*, 13(1), pp. 40–71.

Stromberg, S., 2011. How to Get Republicans to Go Green. *Washington Post*. Available at:

www.washingtonpost.com/opinions/how-to-get-republicans-to-go-green/2011/06/08/
AGx2j9UH_story.html (accessed July 29, 2011).

Wang, M., 2010. Meet the Likely House Committee Chairs Who Promise to Roll Back 'Job Killing' Regs. *Facing South*. Available at: www.nytimes.com/cwire/2010/09/20/20climatewire-climate-change-skeptics-sweeping-gop-senate-75251.html (accessed July 29, 2011).

Wilkinson, K.K., 2010. Climate's Salvation: Why and How American Evangelicals are Engaging with Climate Change. *Environment*, 52(2), pp. 47–57.

Wilkinson, K.K., 2012. *Between God and Green: How evangelicals are cultivating a middle ground on climate change*. Oxford: Oxford University Press.

Williams, D.K., 2011. The GOP's Abortion Strategy: Why Pro-choice Republicans Became Pro-life in the 1970s. *Journal of Policy History*, 23(4), pp. 513–539.

Wunderlich, G., 2004. Evolution of the Stewardship Idea in American Country Life. *Journal of Agricultural and Environmental Ethics*, 17(1), pp. 77–93.

48 Climate change and complexity of gender issues in Ethiopia

Victoria Team and Eyob Hassen

Africa is one of the continents that is predicted (Müller et al., 2014) to be hardest hit by climate change, as is already being experienced (Parenti, 2011). The Ethiopian climate is also changing (Conway and Schipper, 2011). Uncertainty in climate change projections related to inadequate understanding of atmosphere–ocean interactions poses a major challenge for a country's climate change adaptation plans (Simane et al., 2012). Women were identified as one of the vulnerable social groups in a changed climate (Team and Manderson, 2011); those of lower socio-economic classes will also be more vulnerable. Gender inequalities in a changed climate could become increasingly complex. These inequalities occur within the context of the political, economic, and social structures of each individual country, and therefore are not related solely to changed biophysical climatic conditions (Nelson, 2011). The adverse effects of climatic change are already felt in many aspects of Ethiopian women's life, raising gender equalities considerations in the country (Kiros and Hogan, 2001; Abaya et al., 2009). Projected climate change has the potential to exacerbate the existing inequalities further, worsening women's health and well-being indicators and decreasing their survival rate (Kidanu et al., 2009).

In this chapter, we highlight the complexity of climate change-related gender issues in Ethiopia, while also recognizing the role of social class. We discuss the potential climate change issues from a women's health perspective. Our main argument is that the complexity of climate change-related gender issues should be addressed by the relevant changes in social, political, and institutional structures, policies and processes that interact dynamically with gender and class by reducing the levels of inequalities. This multidimensional view of climate–gender interactions can assist the reader in understanding that adaptation will require complex governance processes that address multi-pronged goals and involve multiple stakeholders and constituencies. The described complexity may potentially reduce the effectiveness of climate change interventions. As a country that is already burdened by poverty, inequality, and poor health, it is anticipated that Ethiopia will have limited capacity to deal with the projected climate change issues despite their urgency.

Below we provide a case study from a local private clinic in Ethiopia's capital, Addis-Ababa, demonstrating gender- and class-differentiated impacts of socio-environmental factors, which will no doubt be exacerbated in a changed climate. The proposed case study is concerned with a woman who, with the adoption of the *Environmental Policy* (EPA, 1997) in 1997, was appointed as a community forest guard to patrol the forest and to protect it from illegal logging.

Case study

Lialeese [pseudonym], a 32-year-old Ethiopian woman, was brought in by her relatives to the local private clinic with multiple infected, deep-cut wounds on her face. Lialeese's history was taken from the words of her husband. He said that Lialeese was appointed as a paid forest guard by her local council, *kebele*. She was supplied with a firearm and was requested to control the main road in order to limit the opportunities of unsolicited tree loggers to take the logs to the city, where they could sell them. Her responsibility was that whenever she saw a man carrying a log, she should stop the person, arrest him, and inform the local council. She was also required to go deep into the forest whenever she heard the sound of an axe cutting a tree, and confront that person.

One day Lialeese met her distant relative who was cutting a tree and confronted him with the firearm. Her relative started to negotiate with her, promising to pay her some money for allowing him to cut the tree and not reporting his action to the local council. After Lialeese refused, he left the spot, swearing and threatening to kill her. A few days later he attacked Lialeese in the forest, close to the main road, and cut her face with an axe. No action was taken against the perpetrator by the local council, because Lialeese had no witnesses to support her case. In addition, no support was provided by the local council to Lialeese.

When Lialeese's scarf was removed to allow for examination, her face was disfigured, swollen, and necrotized. Multiple cuts were clearly visible, including a cut through her left eye. Lialeese was in severe pain, restless, and febrile. The referral to the hospital was immediately written and the severity of the situation explained to her husband. However, Lialeese's husband refused to take her to the hospital because he had no money to pay for the hospital services. He also mentioned that she is not eligible for free treatment because she had recent paid work in the local council. The clinic staff informed Lialeese's husband about the potential consequences, including death, that she could face without surgical intervention. He said that he cannot go above God's plans for her, and asked only for pain-relief medication for which he was able to pay.

Lialeese could not speak due to her injury, but she could also not be heard due to the lack of women's voices in Ethiopia. Her voicelessness is a metaphor for the lack of women's environmental activism in Ethiopia. Women have no voice in injury management plans or life- and death-related issues. In this example, we demonstrate gender inequalities in societal values of people's lives, particularly of women of low socio-economic status. Despite the many different ethnic groups in Ethiopia, overall it has a patriarchal system; this system allows men to exercise control over women and to have better access to material and financial resources, including those required for healthcare. Men's lives are valued more than women's. A man is considered the head of the household, the main income earner, and the main decision maker. Considering that the loss will be great if a man dies, women are more procreative in arranging finances to cover healthcare costs required for their husbands. They may rely on borrowing and raising money, asking their relatives to support them financially, and even begging – unlike Lialeese's husband, who decided just to wait and see the outcome. In addition to gender-based personal violence, Lialeese experienced gender-related structural violence in terms of restricted access to healthcare and a failure on the part of the state to provide support for basic needs. Talking about women's disposability, we affirm that gender-based violence is a systematic issue, supported by policy and practice, not just individual decision making on the part of families.

The appropriateness of a forest guard position for untrained and unsupported women is arguable. However, women's poverty, low literacy, and lack of paid positions more suitable for women, especially in rural areas, make them desperate to apply for any available positions, disregarding increased health and safety risks. A man in a forest guard position is less likely to be attacked, particularly by a male relative. This is not a single case where a woman was violently assaulted by a man with the aim to demonstrate his power and to teach other women about what might happen if they do not maintain the societal expectation of their gender roles. Gender accommodations within historically male occupations reinforce the lack of women's power and agency.

In the case described above, which occurred in late 1990s, we highlighted the complexity of existing gender and class issues in Ethiopia. All highlighted issues influencing women's health and well-being were extremely common in Ethiopia at that time and are currently common. We also illustrated the difficulties of 'protecting' forests from deforestation, of implementing environmental policies, and of engaging in climate change mitigation. In a changed climate, these issues will be exacerbated; legal, financial, and social support and healthcare provisions will be limited due to the reduced capacity of state institutes burdened by extreme weather events, population movements, violence, conflicts, and possible wars. In the following sections we discuss the projected effects of climate change, focusing on gender and class.

Extreme weather events

According to recent projections (Setegn et al., 2011b), both frequency and duration of droughts in some areas (for example, in the northern highlands of Ethiopia) have the potential to increase. Floods in Ethiopian lowlands have become intensified due to deforestation and improper land management in the highlands, and this has the potential to increase further (Balcha, 2001; Abaya et al., 2009). The country's preparedness to deal with climate change-induced disasters is low. Despite ongoing international assistance, dealing with drought and related famines remains a challenge (Little, 2008; Tadesse et al., 2008). The country's flood preparedness and emergency response is also insufficient, as the results of the retrospective study investigating local government's coping strategies during the year 2006 floods in Gambella region have shown (Abaya et al., 2009).

Although people from various socio-economic classes can be affected by climate change-induced disasters, poor people are more likely to be exposed to disasters than wealthier people because they live in disaster-prone areas; increased population concentrations in disaster-prone areas is an additional factor that influences disaster exposure (Kim, 2012). Women from low socio-economic classes and their children are less likely to survive natural disasters. For example, in Tigray province, northern Ethiopia during the 1973 to 1991 famines, the reported highest levels of mortality were among children born to illiterate mothers (Kiros and Hogan, 2001). Post-disaster recovery may be particularly difficult in female-headed households in rural areas (Block and Webb, 2001; Abaya et al., 2009). Income diversification, including employment outside of farm work, is key to drought survival and post-famine recovery. Female-headed households are usually less diversified (Block and Webb, 2001) because women are less likely to be involved in paid employment outside of cropping due to their low educational status and family commitments.

Water and food security

In Ethiopia, only 58 percent of the population have access to clean drinking water (Ministry of Agriculture and Rural Development, 2010). Rainfall variability and erratic distribution of rainfall have adversely affected economic growth (Nuru Ali, 2012). Projected climatic changes have the potential to influence rainfall variability further, to decrease the stream flow in main rivers, and to alter the water balance in the Lake Tana Basin, a critical source of fresh water to Ethiopians (Beyene et al., 2010; Legesse et al., 2010; Kim et al., 2008; Setegn et al., 2011a).

The link between water scarcity and food security is a key factor in Ethiopia, where people rely on rain-fed agriculture. Although there were some attempts to improve water harvesting in drought-prone areas, such as building micro-dams, there is growing concern related to an increasing incidence of malaria among people living close to micro-dams (Ghebreyesus et al., 1999; Yewhalaw et al., 2009). Moreover, frequent and prolonged droughts in some areas may impact crop yields and livestock production, contributing to famines (Demeke et al., 2011; Blanc, 2012). Some social issues such as wars and related resettlements (discussed further below), compulsory food provision to armed forces, and guerrilla fighters' reliance on farm products have the potential to intensify food insecurity in the country (White, 2005).

In Ethiopia, men are the primary beneficiaries of international food aid, and women tend to traditionally invest most of the received food aid products in the male members of their families (Broussard, 2012). Some authors (Bonnard and Sheahan, 2009) argue that food aid can negatively affect food prices of locally produced food, and farmers will be required to adjust their prices, generating losses rather than profits, and female-headed households will be the most disadvantaged. However, if the main funders reduce the extent of aid, the poorest families will suffer the most (Little, 2008). The introduction of genetically modified species to increase agricultural productivity and reduce food shortages may be counter-productive and lead to further loss of agricultural biodiversity in Ethiopia and related consequences (Jacobsen et al., 2013).

In Ethiopia, food and water management are women's primary responsibilities; food sharing traditionally comprises part of care for elderly parents, extended relatives, and orphans (Rahmato, 1991). In a changed climate, poor women in particular will be burdened with food preparation, distribution, and cultural expectations related to food management. Again, female-headed households will be particularly insecure (Demeke et al., 2011). With lack of fresh water resources, women will be required to bring water from far away, which may increase their vulnerability to rape (Human Rights Watch, 2005).

Growing girls in poor families will be at increased risk of under-nutrition related to food shortages, which has the potential to delay their physical and psychosocial development and sexual maturation (Belachew et al., 2011) and increase their susceptibility to infectious diseases (Katona and Katona-Apte, 2008). Women's under-nutrition and anemia may increase their chances of dying due to obstetric complications, such as perinatal bleeding, and contribute to low birth weight in full-term infants (Black et al., 2008; Assefa et al., 2012). Limited access to water and lack of food at the time of disasters has the potential to affect breastfeeding. Emergency aid in post-disaster camps usually comprises the distribution of milk substitutes, and lacks breastfeeding support (Gribble et al., 2011). Women's reproductive rights in relation to breastfeeding may be undermined, contributing to infants' under-nutrition and their increased mortality.

Although the lack of water for drinking and cooking may affect women's survival, water use for personal hygiene, including menstrual hygiene, cannot be undermined, particularly for women and girls who underwent female genital cutting. Inadequate menstrual blood passage through the infibulated vagina and retained blood clots – in addition to poor personal hygiene due to lack of water – may increase women's risk of reproductive tract infections (Wagner, 2015).

Population movements

Various patterns of population movements relating to socio-environmental issues have taken place in Ethiopia, including international migration, rural–urban migration, and rural–rural migration (Ezra and Kiros, 2001). Drought-related compulsory resettlements from one geographic area to another were conducted by the military government in 1984/1985 (Baker, 1995). In a changed climate, the country's population mobility can be increased due to increasing levels of extreme poverty, environmental degradation, natural disasters, and social conflicts (Gray and Mueller, 2012; Team and Manderson, 2011; McMichael et al., 2012). Population mobility, particularly uncontrolled urbanization and forced resettlement, may increase the risk of infectious disease transmissions, including malaria (Lynch and Roper, 2011; McMichael et al., 2012).

Both male and female out-migration may take place (Ezra and Kiros, 2001). Male out-migration to cities in order to find jobs usually increases the number of female-headed households in rural areas, placing a disproportionate burden of extreme poverty on women (Gray and Mueller, 2012). To meet basic needs, women in female-headed households would be required to take on activities that were usually conducted by their spouses, in addition to their main household activities and child-rearing tasks (Grawert, 1992). Although wealthier women usually migrate to cities to increase their prospects for education and employment, women from low socio-economic classes move to cities in order to ensure their own and their family's survival (Ezra and Kiros, 2001; Rahmato, 1991). Lacking formal education and having limited job opportunities in cities, many women turn to prostitution (Rahmato, 1991). Limited health-related knowledge increases their vulnerability to unwanted pregnancies and sexually transmitted infections, including HIV/AIDS (Kloos et al., 2007; Kassie et al., 2008).

Children's out-migration may also occur. For example, from 1984 to1986, children residing in famine-affected areas were taken by emergency response aid workers and placed in relief camps (Lothe and Heggen, 2003). Some have never been reunited with their families, including the children who were too small to remember the names of their parents (Lothe and Heggen, 2003). Stress related to resettlement, family separations, and lack of hope for reuniting with family members may considerably influence women's mental health (Kloos, 1990; McMichael and Manderson, 2004).

Infectious diseases

In theory, wetter weather and higher temperatures have the potential to increase the incidence of malaria (Ermert et al., 2011) – which has already reached epidemic proportions in the Ethiopian highlands (Negash et al., 2005) – and diarrheal diseases (Bandyopadhyay et al., 2012). The risk of outbreaks of communicable diseases in warmer climates can be higher among conflict-affected populations, which is related to a breakdown of public health infrastructure (Cascio et al., 2011). As the results of a study investigating post-flood outbreaks in Butajira, Ethiopia in the late 1990s have demonstrated (Emmelin et al., 2009), mortality rates at the time of post-flood epidemics of malaria and diarrheal diseases could be highest among poor people. Poor families in rural areas could be particularly vulnerable to diarrheal diseases due to a lack of sanitation and limited access to clean water (Bartels et al., 2010).

Pregnant women are more likely to suffer from severe forms of malaria than non-pregnant women due to a weaker immune system (Schantz-Dunn and Nour, 2009). The complications of malaria in pregnancy are complex, and may include miscarriage, pre-term labor, low birth weight, congenital malaria, and maternal and infant death due to perinatal bleeding (Schantz-Dunn and Nour, 2009; Uneke, 2011). The prevention and treatment of malaria in pregnant women could be particularly problematic due to side effects of some anti-malarial medications, which include miscarriage and increasing drug resistance to safer medications (Vallely et al., 2007).

Girls' under-nutrition and early marriage predisposes them to HIV infection (Katona and Katona-Apte, 2008). Increased sexual violence accompanying armed conflict, and poverty-related prostitution among displaced women in a changed climate may further increase women's chances of being infected by HIV (McMichael and Woodruff, 2008). HIV-infected women will be particularly vulnerable to food shortages and improper nutrition; under-nutrition further weakens the immune system, contributes to opportunistic infections, and intensifies disease progression (Gwatirisa and Manderson, 2009; Ivers et al., 2009). Most anti-retroviral medications require that they be taken with food at proscribed times – something that is very difficult to achieve when women are running around fetching water, caring for others, dealing with financial situations, and generally trying to survive (Hardon and Dilger, 2011; Mattes, 2011).

The situation could be worsened by a limited capacity of the Ethiopian public health system to deal with an increasing incidence of infectious diseases (Abaya et al., 2009). Even with international support, which is critical for Ethiopia, prognosis for malaria control is pessimistic. Without further increases in funding, the goals of malaria control set up by international aid organizations may not be achieved (Guerra et al., 2008). Recent HIV/AIDS funding cuts from PEPFAR (the US President's Emergency Plan for AIDS Relief), although expected, became a matter of concern for Ethiopian public health officials, who will have trouble sustaining the achievements the country has made in prevention and management of HIV/AIDS without external funding (Reliefweb, 2013).

Wars, conflicts, and violence

The increased occurrence of natural disasters and related population movements may give rise to wars, armed conflicts, and violence at different levels (Raleigh et al., 2008; Parenti, 2011). Border security conflicts may occur as a result of pastoralist movements along and across international borders (Ndaruzaniye, 2011). Conflicts between local and newly resettled people may be related to land redistribution, as in the case of compulsory government resettlements that took place in the 1980s (Balcha, 2001; Milas and Latif, 2000). Fights may occur between pastoralists and the military administration controlling the grazing reserves who may attempt to restrict pastoralists' movements (Rettberg, 2010; Kloos et al., 2010). Lack of water and grazing points may exacerbate inter-tribal conflicts (Balcha, 2001).

Droughts, famines, and extreme poverty were commonly associated with farmers' mental health issues and increased self-harm and family violence

(Deyessa et al., 2008, 2009). Forced government migration, violation of human rights, and lack of social, financial, and health support to newly resettled people are all examples of structural violence (Black, 2003). Women's vulnerability in relation to physical and sexual abuse may increase at the time of war, border and tribal conflicts, and resettlements (McMichael et al., 2012). Women living in refugee camps may confront violence on a daily basis (Feseha et al., 2012).

Final remarks

As with numerous other countries located in Southern Africa, West Africa, and Sub-Saharan Africa (Müller et al., 2014), Ethiopia, and in particular Ethiopian women from lower socio-economic class and rural women, are likely to suffer from climate change-related impacts to a higher extent due to their already vulnerable situation, linked to wars, droughts, and related famines (Teller and Hailemariam, 2011; Ndaruzaniye, 2011). There are several other issues that increase the country's vulnerability and adaptation capacity, including a high level of uncertainty of projected climate change-related impacts, limited local research on the country's adaptive capacity, and the gap between academic research and policy development (Teller et al., 2011; Simane et al., 2012). The Ethiopian government and the governments of other African countries were called to promote sustainable development, acknowledging budgetary constraints related to other priority issues that the countries might experience (International Development Association and International Monetary Fund, 2011). A socially focused, poverty-reduction approach to climate change adaptation may be preferred by some countries, as opposed to a gender-focused approach (Wamukonya and Skutsch, 2002). However, and as the woman in the case study above exemplifies, a lack of gender orientation and women's empowerment in climate adaptation projects may leave women's health and social inequalities unaddressed, limiting poorer women's abilities to cope with climate change impacts and adapt to a changed climate (Terry, 2009). It is important to note that the examined case study does not comprise a sensational report of the unprosecuted assault of the Ethiopian environmental activist and a local government worker. This case study is representative of women's experiences in Ethiopia, and demonstrates the connection between climate change and gender and class inequalities.

African women are already disproportionately carrying the burden of climate change impacts, including water shortage, food insecurity, and related malnutrition and infectious diseases (Magrath and Tesfu, 2006; Ludi, 2009). The development and implementation of gender-sensitive mitigation and adaptation strategies may limit climate change-related impacts on women's health and well-being and reduce the disproportionate injustice (Dankelman, 2010; Haigh and Vallely, 2010; United Nations Secretary-General's High-Level Panel on Global Sustainability, 2012).

References

Abaya, S.W., Mandere, N., and Ewald, G., 2009. Floods and health in Gambella region, Ethiopia: A qualitative assessment of the strengths and weaknesses of coping mechanisms. *Global Health Action*, 2 [online], doi: 10.3402/gha.v2i0.2019.

Assefa, N., Berhane, Y., and Worku, A., 2012. Wealth status, mid upper arm circumference (MUAC) and antenatal care (ANC) are determinants for low birth weight in Kersa, Ethiopia. *PloS one*, 7(6), p. e39957.

Baker, J., 1995. Migration in Ethiopia and the role of the state. In Baker, J. and Aina, T.A. (eds), *The Migration Experience in Africa*. Stockholm, Sweden: Nordiska Afrikainstitutet, pp. 234–256.

Balcha, B.G., 2001. *Environmental, Social and Economic Problems in the Borkena Plain, Ethiopia. WP 2001: 8*. Bergen, Norway: Chr. Michelsen Institute, Development Studies and Human Rights.

Bandyopadhyay, S., Kanji, S., and Wang, L., 2012. The impact of rainfall and temperature variation on diarrheal prevalence in Sub-Saharan Africa. *Applied Geography*, 33(1), pp. 63–72.

Bartels, S.A., Greenough, P.G., Tamar, M., and VanRooyen, M.J., 2010. Investigation of a cholera outbreak in Ethiopia's Oromiya region. *Disaster Medicine and Public Health Preparedness*, 4(4), pp. 312–317.

Belachew, T., Hadley, C., Lindstrom, D., Getachew, Y., Duchateau, L., and Kolsteren, P., 2011. Food insecurity and age at menarche among adolescent girls in Jimma Zone Southwest Ethiopia: A longitudinal study. *Reproductive Biology and Endocrinology*, 9, p. 125.

Beyene, T., Lettenmaier, D.P., and Kabat, P., 2010. Hydrologic impacts of climate change on the Nile River Basin: Implications of the 2007 IPCC scenarios. *Climatic Change*, 100(3–4), pp. 433–461.

Black, R., 2003. Forced migration and sustainable development: Post-conflict opportunities in Ethiopia and Mozambique. *PRAXIS: The Fletcher Journal of International Development*, 18, pp. 31–41.

Black, R.E., Allen, L.H., Bhutta, Z.A., Caulfield, L.E., de Onis, M., Ezzati, M., Mathers, C., and Rivera, J., 2008. Maternal and child undernutrition: Global and regional exposures and health consequences. *The Lancet*, 371(9608), pp. 243–260.

Blanc, E., 2012. The impact of climate change on crop yields in Sub-Saharan Africa. *American Journal of Climate Change*, 1(1), pp. 1–13.

Block, S.A. and Webb, P., 2001. The dynamics of livelihood diversification in post-famine Ethiopia. *Food Policy*, 26(4), pp. 333–350.

Bonnard, P. and Sheahan, M., 2009. *Markets, Food Security and Early Warning Reporting. FEWS NET Markets Guidance, No 6, October*. Washington, DC: FEWS NET, USAID.

Broussard, N.H., 2012. Food aid and adult nutrition in rural Ethiopia. *Agricultural Economics*, 43(1), pp. 45–59.

Cascio, A., Bosilkovski, M., Rodriguez-Morales, A.J. and Pappas, G., 2011. The socioecology of zoonotic infections. *Clinical Microbiology and Infection*, 17(3), pp. 336–342.

Conway, D. and Schipper, E.L.F., 2011. Adaptation to climate change in Africa: Challenges and opportunities identified from Ethiopia. *Global Environmental Change – Human and Policy Dimensions*, 21(1), 227–237. doi: 10.1016/j.gloenvcha.2010.07.013.

Dankelman, I. (ed.), 2010. *Gender and Climate Change: An introduction*. Washington, WA: Earthscan.

Demeke, A.B., Keil, A., and Zeller, M., 2011. Using panel data to estimate the effect of rainfall shocks on smallholders' food security and vulnerability in rural Ethiopia. *Climatic Change*, 108(1–2), pp. 185–206.

Deyessa, N., Berhane, Y., Alem, A., Hogberg, U. and Kullgren, G., 2008. Depression among women in rural Ethiopia as related to socioeconomic factors: A community-based study on women in reproductive age groups. *Scandinavian Journal of Public Health*, 36(6), pp. 589–597.

Deyessa, N., Berhane, Y., Alem, A., Ellsberg, M., Emmelin, M., Hogberg, U., and Kullgren, G., 2009. Intimate partner violence and depression among women in rural Ethiopia: A cross-sectional study. *Clinical Practice and Epidemiology in Mental Health*, 5(1), p. 8.

Emmelin, A., Fantahun, M., Berhane, Y., Wall, S., and Byass, P., 2009. Vulnerability to episodes of extreme weather: Butajira, Ethiopia, 1998–1999. *Global Health Action*, 2, pp. 140–148.

EPA, 1997. *Environmental Policy*, Addis Ababa, April 2. Addis Ababa: Environmental Protection Authority in collaboration with the Ministry of Economic Development and Cooperation.

Ermert, V., Fink, A.H., Morse, A.P., and Paeth, H., 2011. The impact of regional climate change on malaria risk due to greenhouse forcing and land-use changes in tropical Africa. *Environmental Health Perspectives*, 120(1), pp. 77–84.

Ezra, M. and Kiros, G-E., 2001. Rural out-migration in the drought prone areas of Ethiopia: A multilevel analysis. *International Migration Review*, 35(3), pp. 749–771.

Feseha, G.G., Mariam, A., and Gerbaba, M., 2012. Intimate partner physical violence among women in Shimelba refugee camp, northern Ethiopia, *BMC Public Health*, 12, p. 125.

Ghebreyesus, T.A., Haile, M., Witten, K.H., Getachew, A., Yohannes, A.M., Yohannes, M., Teklehaimanot, H.D., Lindsay, S.W., and Byass, P., 1999. Incidence of malaria among children living near dams in northern Ethiopia: Community based incidence survey. *British Medical Journal*, 319(7211), pp. 663–666.

Grawert, E., 1992. Impacts of male outmigration on women: A case study of Kutum/ Northern Darfur/Sudan. *The Ahfad Journal*, 9(2), pp. 37–60.

Gray, C. and Mueller, V., 2012. Drought and population mobility in rural Ethiopia. *World Development*, 40(1), pp. 134–145.

Gribble, K.D., McGrath, M., MacLaine, A., and Lhotska, L., 2011. Supporting breast-feeding in emergencies: Protecting women's reproductive rights and maternal and infant health. *Disasters*, 35(4), pp. 720–738.

Guerra, C.A., Hay, S.I., Mutheu, J.J., and Snow, R.W., 2008. International funding for malaria control in relation to populations at risk of stable Plasmodium falciparum transmission. *PLoS Medicine*, 5(7), pp. 1068–1078.

Gwatirisa, P. and Manderson, L., 2009. Food insecurity and HIV/AIDS in low-income households in urban Zimbabwe. *Human Organization*, 68(1), pp. 103–112.

Haigh, C. and Vallely, B., 2010. *Gender and the Climate Change Agenda: The impacts of climate change on women and public policy.* London: Women's Environmental Network.

Hardon, A. and Dilger, H., 2011. Global AIDS medicines in East African health institutions. *Medical Anthropology*, 30(2), pp. 136–157.

Human Rights Watch, 2005. Sexual Violence and its Consequences among Displaced Persons in Darfur and Chad. A Human Rights Watch Briefing Paper. April 12. Available at: www.cmi.no/sudan/doc/?id=1069 (accessed June 24, 2015).

International Development Association and International Monetary Fund, 2011. *The*

Federal Democratic Republic Of Ethiopia. Joint Staff Advisory Note on Growth and Transformation Plan (GTP) (2010/11–2014/15), August 15. Washington, DC: International Monetary Fund.

Ivers, L.C., Cullen, K.A., Freedberg, K.A., Block, S., Coates, J., Webb, P., and Mayer, K.H., 2009. HIV/AIDS, undernutrition, and food insecurity. *Clinical Infectious Diseases*, 49(7), pp. 1096–1102.

Jacobsen, S.E., Sørensen, M., Pedersen, S.M., and Weiner, J., 2013. Feeding the world: Genetically modified crops versus agricultural biodiversity. *Agronomy for Sustainable Development*, 33(4), pp. 651–662.

Kassie, G.M., Mariam, D.H., and Tsui, A.O., 2008. Patterns of knowledge and condom use among population groups: Results from the 2005 Ethiopian behavioral surveillance surveys on HIV. *BMC Public Health*, 8, p. 429.

Katona, P. and Katona-Apte, J., 2008. The interaction between nutrition and infection. *Clinical Infectious Diseases*, 46(10), pp. 1582–1588.

Kidanu, A., Rovin, K., and Hardee, K., 2009. *Linking Population, Fertility and Family Planning with Adaptation to Climate Change: Views from Ethiopia*. Country study, October. Addis Ababa, Ethiopia and Washington Miz-Hasab Research Center (MHRC) and Population Action International (PAI).

Kim, N., 2012. How much more exposed are the poor to natural disasters? Global and regional measurement. *Disasters*, 36(2), pp. 195–211.

Kim, U., Kaluarachchi, J.J., and Smakhtin, V.U., 2008. Generation of monthly precipitation under climate change for the Upper Blue Nile River Basin, Ethiopia. *Journal of the American Water Resources Association*, 44(5), pp. 1231–1247.

Kiros, G-E. and Hogan, D.P., 2001. War, famine and excess child mortality in Africa: The role of parental education. *International Journal of Epidemiology*, 30(3), pp. 447–455.

Kloos, H., 1990. Health aspects of resettlement in Ethiopia. *Social Science and Medicine*, 30(6), pp. 643–656.

Kloos, H., Haile Mariam, D., and Lindtjorn, B., 2007. The AIDS epidemic in a low-income country: Ethiopia. *Human Ecology Review*, 14(1), pp. 39–55.

Kloos, H., Legesse, W., McFeeters, S., and Turton, D., 2010. Problems for pastoralists in the lowlands: River basin development in the Awash and Omo Valleys. In Kloos, H. and Legesse, W. (eds), *Water Resources Management in Ethiopia: Implications for the Nile Basin*. Amherst, NY: Cambria Press, pp. 253–283.

Legesse, D., Abiye, T.A., Vallet-Coulomb, C., and Abate, H., 2010. Streamflow sensitivity to climate and land cover changes: Meki River, Ethiopia. *Hydrology and Earth System Sciences*, 14(11), pp. 2277–2287.

Little, P.D., 2008. Food aid dependency in northeastern Ethiopia: Myth or reality? *World Development*, 36(5), pp. 860–874.

Lothe, E.A. and Heggen, K., 2003. A study of resilience in young Ethiopian famine survivors. *Journal of Transcultural Nursing: Official Journal of the Transcultural Nursing Society/Transcultural Nursing Society*, 14(4), pp. 313–320.

Ludi, E., 2009. *Climate Change, Water and Food Security*. Background note, March. London: Overseas Development Institute.

Lynch, C. and Roper, C., 2011. The transit phase of migration: Circulation of malaria and its multidrug-resistant forms in Africa. *PLoS Medicine*, 8(5), p. e1001040.

MacGregor, S., 2010. Plus ca (climate) change, plus c'est la meme (masculinist) chose: Gender politics and the discourses of climate change. In Lee, Y.-L. (ed.), *The Politics of Gender*. London: Routledge, pp. 83–101.

Magrath, P. and Tesfu, M., 2006. *Equal Access for All? Meeting the needs for water and sanitation of people living with HIV/AIDS in Addis Ababa, Ethiopia*. Briefing note 6. Addis Ababa, Ethiopia: Water Aid Ethiopia.

Mattes, D., 2011. 'We are just supposed to be quiet': The production of adherence to antiretroviral treatment in urban Tanzania. *Medical Anthropology*, 30(2), pp. 158–182.

McMichael, A.J. and Woodruff, R.E., 2008. Climate change and infectious diseases. In Mayer, K.H. and Pizer, H.F. (eds), *The Social Ecology of Infectious Diseases*. Amsterdam: Academic Press, pp. 378–407.

McMichael, C. and Manderson, L., 2004. Somali women and well-being: Social networks and social capital among immigrant women in Australia. *Human Organization*, 63(1), pp. 88–99.

McMichael, C., Barnett, J., and McMichael, A.J., 2012. An ill wind? Climate change, migration, and health. *Environmental Health Perspectives*, 120(5), pp. 646–654.

Milas, S. and Latif, J.A., 2000. The political economy of complex emergency and recovery in northern Ethiopia. *Disasters*, 24(4), pp. 363–379.

Ministry of Agriculture and Rural Development, 2010. *Ethiopia's Agricultural Sector Policy and Investment Framework (PIF) 2010–2020*. Draft Final Report, September 15, 2010. Addis Ababa: Federal Democratic Republic Of Ethiopia, Ministry of Agriculture and Rural Development.

Müller, C., Waha, K., Bondeau, A., and Heinke, J., 2014. Hotspots of climate change impacts in Sub-Saharan Africa and implications for adaptation and development. *Global Change Biology*, 20(8), pp. 2505–2517.

Ndaruzaniye, V., 2011. *Water Security in Ethiopia: Risks and vulnerabilities' assessment, Issue 2*. Brussels: Global Water Institute.

Negash, K., Kebede, A., Medhin, A., Argaw, D., Babaniyi, O., Guintran, J.O., and Delacollette, C., 2005. Malaria epidemics in the highlands of Ethiopia. *East African Medical Journal*, 82(4), pp. 186–192.

Nelson, V., 2011. *Gender, Generations, Social Protection and Climate Change: A thematic review*. London: Overseas Development Institute.

Nuru Ali, S., 2012. *Climate Change and Economic Growth in a Rain-fed Economy: How much does rainfall variability cost Ethiopia?* Addis Ababa, Ethiopia: Ethiopian Economics Association.

Parenti, C., 2011. *Tropic of Chaos: Climate change and the new geography of violence*. New York: Nation Books.

Rahmato, D., 1991. *Famine and Survival Strategies: A case study from northeast Ethiopia*. Uppsala: Nordic Africa Institute.

Raleigh, C., Jordan, L., and Salehyan, I., 2008. *Assessing the Impact of Climate Change on Migration and Conflict*. Washington, DC: Social Development Department, The World Bank.

Reliefweb, 2013. Concerns over HIV/AIDS funding cuts, January 9. Available at: http://reliefweb.int/report/ethiopia/concerns-over-hivaids-funding-cuts (accessed July 24, 2015).

Rettberg, S., 2010. Contested narratives of pastoral vulnerability and risk in Ethiopia's Afar region. *Pastoralism*, 1(2), pp. 248–273.

Schantz-Dunn, J. and Nour, N.M., 2009. Malaria and pregnancy: A global health perspective. *Reviews in Obstetrics and Gynecology*, 2(3), pp. 186–192.

Setegn, S.G., Rayner, D., Melesse, A.M., Dargahi, B., and Srinivasan, R. 2011a. Impact of climate change on the hydroclimatology of Lake Tana Basin, Ethiopia. *Water Resources Research*, 47(4), p.W04511.

Setegn, S.G., Rayner, D., Melesse, A.M., Dargahi, B., Srinivasan, R., and Worman, A., 2011b. Climate change impact on agricultural water resources variability in the Northern Highlands of Ethiopia. In Melesse, A.M. (ed.), *Nile River Basin: Hydrology, climate and water use*. New York: Springer, pp. 241–266.

Simane, B., Zaitchik, B.F., and Mesfin, D., 2012. Building climate resilience in the Blue Nile/Abay Highlands: A framework for action. *International Journal of Environmental Research and Public Health*, 9(2), pp. 610–631.

Tadesse, T., Haile, M., Senay, G., Wardlow, B.D., and Knutson, C.L., 2008. The need for integration of drought monitoring tools for proactive food security management in sub-Saharan Africa. *Natural Resources Forum*, 32(4), pp. 265–279.

Team, V. and Manderson, L., 2011. Social and public health effects of climate change in the '40 South.' *Wiley Interdisciplinary Reviews: Climate Change*, 2(6), pp. 902–918.

Teller, C. and Hailemariam, A., 2011. The complex nexus between population dynamics and development in Sub-Saharan Africa: A new conceptual framework of demographic response and human adaptation to societal and environmental hazards. In Teller, C. and Hailemariam, A. (eds), *The Demographic Transition and Development in Africa: The unique case of Ethiopia*. New York: Springer, pp. 3–16.

Teller, C., Hailemariam, A., and Teklu, N., 2011. Barriers to access and effective use of data and research for development policy in Ethiopia. In Teller, C. and Hailemariam, A. (eds), *The Demographic Transition and Development in Africa: The unique case of Ethiopia*. New York: Springer, pp. 323–337.

Terry, G., 2009. No climate justice without gender justice: An overview of the issues. *Gender and Development*, 17(1), pp. 5–18.

The World Bank, 2012. *Gender Equality and Development: World Development Report 2012*. Washington, DC: The World Bank.

Uneke, C.J., 2011. Congenital malaria: An overview. *Tanzania Journal of Health Research*, 13(3), pp. 1–18.

United Nations Secretary-General's High-Level Panel on Global Sustainability, 2012. *Resilient People, Resilient Planet: A future worth choosing*. Overview. New York: United Nations.

Vallely, A., Vallely, L., Changalucha, J., Greenwood, B., and Chandramohan, D., 2007. Intermittent preventive treatment for malaria in pregnancy in Africa: What's new, what's needed? *Malaria Journal*, 6(1), p. 16.

Wagner, N., 2015. Female genital cutting and long-term health consequences – Nationally representative estimates across 13 countries. *The Journal of Development Studies*, 51(3), pp. 226–246.

Wamukonya, N. and Skutsch, M., 2002. Gender angle to the climate change negotiations. *Energy and Environment*, 13(1), pp. 115–124.

White, P., 2005. War and food security in Eritrea and Ethiopia, 1998–2000. *Disasters*, 29, pp. S92–S113.

Yewhalaw, D., Legesse, W., Van Bortel, W., Gebre-Selassie, S., Kloos, H., Duchateau, L., and Speybroeck, N., 2009. Malaria and water resource development: The case of Gilgel-Gibe hydroelectric dam in Ethiopia. *Malaria Journal*, 8(21) [online], doi:10.1186/1475-2875-8-21.

49 How climate change makes me feel

Anthony J. Richardson

I feel a maelstrom of emotions

I am *exasperated*. Exasperated no one is listening.
I am *frustrated*. Frustrated we are not solving the problem.
I am *anxious*. Anxious that we start acting now.
I am *perplexed*. Perplexed that the urgency is not appreciated.
I am *dumbfounded*. Dumbfounded by our inaction.
I am *distressed*. Distressed we are changing our planet.
I am *upset*. Upset for what our inaction will mean for all life.
I am *annoyed*. Annoyed with the media's portrayal of the science.
I am *angry*. Angry that vested interests bias the debate.
I am *infuriated*. Infuriated we are destroying our planet.

But most of all I am *apprehensive*. Apprehensive about our children's future.

Index

Page numbers in *italics* denote tables, those in **bold** denote figures.

Printed and bound by CPI Group (UK) Ltd, Croydon, CR0 4YY

17/10/2024

01775686-0016